WAR WINGS

Modern China

WAR WINGS

The United States and Chinese Military Aviation, 1929–1949

Guangqiu Xu

Contributions in Military Studies, Number 211

GREENWOOD PRESS
Westport, Connecticut • London

Library of Congress Cataloging-in-Publication Data

Xu, Guangqiu, 1951–
War wings : the United States and Chinese military aviation, 1929–1949 / Guangqiu Xu.
p. cm.—(Contributions in military studies, ISSN 0883–6884 ; no. 211)
Includes bibliographical references and index.
ISBN 0–313–32004–7 (alk. paper)
1. Aeronautics, Military—China—History—20th century. 2. Zhongguo guo min dang. 3. United States—Military relations—China. 4. China—History—Republic, 1912–1949. 5. China—Military relations—United States. I. Title. II. Series.
UG635.C5X824 2001
358.4′00951′0904—dc21 2001023315

British Library Cataloguing in Publication Data is available.

Library of Congress Catalog Card Number: 2001023315
ISBN: 0–313–32004–7
ISSN: 0883–6884

First published in 2001

Greenwood Press, 88 Post Road West, Westport, CT 06881
An imprint of Greenwood Publishing Group, Inc.
www.greenwood.com

Printed in the United States of America

The paper used in this book complies with the Permanent Paper Standard issued by the National Information Standards Organization (Z39.48–1984).

10 9 8 7 6 5 4 3 2 1

Contents

Preface

Americans played an essential role in the development of Chinese military aviation from 1929 to 1949. From the lifting in 1928 of the arms embargo by the U.S. government on exportation of airplanes to China until the U.S. military aid to China began at the end of 1940, Americans contributed in a variety of ways to the expansion of the Chinese military aviation by selling planes and sending aviation missions to China. With the assistance of Americans, modern aviation schools and aircraft factories were established and the Chinese Air Force was reorganized and enlarged. After 1940, the U.S. government began to extend air aid to the Nationalist government in an effort to defeat Japan. U.S. air aid and air supply to China played an important role in the Sino-Japanese war. After the war, the Truman administration continued to render air aid to the Chinese Nationalists during the civil war until 1949.

Although historians have written a number of books and articles on the Chinese commercial aviation, little has been written about American assistance to the Chinese military aviation between 1929 and 1949. Why did the Nationalists promote the development of military aviation in China? How did Americans help the Chinese develop military aviation before 1940? How did the U.S. government support the Nationalist government in the air under the Lend-Lease program between 1941 and 1949? What was the impact of the expansion of Chinese military aviation on the Chinese political situation, on the military campaigns against warlords and Communists from 1929 to 1936, on the Sino-Japanese war of 1937–1945, and on the civil war of 1946–1949? How did the U.S. air assistance and aid to China affect the relations between Japan and the United States and the relations between the Washington and London governments during this period? Why were the Nationalists who received American air assistance and aid for 20 years defeated by the Communists who had no air power? These engaging questions, none of the answers to which are readily at hand in existing literature, encouraged this writer to plunge into government documents and manuscript materials in the United States, the United Kingdom, and the People's Republic of China.

Fortunately, key materials particularly for the U.S. side of things have been preserved in the National Archives. For example, the records of the Office of the Chief of Naval Operations, the United States Senate, the Department of State, the War Department, both General and Special Staffs, and the Department of Commerce preserve many documents with regard to American assistance to Chinese military aviation. In addition, the U.S. Air Force Archives at Maxwell, Alabama, the Hoover Institution on War, Revolution, and Peace (located on the campus of Stanford University), the George C. Marshall Foundation Library (Lexington, Virginia), the Division of Manuscripts of the Library of Congress, the Second Historical Archives of China (Nanjing), the Archives of Guangdong province (Canton), and the Public Record Office (London) also preserve some documents, manuscripts, and personal papers with regard to this topic. I am grateful to those archivists and their staffs who work in the above institutions.

Dr. Wayne S. Cole, master historian of American isolationism in the 1930s, helped me immensely in exploring byways of this subject and in locating material sources. His ideas, suggestions, and comments kept me from straying from the right track. I am deeply indebted to him for his kindness in reading, criticizing, and offering suggestions regarding various parts of the manuscript at various stages. I am also very grateful to Dr. Clifford Foust, who read the manuscript and guided me on points of detail. I owe special appreciation to Dr. William Leary, who read the entire manuscript and criticized style and substance. The Department of History of the University of Maryland generously underwrote travel to the Hoover Institution and the Public Record Office. Finally, I should like to thank my wife, who cared for our children while I concentrated on research and writing; she succeeded in keeping the noise level of a pair of busy boys tolerable.

THE USE OF PINYIN

This book uses pinyin romanization throughout, with some exceptions for place names and personal names that are long familiar in the West or difficult to recognize in pinyin. Thus Peking and Canton are retained in preference to Beijing and Guangzhou, and Chiang Kai-shek is used rather than Jiang Jieshi.

List of Abbreviations

AVG	American Volunteer Group
CAF	Chinese Air Force
CAT	Civil Air Transport
CATF	China Air Task Force
CCP	Chinese Communist Party
CACW	Chinese-American Composite Wing
CNAC	China National Aviation Corporation
GMD	Guomindang (Nationalist Party)
JUSMAG	Joint United States Military Advisory Group
SWNCC	State-War-Navy Coordinating Committee
UNRRA	United Nations Relief and Rehabilitation Administration

Introduction

In 1926, the Nationalist Party (Guomindang or GMD) in Canton launched the Northern Expedition against the local warlords in order to reunify China. By early 1928, under the leadership of Chiang Kai-shek the Nationalist Party had concluded deals with regional warriors, bringing China under the GMD banner. In late 1928, the Nationalists established a central government in Nanjing; the most important feature of the Nanjing government was its militaristic basis. The GMD government, led by Chiang Kai-shek, actually controlled only a small part of China, mainly in the lower Yangzi valley, and was still politically shaky, facing the opposition of the warlords who tried to organize another "National government." Local warlords, who strove to maintain their own bases and armies, put up strong resistance against the Nationalist Army's attacks. The warlords were not the only problems confronting Nanjing. The Chinese Communist Party (CCP), following the collapse of the first GMD-CCP United Front in 1927, lost no time in initiating guerrilla wars against the Nationalists. They established bases in South China and posed a threat to the Nanjing government.

With the stress on the military, Chiang endeavored to unify China by force. Nanjing's army was inadequate to overcome those powerful warriors and Communists. To augment its military power, Nanjing decided to develop a modern air force which would help the army defeat the internal enemies while the warlords were also interested in air power.

Military aviation proved disappointing in China in the late 1920s for both the Nationalists and the regional warlords. The regional warlords used a few out-of-date aircraft made in the1910s for maintaining power and keeping territories, and the Nationalists also were using outdated aircraft. Obviously, these old airplanes produced insignificant results; the Chinese seemed not to have enjoyed the technology advances in military aviation after World War I.

The Chinese had to purchase aircraft from other countries because China did not have any aeronautical industry at that time. The same year that the Nanjing government was created, the United States unilaterally lifted its embargo on the export of civilian airplanes to China. When the major world powers, including the United States, Great Britain, and France, agreed collectively in 1929 to lift the China arms embargo, foreign aircraft manufacturers thereafter began to sell a large number of their planes to China.

The Japanese attack of September 18, 1931, which led to Japan's occupation of Manchuria, quickly convinced the Nanjing government of the need to develop its air force quickly and regardless of cost; the enemies were now both domestic warlords and the Japanese. To help China speed its military aviation program in the years thereafter, American business enterprise took the lead in selling military airplanes and dispatching air missions to China. With American assistance, the Nationalist government established a Central Aviation School in 1933, built a Central Aircraft Factory in 1934, and thoroughly reorganized the Nationalist Air Force. It soon became apparent that Americans had come to dominate Chinese military aviation, and nationalist military aviation reached a new and higher stage in both size and technical skills before the Sino-Japanese war broke out in July 1937.

In the meantime, American businessmen were also interested in the aircraft market with the regional warlords in South China. In spite of the protest of the Nanjing government, Americans sold military aircraft to the Canton government which was independent of the control of the Nanjing government. Americans also helped to establish a modern aircraft factory in Canton and to train the Cantonese aviators.

The U.S. government was cautious in helping the Chinese develop military aviation in order not to provoke the Japanese between 1929 and 1936. After the Sino-Japanese war erupted in July 1937, the Roosevelt administration started to impose a partial arms embargo on China and to prohibit American aviators from serving in the Chinese Air Force in an effort to keep the United States out of the war in East Asia. In the next several years, the Soviet Union, hitherto absent from the picture, began to assist China in its fight against Japan by sending pilots and airplanes. Although the Soviet Union had considerable influence on the Chinese Air Force between 1938 and 1941, Americans encouraged by their government still sold many aircraft to China and continued to help the Chinese manage their aircraft factories and supervise their aviation schools.

When the Russians turned increasingly to preparing to defend their own country from attack by Germany, the Chinese government turned to the U.S. government again for pilots and airplanes, and President Franklin D. Roosevelt approved broad military aid to China on December 19, 1940. American governmental military aid to China emphasized aircraft and pilots, and Americans dominated Chinese military aviation again after 1940 as they had done before 1938. Under the Lend-Lease program, China received more U.S. government air aid, especially after the Pearl Harbor attack. American air aid and air supply to the Chinese played an important role in the war. After the surrender of Japan, the Truman administration continued to expand and strengthen the Nationalist Air Force. In spite of American air aid, the Chinese

Communists, who did not have any air power, defeated the Nationalists and forced them to flee to Taiwan in 1949.

Study of the U.S. air support for China from 1929 to 1949 has significance for the following reasons.[1] First, this topic is essential to understanding the development and expansion of Chinese military aviation and its impact on the political situation in China during this period. Although some books and articles relating to this topic have been published, this subject has to be thoroughly studied. How did the Chinese develop military aviation with American assistance? What role did the Nationalist Air Force play in the military campaigns against the local warlords and the Chinese Communists in the unification movement between 1929 and 1936, in the Sino-Japanese war of 1937–1945, and in the civil war of 1946–1949? This book will try to paint a picture of the development and expansion of the Nationalist Air Force assisted by the Americans and the U.S. government during the period of 1929–1949 and discuss the impact of this air force on the Chinese political situation and wars in China.

Second, study of this subject is also useful to any examination of the relations between the United States and Japan. Japan was extremely concerned over American assistance to the Chinese Air Force and disliked American participation in the development of Chinese military aviation. The Japanese government regarded the Chinese Air Force under American influence and control as an attempt by the United States to prevent Japan's expansion and to extend American military influence in East Asia. The Soviet Union came to control the Chinese Air Force between 1938 and 1940, but the U.S. Department of State continued to encourage the sale of military airplanes to China and advocated a moral embargo on aircraft sale to Japan. In 1941, Washington began providing China with many of the latest aircraft and began allowing American pilots to serve in the Chinese Air Force. At this time, the U.S. government not only helped the Chinese Air Force to resist Japanese air assault, but also encouraged them to bomb the Japanese homeland.

How did the Japanese respond to the American air assistance to China?[2] What policy did the U.S. government adopt in the face of the Japanese pressure? What was the impact of U.S. government air support of China on the Pearl Harbor attack? [3] Numerous books and articles have been published on American–East Asian relations between 1931 and 1941, but the issue of American air assistance to China is not thoroughly studied. Analysis of American air assistance to China will cast new light on the study of U.S.–East Asian relations from a different angle.[4] This book will examine how U.S. air assistance to the Chinese became a problem in the U.S. –Japan relations and study the impact of U.S. government air support of China on the Pearl Harbor attack.

Third, the study of American assistance to Chinese military aviation contributes to an understanding of relations between Great Britain and the United States in the 1920s and 1930s. After World War I, Chinese warlords remained locked in a war for power and bases. As the Chinese civil struggle became more severe and widespread, the warlords needed more arms and ammunition as well as airplanes. To end the fighting and to prevent a further disintegration of China, the United States initiated a proposal in May 1919 for a China arms embargo, which quickly received the support of Great Britain as well as other powers. The major powers imposed the China arms embargo between 1919 and 1929. After the embargo was lifted, China was still racked by civil strife. When the Nanjing government endeavored to unify China by force, the local warlords, who strove to maintain their own armies and bases, resisted this Central government. Political divisions and tensions persisted between Nanjing and the local governments until 1936.[5]

To strengthen their military powers, the provincial warlords purchased a large quantity of arms and munitions, especially airplanes, from the Western powers because the air forces became increasingly important in settling political and military differences in China. The provincial governments in China strove to develop air power not only to maintain local control, but also to protect against Nanjing's attempts to unify and to dominate China. The local authorities sought foreign assistance in developing military aviation programs; foreign businessmen took the opportunity to extend their influence in China by sending air missions, building aircraft factories, and establishing flying schools. American, British, French, Italian, and German aircraft businessmen were scrambling for the Chinese aircraft market; American and British manufacturers soon became major competitors for aircraft orders in South China, which led to the American domination of military aviation in Guangdong province and the British in Guangxi.[6]

Several Western countries, such as the United States, Great Britain, and France, continued to regulate their arms exports to China after 1929 by requiring licenses for these exports and issuing them only for shipments destined for and approved by the Nationalist government in order to prevent arms from falling into the hands of Nationalist government's opponents. The China arms embargo and arms control implemented by the major powers encountered great pressure from their aircraft business when the Chinese warlords became intrigued with the military use of the airplanes. The aircraft businessmen's rivalry embroiled their governments in a major dispute over the China arms embargo and over the regulation of arms sales to China. American and British airplane businessmen were concerned with the problems of how to dominate the Chinese market for their aircraft industries hit greatly by the Great Depression while the officials of the State Department and Foreign Office were more concerned with the broader diplomatic objectives in China.[7] Therefore, American-British aircraft competition in South China constituted one of the challenges to the China arms embargo and international cooperation on control of arms sales to China in the 1920s and 1930s.

The study of Anglo-American aircraft competition highlights the issue of Anglo-American cooperation in China in the 1920s and 1930s.[8] During that time American and British interests in China were identical—a unified China without Japanese dominance. Common interests involved cooperation on the part of Washington and London.[9] This common goal, which characterized Anglo-American relations in East Asia, established a foundation of the Anglo-

American cooperation in China. Anglo-American cooperation was important in helping the Chinese establish a strong unified China and in opposing the Japanese military expansion in China. The China arms embargo and the international regulation of arms sales to China were part of the Anglo-American effort to adopt a cooperative policy in East Asia.[10] Anglo-American cooperation was crucial in implementing the embargo agreement from 1919 to 1929 and in regulating their arms sales to China from 1930 to 1936. However, neither the United States nor Great Britain wanted to sacrifice its own aircraft interests for the stability of China when other countries were exporting aircraft to the Chinese militarists or to the Nationalist government's internal opponents. Anglo-American competition in the South China aircraft market tested the policy of Anglo-American cooperation toward China in general and on arms sales to China in particular throughout the years of 1926–1936.

Fourth, this subject is valuable for understanding the U.S. China policy in the 1930s. During the 1930s, the United States exported arms, munitions, and war materials to China. In comparison with the export of airplanes to China, however, those sales were insignificant. The most important military shipments from the United States to China at the time were aviation-related. In the 1930s, Americans helped the Chinese develop not only military aviation but commercial aviation as well, but most of the airplanes exported from the United States to China were military. The aircraft factories established by Americans in China manufactured mainly military airplanes and the aviation schools established by Americans principally trained pilots for the Chinese Air Force. This study covers all imported airplanes, airplane factories, and aviation schools within the scope of military aviation because aviation schools and aircraft factories were under control of the Chinese military, and imported commercial airplanes often were used militarily. The U.S. China arms policy in the 1930s was in fact the China airplane policy of the American government.

Aircraft are not ordinary commodities of international trade. The sales of airplanes, especially military aircraft, to China during the 1930s changed the balance of power not only within that country but also in East Asia more broadly. The China arms policy, an integral and increasingly important part of the broad political policy of the American government toward China, was based on two general goals: to discourage civil strife in China, and to refrain from placing American exporters at an unjustifiable disadvantage in competing with other exporter countries. The first principle was essentially humanitarian, although it had obvious political implications; the second was economic. The U.S. State Department consequently maintained a license system to control the sale of airplanes, especially warplanes, from the United States to China. But U.S. policy in these matters was not always easily set; there were conflicting opinions and interests within the Washington establishment. The China arms policy was not only political but also commercial and economic. The Departments of Commerce and War were anxious to stimulate the sale of military airplanes to China without consistently considering the political consequences of those sales within and outside China. The Department of State, on the other hand, sometimes opposed selling too many airplanes to the Nationalist government or the local governments in South China, fearing the aggravation of diplomatic problems. To study the motives of the Department of State in amending or revising the China arms policy again and again in the 1930s contributes to an understanding of how the Department of State formulated its China policy. This book will examine particularly the differing attitudes and policies of the State, War, and Commerce Departments on the sale of airplanes to China to determine how they developed in the 1930s.

Fifth, this subject helps the study of the Sino-U.S. diplomacy and military cooperation during the wartime between 1942 and 1945. After the outbreak of the Pacific War in December 1941, at the request of China, President Roosevelt in January 1942 appointed General Joseph W. Stilwell chief of staff to Chiang Kai-shek, the supreme commander of the China theater. This appointment signaled the beginning of close military cooperation between the United States and China, and large-scale American military aid to the Chinese. From 1942 to 1945, in aviation matters, the U.S. War Department established an American air force in China, provided the Chinese Air Force with American aircraft, and trained the Chinese pilots and technicians. In the meantime, the United States set up the air supply route over the Hump in the Himalayas after Japan penetrated into Burma so as to continue delivering to the Chinese the most urgently needed war materials. American air support for the Chinese vastly enhanced the combat effectiveness of the Chinese military forces and contributed greatly to the defeat of Japan in China.

Between 1942 and 1945, however, serious differences existed between the U.S. and the Chinese governments not only on political goals but also on military strategies. The conflicts on military goals and strategies between China and the United States continued until the end of 1944. The rise and development of United States–Chiang contradictions were evident in a conflict between Stilwell and Chiang. Historians differ on the sources and causes of the conflict between the Washington and Chongqing governments as well as between Stilwell and Chiang Kai-shek. Some scholars maintain that the reason for the wartime friction and frustrations between China and the United States was that Chiang refused to employ his best-equipped armies against the Japanese and to commit forces to offensive action while demanding more U.S. military aid.[11] Stilwell utilized everything in the fight against the Japanese while Chiang preferred hoarding the Lend-Lease materials and military forces under Stilwell's control for a future fight against the Communists.[12] In defense of Chiang, some historians argue that Chiang and his forces did actively resist the Japanese after the United States entered the war and that Stilwell, despite his military credentials, was essentially miscast in the China theater. He was not really well suited for this job because he knew very few Chinese leaders personally and lacked appreciation of the immense complexity of modern Chinese military politics.[13] Washington and Chongqing disagreed on a number of issues, but the issue of American air aid and air supply to China was one of the most serious conflicts between these governments during that time. Study of this subject helps to examine the U.S.-China diplomatic relations and military cooperation, and to shed new light on the causes of Sino-U.S. conflicts during most critical years of the war.

Sixth, the attitude and policy of the Chinese Communist Party toward the United States from 1945 to 1949 has been an interesting subject of historians and remains a source of controversy.[14] Recently released documents in the Soviet Union and China produce new study on the CCP's attitude toward the United States in the late 1940s. These new articles and books raise an important question: Did the Truman administration have a chance or not have a chance to avoid the CCP-U.S. confrontation in the late 1940s? Historians differ on whether the Chinese Communists did really seek U.S. support, wanted to develop trade relationship with the United States, and tried to pursue CCP-U.S. friendship during that time. The debate continues today.[15]

Numerous articles and books have been published on the Chinese Communist policy toward the United States during the postwar period, but the study of U.S. military aid, especially U.S. air aid, to the Guomindang during the civil war of 1945–1949 still has received little attention.[16] In the immediate aftermath of the Pacific War, the Truman administration continued military aid to Chiang Kai-shek. Under the Lend-Lease program, the U.S. government transferred many aircraft and vessels, and much military equipment, and sold large quantities of military and civilian surplus property to the Nationalist government. Washington continued the training program of Chinese military personnel in both China and the United States and sent military advisers to China in an effort to modernize the Nationalist military forces. The U.S Army Air Forces (USAAF) and the Navy played a significant role in rushing hundreds of thousands of Nationalist ground forces to Japanese-occupied areas to prevent the Chinese Communists from conquering these regions.

American air aid was an important part of U.S. military aid to the GMD government after the war. What was the policy of the Truman administration toward the Nationalist military and civilian aviation after the war? How did Americans strengthen the power of the Chinese Air Force? What role did the Chinese military and civilian aviation play during the civil war? What was the attitude of the Chinese Communists toward U.S. air aid and air assistance to the Nationalists? The answers to the above questions help us to understand the origins of the CCP-U.S. confrontation. This book examines the U.S. policies toward the Chinese military and civilian aviation during the postwar period, American air aid and assistance to the Nationalist government, and the impact of U.S. air aid on the attitude of the Chinese Communists toward the Washington government.

Finally, to survey and analyze the aviation aspect of the Sino-U.S. relationship from the establishment of the Nanjing government to the end of the Nationalist regime in China situates the study on the effectiveness of air power as an advanced weapon. The Nationalists received American air assistance from 1929 to 1940 and U.S. government air aid from 1941 to 1949. With strength in the air, Chiang Kai-shek's military power gained domination over his internal enemies and won some victories over the warlords and the Communists.

Although the Nationalists owned the modern military weapons, they were finally defeated by the Communists. Why was the modern military weapon unable to save the Nationalists? Why did the Nationalist Air Force fail to become an effective fighting force after 20 years of American assistance and aid? This book tries to answer these questions.

A study of the history of American air assistance and aid to the Chinese military aviation has contemporary significance as well. History is a mirror; lessons may be drawn from it. The crucial role of air power in modern warfare has been readily apparent since World War II, if not earlier. At present the United States is one of the major exporters of military planes to the Republic of China in Taiwan. These exports to Taiwan are an aspect of the overall China policy of the United States, and these airplanes greatly strengthen the Taiwanese Air Force. Indeed, the sale of American warplanes to Taiwan has been a thorny diplomatic problem between Peking and Washington since 1949 and continues to be.

This book is divided into seven chapters. It chronically examines how Americans and the U.S. government became involved in Chinese military aviation, discusses the American government's attitudes and actions toward that involvement, and analyzes the impact of the development of Chinese military aviation on China and on the international relations relating to China. In 1928, as noted above, the American government revoked the ban on exporting commercial planes to China and the first modern commercial airplane was exported from the United States to China in that year. Twelve years later, in December of 1940, President Roosevelt approved a broad military aid program to China. This was the first time that the American government, rather than American businessmen, provided aviation supplies to China. After 1940, the U.S. government played increasingly important roles in developing Chinese military aviation until 1949.

NOTES

1. Several books, articles, and dissertations relate to this topic in part: William Leary's book, *The Dragon's Wing: The China National Aviation Corporation and Development of Commercial Aviation in China* (Athens: University of Georgia, 1976), treats the development of commercial aviation in China from 1929 to 1945 and is drawn on here. Gordon K. Pickler's Ph.D. dissertation, "United States Aid to Chinese Nationalist Air Force, 1931–1949" (Florida State University, 1971) and Daniel Ford's *Flying Tigers: Claire Chennault and the American Volunteer Group* (Washington, D.C.: Smithsonian Institution Press, 1991) provide important background material. William Leary's article, "Wings for China: The Jouett Mission, 1932–1935," *Pacific Historical Review*, 38 (1969): 447–62, studies the role of the Jouett Mission in developing the Nationalist Air Force; some of his ideas are adapted here. In Elton Atwater's study, *American Regulation of Arms Exports* (Washington, D.C.: Carnegie Endowment for International Peace, 1941), some chapters discussing the U.S. arms sales to China in the context of United States regulation of arms exports to China before 1940 also provide some new sources for this book. In Michael Schaller's book, *The U.S. Crusade in China, 1938–1945* (New York: Columbia University Press, 1979), the American

decision to reverse the trend and to come to aid China is ably dealt with. Arthur N. Young's *China and the Helping Hand, 1937–1945* (Cambridge, Mass.: Harvard University Press, 1963) is also helpful. As for books in the Chinese language and from the Chinese point of view, this writer is aware of only a few books relating in part to this subject—Jin Yao, *Zhongguo hanghong shi* [A history of China's aviation] (Zhengzhou: Daxian chubanshe, 1998); Shushan Li, *Zhongguo hongjun zhanshi* [The fighting history of the Chinese Air Force] (Guangzhou: Guangdong renmin chubanshe, 1996); Changying Jing, *Zhongguo hanghong shi* [The history of Chinese aviation] (Taipei: Zhongguo zhiyi chubanshe, 1993); and Minfu Ma, *Zhongguo junshi hangkong, 1908–1949* [China's military aviation, 1908–1949] (Beijing: Hanghong gongye chubanshe, 1994).

2. For a scholarly study of the Japanese-American relations in the 1930s, see *Pearl Harbor as History: Japanese-American Relations, 1931–1941*, ed. Dorothy Borg and Shumpei Okamoto (New York: Columbia University Press, 1973), which focuses on the decision-making process within the respective foreign elites, presses, and private organizations; *Pearl Harbor Reexamined: Prologue to the Pacific War*, ed. Hilary Conroy and Harry Wray (Honolulu: University of Hawaii Press, 1990); James H. Herzog, *Closing the Open Door: American-Japanese Diplomatic Negotiations, 1936–1941* (Annapolis, M.D.: Naval Institute Press, 1973); and Walter LaFeber, *The Clash: A History of U.S.-Japan Relations* (New York: W. W. Norton, 1997).

3. Historians have different opinions on the major reasons for the Pearl Harbor attack. Some scholars accuse President Franklin D. Roosevelt of provoking the Japanese attack by imposing the oil embargo and of maneuvering the nation into the war. Intervention served to cloak domestic political failure. For this school, see Charles C. Tansill, *Back Door to War: The Roosevelt Foreign Policy, 1933–1941* (Chicago: Regnery, 1952) and Charles A. Beard, *President Roosevelt and the Coming of the War, 1941* (New Haven, Conn.: Yale University Press, 1948). Most historians, however, in defending President Roosevelt's policies and rejecting the conspiracy theory, argue that the aggressive ambitions of the Japanese militarists had caused the attack. See, for example, Basil Rauch, *Roosevelt From Munich to Pearl Harbor* (New York: Creative Age, 1950); Gordon Prange, *At Dawn We Slept: The Untold Story of Pearl Harbor* (New York: McGraw-Hill, 1981), which is the most detailed and reliable; and Waldo Heinrich, *Threshold of War: Franklin D. Roosevelt and American Entry into World War II* (New York: Oxford University Press, 1988), the best book on the American-Japanese crisis in 1941, focusing on how Roosevelt took a tough policy stance toward Japan because he was concerned about the collapse of the Soviet Union.

4. Scholars differ on what policies the U.S. government established toward East Asia in the 1930s and why the policy makers pursued such policies. Some historians hold that U.S. Japan policy was too soft in the 1930s and too hard in 1941. Some scholars argue that the U.S. government tried its best to help China against Japan after the Manchurian incident, but American aid was limited because of the Great Depression, the pressure of isolationism, and the American people's lack of interest in East Asia. Whitney Griswold's *The Far Eastern Policy of the United States* (New Haven, Conn.: Yale University Press, 1938) is still the only synthesis of U.S. East Asia policy. Warren Cohen, *America's Response to China: A History of Sino-American Relations*, 3d ed. (New York: Columbia University Press, 1990), disagrees with Griswold's opinions, arguing that the destiny of China was a matter of indifference to American policy makers who treated China not as a nation but as an object to be used in war and diplomacy. Akira Iriye's *Across the Pacific: An Inner History of American–East Asian Relations*

(New York: Harcourt, Brace and World, 1967) which is the broadest treatment of American–East Asian relations, stressing misunderstanding, misperception, and miscalculation in China, Japan, and the United States. The best study on the East Asian crisis is Dorothy Borg's *The United States and the Far Eastern Crisis of 1933–1938* (Cambridge, Mass.: Harvard University Press, 1964) which chronicles America's failure to aid China until late in the 1930s. This book shows the continuity of American passivity toward changing Japanese tactics, emphasizing American appeasement of Japan. However, Herbert Feis, *The Road to Pearl Harbor: The Coming of the War between the United States and Japan* (Princeton, N.J.: Princeton University Press, 1950), stresses the patience in American policy. James Thomson, *While China Faced West* (Cambridge, Mass.: Harvard University Press, 1969), argues that the United States had difficulties in assisting China due to the Great Depression.

5. For the study of the Chinese warlords in the 1920s and 1930s, see Lucian Pye, *Warlord Politics: Conflict and Coalition in the Modernization of Republican China* (New York: Praeger, 1971); Lloyd E. Eastman, *The Abortive Revolution: China under Nationalist Rule, 1927–1937* (Cambridge, Mass.: Harvard University Press, 1974); Diana Lary, *Region and Nation: The Kwangsi Clique in China Politics, 1925–1937* (London: Cambridge University Press, 1974); James Sheridan, *China in Disintegration: The Republican Era in Chinese History, 1912–1949* (New York: Free Press, 1975); and Wai-chor So, *The Kuomintang Left in the National Revolution, 1924–1931* (Hong Kong: Oxford University Press, 1991).

6. For the study of Anglo-American economic competition, see Stephen J. Rabe, "Anglo-American Rivalry for Venezuelan Oil, 1919–1929," *Mid-America* 58 (1976): 97–109; Frank C. Costigliola, "Anglo-American Financial Rivalry in the 1920s," *Journal of Economic History* 37 (1977): 911–34; M. Ruth Megaw, "The Scramble for the Pacific Anglo-United States Rivalry in the 1930s," *Historical Studies* 17 (1977): 458–73; Emily S. Rosenburg, "Anglo-American Economic Rivalry in Brazil during World War I," *Diplomatic History* 2 (1978): 131–52. For the study of Anglo-American competition in China, see Kwang-ching Liu, *Anglo-American Steamship Rivalry in China, 1862–1874* (Cambridge, Mass.: Harvard University Press, 1962) and Roberta Dayer, *Bankers and Diplomats in China, 1917–1925: The Anglo-American Relationship* (Bournemouth, England: Bourne Press, 1981).

7. For the background on the U.S. aviation industry, see Gene Simonson, *The History of the American Aircraft Industry: An Anthology* (Cambridge, Mass.; MIT Press, 1968); John B. Rae, *Climb to Greatness: The American Aircraft Industry, 1920–1960* (Cambridge, Mass.: MIT Press, 1969); Barry Bluestone, *Aircraft Industry Dynamic: An Analysis of Competition, Capital, and Labor* (Boston: Auburn House Company, 1981). For an account of the history of the aeronautic industry in England, see Keith Hayward, *Government and British Civil Aerospace* (Manchester, England: Manchester University Press, 1983), passim, especially pp. 12–18; Sebastian Ritchie, *Industry and Air Power: The Expansion of British Military Aircraft Production, 1935–1941* (London: Frank Cass, 1997).

8. As to relations between Great Britain and the United States with respect to East Asia in the 1920 and 1930s, some historians see positive elements in the cooperation between those two countries. They stress instances of British and American cooperation but neglect important disagreements between Washington and London. For example, see Roberta Dayer, "Struggle for China: The Anglo-American Relationship, 1917–1925" (Ph.D. dissertation, State University of New York at Buffalo, 1972); Gerald E. Wheeler, "Isolated Japan: Anglo-American Diplomatic Cooperation, 1927–1936," *Pacific Historical Review* 30 (1961): 165–78. Other historians, however, contend that

competition prevailed over cooperation between those two countries. See Max Beloff's article, "The Special Relationship: An Anglo-American Myth," in *A Century of Conflict, 1850–1950: Essays for A. J. Taylor* (ed. Martin Gilbert, London, 1966); Noel Pugach, "Anglo-American Aircraft Competition and the China Arms Embargo, 1919–1921," *Diplomatic History* 2 (1978): 351–71.

9. For the study of East Asian policy of Great Britain, see Roger Louis, *British Strategy in Far East, 1919–1939* (Oxford: Clarendon Press, 1971); G. Bennett, "British Policy in the Far East, 1933–1936: Treasury and Foreign Office," *Modern Asian Studies* 26 (1992): 545–68; Antony Best, *Britain, Japan and Pearl Harbor: Avoiding War in East Asia, 1936–1941* (London: Routledge, 1995).

10. For a scholarly study of the China arms embargo, see Zunkong Chen, *Lieqiang tui zhongguo de junhuo jinyun, 1919–1929* [China arms embargo, 1919–1929] (Taipei, Taiwan: Institute of Modern History, Academia Sinica, 1983); Anthony Chan, *Arming the Chinese: The Western Armaments Trade in Warlord China, 1920–1928* (Vancouver: University of British Columbia Press, 1982); Stephen Valone, *"A Policy Calculated to Benefit China": The United States and the China Arms Embargo, 1919–1929* (New York: Greenwood Press, 1991); Xiaoming Zhang, "Toward Arming China: United States Arms Sales and Military Assistance, 1921–1941," (Ph.D. dissertation, University of Iowa, 1994). Scholars differ on whether the China arms embargo was a failure or a success. Most scholars conclude that the embargo was an unqualified failure, but Valone argues that the embargo was successful in counteracting the Japanese attempt to dominate China and in providing the Chinese with humanitarian protection from civil strife.

11. See Riley Sunderland, "The Secret Embargo," *Pacific History Review* 29:1 (1960), pp. 75–80; Herbert Feis, *The Chinese Tangle: The American Effort in China from Pearl Harbor to the Marshall Mission* (New York: Atheneum, 1965); Warren Cohen, "Who Fought the Japanese in Hunan? Some Views of China's War Effort," *Journal of Asian Studies,* 27 (Nov. 1967): 111–115; John E. Shephard, "Warriors and Politics: The Bitter Lesson of Stilwell in China," *Parameters* 19 (1989): 61–75; and Michael E. Haith, " 'Vinegar Joe' Stilwell in the CBI," *Military Review* 72 (1992): 73–83.

12. See Shuoqi Wu, "Some Remarks on Joseph Stilwell's Relationship with Jiang Jieshi," *Republican China* 14 (1989): 37–55; Qifu Zhang, "On the Stilwell Incident and the Contradiction Between the United States and Jiang Jieshi," ibid., pp. 21–36; and Wandi Jiang, "Old Soldiers Never Die," *Beijing Review* (Nov. 11–17, 1991): 33–35.

13. See Chin-tung Liang, *General Stilwell in China, 1942–1944: The Full Story* (New York: St. John's University Press, 1972); Jui-te Chang, "Nationalist Army Officers during the Sino-Japanese War, 1937–1945," *Modern Asian Studies,* 30 (1996): 1033–56; and Hsi-sheng Chi, "General Stilwell's Approach to China and the Burma Campaign," *Sino-American Relations* 24 (1998): 65–84.

14. Some historians minimize ideological constraints on the Chinese Communist policy toward the United States, arguing that the Truman administration lost its chance at the end of the Sino-Japanese war when the Chinese Communist leaders were frustrated in their attempt to avoid CCP-U.S. confrontation and to reduce their dependence on the Russians. They see Mao Zedong as attempting to create an independent foreign policy rather than committing himself to the Russians or the Americans. Mao was flexible in his view of the Truman administration but seriously influenced by U.S. military and economic aid to Chiang Kai-shek. For this interpretation, see James Reardon-Anderson, *Yenan and the Great Powers: The Origins of the Chinese Communist Foreign Policy,*

1944-1946 (New York: Columbia University Press, 1980), focusing on the disappointment of the Chinese Communist leaders in their effort to avoid CCP-U.S. conflict and the CCP's dependence on Soviet Communism after the war. William Stueck supports this argument in his book, *Road to Confrontation: American Policy toward China and Korea, 1947–1950* (Chapel Hill: University of North Carolina Press, 1981). Michael Hunt also argues in a similar vein in his article, "Mao Tse-tung and the Issue of Accommodation with the United States, 1948–1950," in *Uncertain Years: Chinese-American Relations, 1947–1950* (New York: Columbia University Press, 1980), edited by Dorothy Borg and Waldo Heinrichs. Odd Arne Westad's *Cold War and Resolution: Soviet-American Rivalry and the Origins of the Chinese Civil War, 1944–1946* (New York: Columbia University Press, 1993) emphasizes that Mao's dream of establishing an independent China without control by external forces was shattered by the Truman administration in the immediate aftermath of the Second World War. But some historians developed contrary arguments. Russell D. Buhite's book *Soviet-American Relations in Northeast Asia, 1945–1954* (Norman: University of Oklahoma Press, 1981) tries to uphold this interpretation by underlining American concern over Soviet expansion in Northeast Asia, which posed a challenge to American security interest in Asia. John Hansen Feaver's work, "The Truman Administration and China, 1945–1950: The Policy of Restrained Intervention" (Ph.D. Dissertation, University of Oklahoma, 1980), holds that President Truman regarded the rising of the Chinese Communists as a threat to the American interest in China and justified his support of the Nationalist government. Okabe Tatsumi, "The Cold War and China" in *The Origins of the Cold War in Asia* (New York: Columbia University Press, 1977), edited by Yon Osuke Nagai and Akira Iriye supports this argument. See also Steven M. Goldstein, "Chinese Communist Policy Toward the United States: Opportunities and Constraints, 1944–1950," in *Uncertain Years: Chinese-American Relations, 1947–1950.*

15. For these arguments, see John W. Garver, "The Origins of the Second United Front: The Comintern and the Chinese Communist Party," *China Quarterly*, no. 113 (March 1988); Steven Goldstein, "Sino-American Relations, 1948–1950: Lost Chance or No Chance?" in *Sino-American Relations, 1945–1955, A Joint Reassessment of a Critical Decade* (Wilmington, Del.: Scholarly Resources Imprint, 1989), edited by Harry Harding & Yuan Ming; Michael M. Sheng, "Chinese Communist Policy Toward the United States and the Myth of the 'Lost Chance,' 1948–1950," *Modern Asian Studies,* 28 (1994); Jian Chen, *China's Road to the Korean War: The Making of the Sino-American Confrontation* (New York: Columbia University Press, 1994); and Michael Hunt, *The Genesis of Chinese Communist Foreign Policy* (New York: Columbia University Press, 1996).

16. As far as the author knows, only a few historians discuss this issue in their studies. See, for example, Chonghal Petey Shaw, *The Role of the United States in Chinese Civil Conflicts, 1944–1949* (Salt Lake City, Utah: Charles Schlacks, Jr., 1991); and William M. Leary, Jr., "Aircraft and Anti-Communists: CAT in Action, 1949–52," *China Quarterly*, no. 52 (1972).

Chapter 1

Aircraft Sales and the China Arms Embargo, 1919–1929

After the First World War, the Chinese became more interested in aviation as the former Allies were anxious to sell their large stocks of surplus aircraft. The British were particularly interested in the potential market in China and lost no time in sending their agents to Peking. The Peking government asked the British to recommend an adviser to China to develop commercial air service. After the British manufacturers, Handley-Page and Vickers, delivered their aircraft to China and were prepared for the inauguration of China's first commercial air service, a civil war broke out in 1920 between the Peking government and the warlords in Manchuria. The Peking authorities immediately commandeered all commercial airplanes for military use. Since the British did not want to be involved in any fighting, they withdrew their aircraft from the Peking airdrome in order to prevent the military authorities from using them for bombing. Due to the escalating civil war in China and Peking's financial problems, the British effort to introduce commercial air service into North China failed in the early 1920s.[1]

The Chinese did not successfully develop commercial aviation in the early 1920s, but they had already used aircraft in combat. As early as in 1914, the Peking government first used airplanes for observation and bombing in suppressing the Bailang rebellion. Later, various provincial warlords, realizing the military role of aircraft, began to organize their own air forces. In 1921 the Baodingfu Air Corps was formed in Hebei province, and the Northeastern Aviation Department began in Manchuria. The next year a Northeastern Aviation Bureau was formed.[2]

In April 1922, in North China aircraft were employed in the warlords' war between the Zhi and Feng cliques, marking the first time that planes were used in a major skirmish in China. During that fighting, the warlords of the Zhi clique took advantage of their superiority to send airplanes to bomb intensively the defense works and strongholds of the Feng clique in North China; the latter warlord clique suffered such a heavy blow as to contribute directly and immediately to its defeat. The army of the Feng clique was forced to flee northward to Manchuria.[3]

After that war, the Feng clique examined the causes of its failure. Lacking airplanes, these warlords determined to develop their own air force by purchasing French and other countries' airplanes. During that period, the Fengtian officials at Shenyang, Manchuria, initiated an aviation bureau. Marshal Zhang Xueliang assumed the directorship of the Northeastern Aviation Bureau on September 4, 1923. Several air squadrons, each consisting of 10 or more planes, were formed and carried colorful names, such as the "Flying Dragons," "Flying Tigers," and "Flying Eagles." The Northeastern Army planes consisted of English Handley-Pages and Vickers, French Breguets, American Curtisses, Vought Corsairs and Fords, and German Junkers. Many foreigners, mainly French, served as instructors and technicians to the Northeastern Aviation Bureau. France's influence among aviation personnel in Shenyang exceeded that of all other countries. Chinese cadets and aviation students were trained, and a nucleus of Chinese experts came into existence.[4]

By the time a second phase of fighting between the Feng and the Zhi cliques erupted in September1924, the Feng clique had a decidedly stronger air corps than the Zhi clique. While its infantry was marching toward Peking, the Feng clique's leader, Zhang Zuolin, sent 50 airplanes to the front. At the same time, General Wu Peifu of the Zhi clique challenged them with only 30 or so airplanes. This battle was the first time planes appeared in large numbers in a Chinese war. The Feng's airplanes observed the operation of the Zhi clique's infantry and bombed their defense works and railroads. The bombing was so severe that the Zhi clique could not resist the attacks and was defeated. Afterward, in the autumn of 1925, the Air Force Command of the North East was instituted under the control of Zhang Xueliang, who was a son of Zhang Zuolin, comprising five squadrons and 100 airplanes, the strongest air force in China at the time. During that period, other warlords established a number of aviation schools and airplane repair shops in North China.[5]

In South China, aviation was also being developed. As early as 1920, Dr. Sun Yat-sen's Revolutionary Army began to use airplanes for bombing and observation in wars against the local warlords. In November of that year, he returned to Canton to establish a Revolutionary government, and, subsequently, a Canton Aviation Bureau was formed in February 1922. In the campaign against warlord Chen Jiongming, the Canton Air Corps played a limited role in defeating Chen and defending the Canton Revolutionary government. In September 1924, the Guangdong Military Aviation School was begun under Sun Yat-sen's government.[6]

After the beginning of the Nationalist Revolution in 1924, Russian military advisers to the Nationalist Army organized a new military aviation unit at Canton. Russian military aviators taught Chinese pilots to fly the obsolete planes. By the onset of the Great Northern Expedition of Sun's Nationalism in the summer of 1926, the Russian fliers were admitted to the Canton Aviation Corps. The Chinese students learned rapidly under their Russian teachers, and during the fighting along the Yangzi valley, they eventually replaced their teachers in the air and flew planes by themselves.[7]

In the Northern Expedition, the airplanes of the Nationalist Aviation Corps, which evolved from the Canton Corps, played a highly visible but marginal role. They took photographs of the defense works of the army of Wu Peifu, directed artillery fire of the Nationalist Army, and dropped bombs on the strongholds of the enemy. In the battles of Tingxi Bridge and Hesheng Bridge near Hankou, the air attack prior to the infantry assault was so heavy that when the infantry of the Nationalist advanced, warlord Wu's army escaped. Although the successes of the Northern Expedition may have contributed in small measure to air warfare, the military use of airplanes consisted largely of lavish displays of equipment with only occasional bombing and observation missions.[8]

In spite of aircraft's limited effectiveness in the Northern Expedition, Chinese Nationalist military leaders increasingly realized the importance of air in future warfare and took steps to develop their air forces. Because China had not yet established an aeronautic industry, it had to purchase airplanes from abroad. The China arms embargo, however, seemed to be an obstacle to the import of airplanes. That embargo now needs attention.

THE CHINA ARMS EMBARGO, 1919–1928

The origins of the China arms embargo came from the civil wars in China after the First World War when the major imperial powers returned to China to pursue their commercial interests, and Chinese warlords were still locked in a war for power and bases. Those civil wars in China provided a welcome market for the arms and munitions manufacturers of the Great War. Orders for arms and munitions had diminished quickly after fighting subsided in Europe, and surplus war materials piled up in warehouses in the arms exporting countries. As the Chinese civil struggles became more severe and frequent, the warlords needed more arms and munitions, including airplanes. Thus, the traders and dealers from Western countries competed for that big market which resulted in hundreds of thousands of tons of arms and munitions from all over the world flowing into China.[9]

When the civil war among Chinese warlords steadily grew more severe in 1918, the situation in China deteriorated. To attempt to end the fighting and prevent a further disintegration of China, at the beginning of 1919 the United States urged the other powers to come to an agreement regarding the arms traffic. The British government welcomed this initiative because, as London saw it, British trade would benefit from a stable China. As a result, on instruction from London, on May 5, 1919, British Minister Sir John Jordan, as the dean of the diplomatic corps, presented a document to the Chinese government that stated:

> The governments of Great Britain, Spain, Portugal, the United States, Russia, Brazil, France and Japan have agreed effectively to restrain their subjects and citizens from exporting to or importing into China arms and munitions of war material destined exclusively for their manufacture until the establishment of a government whose authority is recognized throughout the whole country and also to prohibit during the above period the delivery of arms and munitions for which contracts have already been made but not executed.[10]

The China arms embargo was, on the surface, an attempt of the major powers to undertake a cooperative policy dedicated to humanitarian and moralistic ends. Notably, this policy marks the first time that the United States took the lead in organizing an international arms embargo effective against a single and particular country.[11] After the arms embargo was imposed, the U.S. government, in cooperation with the British government, tried its best to implement the embargo and encouraged other countries to implement it.

The embargo, however, failed to define the scope of the munitions trade that was proscribed. The U.S. Department of State considered all types of airplanes, both military and commercial, to be included within the embargo, but other countries did not. On October 1, 1919, only a few months after the powers signed the China Arms Embargo Agreement, the Peking government entered into a contract with Messrs. Vickers, Limited, of London, for the purchase of 100 Vimy aircraft.[12] The British government granted a permit to British businessmen for sale of those airplanes to China because they considered them "commercial airplanes," falling outside the scope of embargo. In fact, when those airplanes were later shipped to China, they were used for military purposes.

Regarding commercial aircraft exempt from the embargo, other countries, France, for example, continued to export a considerable number of airplanes to China after 1919. The consequence was that American firms that desired to sell airplanes to China complained the Department of State was disposed to include airplanes among the arms and munitions while other countries were not. Provoked by these complaints, the Department of State on September 17, 1920, instructed the U.S. Legation in China that since other governments had permitted their nationals to enter into contracts for commercial airplanes, the U.S. government did not feel warranted in interposing objections to its nationals doing likewise when the airplanes ostensibly were destined and built strictly for commercial use.[13]

All this notwithstanding, the Department of State's control of arms shipment to China was without legal basis until January 31, 1922, when the 67th Congress of the United States approved a joint resolution entitled, "To Prohibit the Exportation of Arms or Munitions of War from the United States to Certain Countries, and for Other Purposes."[14]

President Warren G. Harding fulfilled the resolution by issuing, on March 4, his Proclamation No. 1621 declaring it unlawful to export arms or munitions of war to China.[15] Under the joint resolution and the president's proclamation, not only arms and munitions but also airplanes, all airplanes, and aeronautical equipment and parts were not permitted to be exported from the United States to China. All aircraft were defined as arms. Thus instructed, the Department of State reverted two weeks later to its original position regarding all aircraft as within the definition of arms and munitions, and therefore unexportable without department permission. It informed the Department of Treasury to that effect.[16]

Since other countries were still exporting airplanes to China, the Department of State was reluctant to forbid exporting commercial airplanes from the United States. The Slevin case was one of the examples. On May 10, 1922, James Slevin, a U.S. citizen and a representative of the Curtiss Aeroplane Company, concluded an agreement with Chinese businessman Lin Jingjin to operate a commercial airline in the territory around Fuzhou. By this agreement, Curtiss would sell to the Fuzhou Motor Car Company 12 aging J.N.4 D 2 planes equipped with OX engines, which were war surplus, and the Great China Air-Way Company would use those planes in establishing air routes.[17] Informed that these 12 aircraft were intended solely for legitimate commercial purposes, the State Department quickly issued a permit authorizing the Curtiss Company to ship to Slevin those airplanes with OX engines. The American Legation in China opposed the permit on the grounds that the airplanes fell under the prohibitions of the president's proclamation. The State Department informed Curtiss that it had decided to cancel the permit, in spite of the fact that half of the planes had already been shipped to China. The State Department subsequently asked the Curtiss Company not to deliver the remainder of the planes. When the initial six airplanes were delivered to Shanghai, they attracted the attention of other governments, especially Great Britain and Japan. Beginning in 1921, those two countries had refused to approve applications of their nationals for the export of aircraft to China.[18]

The upshot of this dilemma came in October 1922, when foreign diplomats met in Peking to modify the embargo agreement and to discuss problems of shipping aircraft to China. The British and Japanese ministers announced that their governments approved the inclusion of aircraft in the embargo. The American minister supported this view, but the French representative insisted that commercial aircraft be exempt. In the end, the diplomats could agree only on the following interpretation to be added to the China Arms Embargo Agreement: "This is understood to include aircraft other than commercial aircraft."[19]

After 1922, the Department of State declined to approve any application to export any airplanes to China. James Slevin applied again in 1923 for a license to send 24 Curtiss airplanes to China. Secretary of State Charles Evans Hughes declined his request after receiving a telegram from Jacob Gould Schurman, American minister in Peking, who opposed granting the permit.[20]

Although the Department of State in cooperation with the British government tried its best, it seems, to execute the embargo agreement and nagged the other countries to implement it, the embargo had little effect. Large quantities of arms and munitions, including airplanes, were still delivered to China, openly and secretly, from a variety of countries, including the United States. The civil wars continued in spite of the establishment of the Nationalist government in Nanjing. The failure of the China arms embargo may be attributed to the powers' competition for the Chinese market, the large profits derived from the sale of munitions, and the Chinese urgent demand for aircraft.

REMOVAL OF THE EMBARGO ON AIRCRAFT

By the later 1920s, the overall Chinese demand for airplanes steadily increased. The first stimulus was that China began developing commercial aviation. Although China was a country of vast distances between great cities and a large population center, there were only about 6,000 miles of railways, and those were in a sadly depleted condition. Roads were still very poor, and paved interurban highways simply did not exist. It would have required millions of dollars and taken many years to develop adequate railroad and highway networks. Airplanes were a promising solution to China's transportation difficulties because the terrain and climate, in general, were highly favorable to cross-country flying, and the investment seemed less than the alternatives and the lead-time relatively short.

As early as the summer of 1921, the China Aviation Bureau, a government agency, opened two short commercial routes (Peking-Jinanfu and Peking-Beidahe), in which passengers, mail, and cargo were carried. Both routes, however, had to be abandoned within several years because of financial problems. Nonetheless, the successful development of commercial aviation in other countries led Chinese governmental leaders to remain convinced of its aviation's prospects in China. Since China had not yet established an aeronautic industry, importation of foreign airplanes continued to be seen as an urgent need.[21]

A second reason the Chinese craved airplanes was the civil war. The Nationalist government, actually controlling only a small part of the immense country of China, attempted to unify China by force of arms, which only served immediately to provoke great resistance from local warlords who were determined to maintain their own bases and armies. A series of civil wars between the central and local governments took place in 1928 and 1929. By that time, not only had the Nationalist government come to appreciate the important role of aircraft in unifying China, but also the warlords had perceived air power as critical to their survival. While the China arms embargo was still in force abroad, military airplanes were not easily available from other countries, so both the Central government and the warlords yielded to the ruse of purchasing commercial airplanes and adapting them for military purposes. Not surprisingly, a sudden and great interest on the part of the new Chinese authorities in purchasing airplanes occurred with the renewal of the civil wars.[22]

With the increased demand for military and commercial airplanes in China, foreign aircraft manufacturers and their governments were attracted to the market. The British government was the first to remove commercial airplanes from articles listed in the arms embargo in the summer of 1926. The American minister then in China, John MacMurray, reported that information to the Department of State at the early July,[23] and on August 5, the British Legation formally advised him that the following instruction had been issued to the British consular officers: "It has been decided by His Majesty's Principal Secretary of State for Foreign Affairs that the importation of aeroplane for commercial purposes should not be regarded as coming within the scope of the 1919 arms embargo of the King's regents [regulation]."[24]

Representatives of British aeronautic interests soon began to sell airplanes to China. The British De Havilland Aircraft Company opened a showroom in Shanghai with a Gypsy Moth airplane on exhibition. In late 1928, the local representatives of De Havilland secured an order from the Aviation Bureau of Nanjing for four Gypsy Moth training aircraft, marking the first time in seven years that the Chinese ordered British aircraft.[25]

Similarly, in the summer of 1928 the Messrs. A. V. Roe and Co. Ltd. of England began to negotiate with the Canton Aviation Bureau for the purchase of its airplanes. After months of exhaustive negotiations, a contract was signed in January of the next year for 14 Avro two-seater light planes, completely supplied with spare engines. This was the first time that British airplanes were sold to the Canton government. Because the Canton Aviation Bureau was a military unit, there could have been little doubt those planes were intended for military purposes.[26]

France was not to be left behind. Particularly, French businessmen increased their sale of airplanes for military use to provincial governments. The French had no difficulty in delivering airplanes to China because Paris, after signing the 1919 agreement, had been pressing for the exemption of commercial aircraft from the China arms embargo. Before France's airplanes were delivered to Canton, however, John MacMurray had a conversation at the beginning of 1928 with the head of the Canton Aviation Department. The Canton official confirmed to him that about 14 French planes would arrive in Canton within the month, but that he would rather have purchased those airplanes from the United States had that been possible. In a telegram to his secretary of state reporting his conversation with the provincial official, MacMurray asked: "How do the French get away with this, to disregard the present arms embargo?"[27]

The American consul general at Canton confirmed later to the Department of State that seven of the 17 French airplanes purchased by the Canton Aviation Bureau had arrived on April 26. The local French consul assured the American that he had known nothing about the order for the airplanes until he was called upon a few days before to send an officer down to survey some of the machines. The Frenchman insisted that those machines were commercial types only and that they were to be used solely for commercial purposes. Credulous though the French consul may have been, he did not believe that those airplanes could be regarded as infringing upon the arms embargo agreement. The U.S. consul general, Douglas Jenkins, insisted that those planes were intended all along for military purposes because the numbers were too large for commercial use at Canton without very serious financial losses. He supported his suspicions with a Canton newspaper report that regarded those airplanes as having been purchased by the Canton government for military purposes.[28]

The influx of these British and French craft into China aroused the attention of American airplane businessmen, and they sharply attacked the China arms policy of the Department of State. In its *Aircraft Yearbook* for 1928, the Aeronautical Chamber of Commerce of America held that:

> The American embargo is severe enough to particularly close the market to American aircraft industry. . . . Under these conditions American manufacturers are severely handicapped in competition with European constructors whose governments place no such restrictions upon their activities, but go far towards facilitating successful negotiations through direct aviation missions with aircraft and demonstrating personnel paid for by their governments.[29]

Stimulated by the aircraft deliveries, the complaints of its agents in China, and remonstrance from the U.S. industry leaders, the State Department undertook to reexamine its stand. Consul Jenkins first drew the attention of the American Legation in China to the possibility of lifting the arms embargo on commercial aircraft, holding up the case of the 17 French planes that had been purchased by the Canton Aviation Bureau. He also stated that the British government no longer opposed the exportation to China of nonmilitary aircraft. The American Legation supported that suggestion and, in turn, brought the matter to the attention of the Department of State.[30]

Minister MacMurray then asked the Department of State to lift the arms embargo on commercial aircraft. On June 20, he telegraphed Frank B. Kellogg, secretary of state, stating: "I venture under present circumstances to request the Department's reconsideration of the question of including commercial airships under our interpretation of the arms embargo agreement." Several days later, Kellogg, on his own authority, decided that the Department of State would grant permits to export commercial airplanes from the United States to China.[31]

Kellogg clarified the new policy of the State Department in a letter to influential Senator Hiram Bingham.[32] In order to fulfill the policies and duties arising from the arms embargo understanding of 1919 and the President's Proclamation of 1922, the secretary of state had required that persons or firms desiring to export airplanes to China should explain the circumstances of the transaction and should apply for an export permit. This precaution had been rendered necessary by the fact that in the recent civil wars in China commercial airplanes were adaptable, after little or no alteration, to warlike uses.[33] Kellogg further indicated that in the case of commercial-type airplanes and in the absence of circumstances necessitating caution, permits for export would be granted immediately following the submission of the required application.

Under that policy, American persons or firms desiring to export commercial airplanes to China who explained the circumstances of the transaction could apply for export permits, the paperwork still regarded as necessary to prevent the export of airplanes clearly intended for military use. In other words, when the planes were classified as "munitions of war," export was forbidden according to the President's Proclamation of 1922, but when they were allegedly for nonmilitary use, presumably they could be exported.

The State Department was keenly aware that commercial aviation in China was just beginning, and the country did not immediately need many commercial airplanes. There was a great likelihood that most of the aircraft imported would be used in military or military-support roles in the civil wars. The State Department was conflicted because it did not want to see American businessmen lose profits from the China market. Therefore, it felt compelled to lift the embargo on airplanes. Business realities triumphed over the political goal of fostering a united and strengthened China.

AMERICAN AIRPLANES OVER CHINA

The unilateral revocation of the ban on the export of commercial planes by the U.S. government in June of 1928 opened a new market for American aircraft manufacturers and marked a new epoch in the history of Chinese aviation. Oddly, Canton, rather than Nanjing, served as the venue for the earliest sales of U.S. craft in China after the embargo lifting. The first modern American airplane delivered to China since 1922 reached Canton on October 9, 1928, a plane purchased by General Zhang Weizhang. Zhang, director of the Canton Aviation Bureau, had received his aviation experience in the United States. Inspired by the news of Charles A. Lindbergh's solo trans-Atlantic flight in 1927, Zhang was anxious to acquire an American plane similar to the *Spirit of St. Louis*, and approached an official of the Department of Commerce. After considerable cabling back and forth between China and the United States, Zhang placed an order with the B. F. Mahoney Aircraft Corporation of San Diego for a Ryan Brougham plane equipped with a Wright Whirlwind engine similar to the *Spirit of St. Louis*. When that plane was shipped to Canton, Zhang appropriately named it the *Spirit of Canton*, and assembled it with the help of a fellow officer who had also received some flying training in the United States. A second Ryan Brougham, identical to the first monoplane, arrived in Canton a little over a month later.[34]

The general took his Ryan plane on a dramatic tour of China, northward from Canton to Hankou, east to Nanjing, north again to Peking and Shenyang, and then southward, by way of Tientsin and Shanghai, back to Canton by November 20. That successful flight established Zhang as the "Lindbergh of China," and, subsequently, he became the director of the Central Government Aviation Bureau in Nanjing. The flight also awakened much wider interest among the Chinese in aviation and in American aircraft. China's aircraft market was now opened for American airplanes.[35]

To sell more Ryan airplanes to the China market, Mahoney-Ryan appointed the Lloyd E. Gale Company as agent in China. Gale first visited China in 1914 as a civil engineer with the Standard Oil Company, and later joined the American Trading Company, managing its Hankou branch for five years. When that organization closed the office, Gale took over many of its principal agencies and began his own business. Later, he specialized in motor vehicles; he came to own the finest motor car service station in China. The Gale Company was the first American trading company in China to represent directly a U.S. airplane manufacturer.[36]

In early December 1928, Lloyd Gale contracted with the Wuhan Civil Aviation Company, Ltd., Hankou, to purchase five Ryan airplanes manufactured by Mahoney Air Corporation.[37] These five Ryan planes consigned to the Wuhan military authority arrived in Wuhan in mid-February of the next year. On the first test flight, Mrs. Gale entered Chinese history as the first female to make a flight in China.[38] Shortly, the Gale Company succeeded in signing a second contract, this one with the Canton Aviation Bureau to purchase five Waco airplanes at a total cost of $55,125. About the same time, the local military authority of Yunnan province also purchased two Ryan-Mahoney airplanes for $34,000. And finally, nine planes, representing an expenditure of $118,000, were purchased by the Chinese authorities in South China during that spring.[39] The hopes of American manufacturers were quickly fulfilled.

Americans were successful in opening the aircraft market in South China because a handful of Cantonese pilots were familiar with American aircraft. The United States had previously sold 11 American-made seaplanes to M. Ricou, French financier in Macao in 1919. Those planes were used briefly for the short run between Macao and Hong Kong. Later, the American pilot and former naval officer Walter D. Bonner arrived in Hong Kong to take charge of operations. Owing to a shortage of financial support for an airline at that time, the project collapsed a year after its inauguration; the airplanes were purchased by Chu Chewen, director of the Canton Aviation Bureau, and moved to that city up the Pearl River. In Canton, the airplanes were converted from seaplanes to land and absorbed eventually into the Canton Flying School.[40]

In addition, about this time four Curtiss JN-6 planes, formerly used by the U.S. Army, were smuggled into Canton from Manila. Those planes were purchased by the Cantonese in Manila and entered Canton through the port of Guangzhouwan, a

French-leased territory on the Leizhou peninsula in Guangdong province. In 1928, the Canton authorities were still using those twelve-year-old planes equipped with OX engines.[41]

The second reason for the leading role of Canton was that most of the Cantonese aviators received their training in the United States. General Zhang Weizhang was trained in America. General Wang Binghang learned to fly in Emeryville, California, just north of Oakland. Wang Yupei and Wang Yuquan, both of whom were born in nearby Marysville, trained at a commercial school in Chicago in 1921. Captain Luo Ping learned to fly in Chicago in 1921 and returned to China in 1922 with Wang Yupei and Wang Yuquan.[42] Because of their training in the United States, the Cantonese aviators nurtured natural preferences for American airplanes with which they were familiar. Furthermore, American diplomatic officers in Canton proved extremely solicitous in assisting American businessmen to open the China market. Consul General Jenkins was one of these. By 1928 he took a keen interest in the potential of the South China market for American commercial planes and used every opportunity to establish friendly contacts with General Zhang Weizhang and other officers connected with the Canton Aviation Bureau. The sale of the five Waco planes to the local authorities was but one example of American diplomatic officers assisting American businessmen.

In November 1928, the Canton Aviation Bureau selected the Advance Aircraft Company of Troy, Ohio, to supply 10 Waco-10 planes, intended for training commercial and military pilots in connection with the Canton Flying School operated by the Canton Aviation Bureau. It was Consul Jenkins who cabled the Department of State asking for a bid on those planes. After receiving Jenkins' cable, Wilbur J. Carr, assistant secretary of state, passed it on to the Department of Commerce, which in turn contacted the Advance Company. Only a few days before Christmas Jenkins received a telegram of quotations and transmitted it promptly to the Canton Aviation Bureau.[43]

Several months later, in March 1929, Earl Baskey, as a representative of the L. E. Gale Company, arrived in Canton with a letter of introduction from the American consul general at Hankou, Walter A. Adams. Baskey, an erstwhile squadron commander with the U.S. forces during World War I, associated himself with the Mahoney Aircraft Corporation after the Great War.[44] During the negotiations, Jenkins transmitted Baskey's telegram in the gray consular code through the Navy Department to his superior principal at Hankou when Baskey failed to communicate in the normal way. The Canton Consulate also put Baskey in contact with the local branch of the National City Bank of New York through which agreements were finally made for the issuance of letters of credit to facilitate payment. Finally, on April 2, the Canton Aviation Bureau entered into an agreement with the L. E. Gale Company of Hankou for the purchase of five Waco airplanes. The assistance of U.S. diplomatic officers in China clearly proved highly timely in helping establish the China market for American aircraft.

U.S. airframe companies did not, however, neglect the Central government in Nanjing. Clement M. Keys, the head of the Curtiss-Wright complex, sent his associate R. O. Hayward to China in December 1928 to discuss the development of commercial aviation in China with Minister of Railways Sun Fo. During the negotiation, Hayward informed Fo that funds would be available for Nanjing in return for a monopoly on all mail and passenger service in China. After the Nanjing government promised not to contract with any other party for commercial air service before April 1, 1929, Keys decided to send Major William B. Robertson to China. As the president of Curtiss-Robertson Airplane Corporation, Robertson led a party of aviation experts to Shanghai in February 1929 to study China as a market for American airplanes. Robertson had given Charles Lindbergh his first job as a pilot and had been instrumental in raising funds for Lindbergh's historic flight across the Atlantic. His party, which labeled itself the Aviation Exploration Corporation, was a subsidiary of the Curtiss group of companies, the largest combination of aviation interests in the United States. The corporation represented a number of companies, such as the Curtiss Aeroplane and Motor Corporation (New York); the North-American Aviation Corporation (New York); the Curtiss-Robertson Airplane Corporation (St. Louis); the Curtiss Caproin, Incorporated (New York); and others.[45]

The Aviation Exploration Corporation, backed by a powerful group of American bankers, included aeronautic pioneers, such as Commander Roland R. Riggs, president of the Transcontinental Air Transport; former United States naval officer and legal adviser to the Curtiss companies James Wilson, who consulted with the Chinese government for more than 15 years; and E. L. Sloniger, former air mail pilot. They set up their offices in Shanghai. When Major Robertson was asked about the nature and extent of the party's inquiries in China, he replied to a representative of the *China Weekly Review*, "We shall probably be here for several weeks. During this time we expect to visit as many Chinese as possible who are interested in aeronautical industry and to determine how American airplanes can best satisfy China's needs in this respect."[46]

The results of the market probing developed quickly. On April 17, 1929, the Aviation Exploration Corporation signed two contracts with Sun Fo, minister of railways and president of the China National Aviation Corporation. The first contract, the so-called "Air Mail Contract," provided that the American company agree to carry mail for the Nanjing government on three trunk lines, with service to begin within six months. The proposed three lines would connect Nanjing with Peking, Canton with Hankou, and Shanghai with Hankou via Nanjing. The second contract, "Contract for the Establishment and Operation of Flying Schools, Factories, and Aero-Transportation," was "to establish and operate flying schools at appropriate times and places throughout China, subject to the previous written authorization of the Corporation. . . . The Company offers to co-operate with the Corporation in every way in the establishment of aircraft factories in China."[47]

Robertson then announced that the Aviation Exploration had received the additional privilege of engaging in the transport of passengers and freight on its own account. Americans would open schools immediately to train Chinese pilots and manufacture airplanes and equipment in China initially. The success of the Curtiss interest in commercial aviation in China paved the way and provided critical experience for its later involvement in Chinese military aviation. To facilitate these extensive investments, Ernest B. Price, former United States consul in Nanjing, organized a new corporation, the China Aircraft Company, for the Curtiss-Wright interests.[48]

Although all of the airplanes imported from the United States and other countries were civil or commercial planes, Chinese military authorities converted many of them directly and immediately for military purposes and employed them in fighting during that time. The American consulates general at Canton, Hankou, and Tienjing duly reported those facts to the Department of State. Douglas Jenkins reported to the Department of State that the first American commercial plane purchased by the Canton government had been used for semimilitary activities, such as the dissemination of propaganda and the transmission of dispatches.[49] In April of 1929, Jenkins further reported that the Canton Aviation Bureau was remodeling many of the commercial airplanes recently purchased with a view to using them in the event of hostilities between the Cantonese troops and the Nationalist Army. The Ryan-Mahoney planes acquired only a few months earlier from the United States had been fitted with machine guns.[50]

The Department of State, free from its principle of suppressing civil strife in China, was concerned about the conversion of commercial airplanes to military use and attempted to dissuade American businessmen from selling airplanes directly to the Chinese military authorities. Whether their representatives were halfhearted and unpersuasive or whether the aircraft assembling firms were so marginal and fragile in the postwar period is uncertain; what is certain is that those companies persisted in expanding their opportunities in China. The case of the Challenger Robin and Falcon airplanes is instructive.

In March 1929, the Aviation Exploration Cooperation sent four different Curtiss planes, the Challenger Robin, the Falcon, the OX-5 Robin, and a flying boat, from San Francisco to Shanghai consigned to the Chinese Ministry of Communication. These were to be utilized for "demonstration" purposes. Sun Fo asked Robertson at the end of the month to sell or to lend the planes to the Chinese government, to which Robertson replied that he had no objection if C. M. Key, president of the Aviation Exploration Corporation, permitted that arrangement. Immediately, Robertson dispatched a cable to Key. On the next day, Robertson received an answer: “Department of State permission to ship airplane was on the basis of commercial operations with our control. Cannot change without recommendation of MacMurray American minister.” On the same day, the company dispatched a telegram to MacMurray, soliciting his recommendation.[51]

R. S. Norman, American adviser to the minister of communications, arranged a conference on March 31 with Sun Fo; Col. Zhang Qingyu, director of the Aeronautical Bureau of the Chinese Army; and Robertson. At the conference, Colonel Zhang proposed to purchase the four planes for the sum of $51,000, paying 25 percent down and the balance in 90 days. Sun Fo agreed on the basis that he assumed full responsibility for the machines and authorized Colonel Zhang to sign his name to the agreement and take delivery in Sun Fo’s name.

In the course of the conference, Robertson was told that if his company assisted the Chinese government with the sale or use of those planes, it would materially strengthen the company’s position with the Chinese government. Therefore, Robertson agreed to the proposal, but he said that it was still contingent on the approval of C. M. Key. Robertson clearly understood that the situation might lead to a diplomatic controversy if the Chinese government used those planes for war purposes since the titles of those four planes would still be in the Curtiss

Company's name. On April 1, Robertson received an answer from MacMurray through the American consul at Shanghai, saying that MacMurray had no objection to the sale of the airplanes to Sun Fo. The next day, Minister of Finance T. V. Soong, Sun Yat-sen's brother-in-law and close associate of Chiang Kai-shek, called Robertson to come to his home. He explained that instruction had been received from President Chiang Kai-shek to pay Robertson $25,000 in gold as soon as the papers were drawn up and completed. Robertson said that when he heard from Key, the papers would be executed. A conference was held on the next day in R. S. Horman's office. In the course of that conference, Colonel Zhang dispatched a telegram that he had received from President Chiang Kai-shek authorizing the expenditure of $51,000 in gold to buy the Curtiss airplanes. Zhang further told Robertson that if he were to be of any service to the Chinese government, he had to make a decision immediately because the war would end in a few days, and the planes would not be needed after that time. It appeared that the Nationalist government wanted to use those new modern planes for the fighting in Hankou.

Robertson, however, did not receive an answer from Key until April 3. To receive the benefits desired from cooperating with the Chinese government, Robertson decided to lend two airplanes that had been test-flown to the Nationalist government for military purposes. On the same day, E. L. Sloniger piloted the Falcon and a Chinese piloted the Challenger Robin—with Colonel Zhang as a passenger—to Nanjing. On April 6, in response to a request by the Curtiss Company for selling or renting the planes to the Chinese government for military purposes, Henry L. Stimson, secretary of the Department of State, instructed Minister MacMurray in a telegram that "the Department can give no approval and cannot allow itself to be associated with this particular proposed transaction."[52] Several days later, Robertson received information from the American consul general's office that the State Department had refused permission for Robertson's company to sell or rent the planes to the Chinese government. The Chinese military authorities, however, already had two airplanes in hand for the war.[53]

From this case, it is clear that the Department of State tried to use its office to prevent U.S. commercial planes from being shunted to military use once having arrived in China. It is equally clear that the Department was caught on the horns of a dilemma. The State Department could not resolve the dilemma partly because some of its officers were not prepared to choose between the horns, as revealed in a contemporaneous memorandum of the Division of the Far Eastern Affairs: "airplanes, even of commercial type, will in all likelihood be used in warfare, if war breaks out again. However, in spite of the possible risks and disadvantages that are inherent in this branch of our export trade to China, it seems desirable, on the whole, to encourage it as far as possible," because "if the Chinese do not obtain their planes from the United States, they will doubtlessly obtain them from other countries."[54] Here we see exemplified a fundamental predicament not only of the State Department but other diplomatic serves as well: There is no easy resolution when a general political principle conflicts with an immediate and genuine need to advance the cause of the business establishment, a cause to which the department was committed.

The United States took the lead in 1919 in urging other countries to organize the China arms embargo for the purpose of discouraging the civil wars in China, but it shrank from taking a similar role in 1929. As a great power and as a democratic country, the U.S. government accepted the moral responsibility to help the Chinese establish a stable and democratic government and suppress civil warfare. But at the same time economic considerations intruded on the scene. By 1929, the tilt in the State Department appeared to have shifted from primarily moral grounds to practical business considerations. To sell more aircraft to China, the State Department had to consider lifting the China arms embargo.

THE LIFTING OF THE CHINA ARMS EMBARGO IN 1929

The Nanjing government was trying to encourage the great powers to lift the arms embargo on China because it was also in urgent need of Western arms and airplanes while the China arms embargo still prohibited Nanjing from purchasing arms as well as airplanes. The hard reality was that the China Arms Embargo Agreement had never been enforced effectively in its ten-year history. Some signatory governments proved unlikely to prohibit completely their citizens from selling arms and munitions to the Chinese warlords. Nonsignatory countries, of course, remained free to export arms and munitions to China as they chose. Russia, originally a signatory power but later unwilling to recognize that agreement, attempted to expand its influence in China by providing the Nationalist Army with increasing quantities of arms and munitions after 1926. Germany also became a major exporter of arms and munitions to China.[55]

By 1928, the problem of overall arms sales, including military aircraft, to China from other countries had become more serious. Chinese Maritime Customs records show that the total value of arms and munitions of war imported into China increased:

Year	Value in Haiguan Taels
1923	805,036
1924	1,615,711
1925	7,311,584
1926	2,120,476
1927	6,897,716
1928	11,400,315[56]

The major exporting countries in 1928 were as follows:

Countries	Value in Haiguan Taels
Norway	4,868,550
Germany	3,208,897
Poland	2,536,646
Japan	393,005
United States	49,216
Great Britain	34,577
France	19,398[57]

Some of the obvious lessons of these figures were not lost on American businessmen and commercial officials in China. Even prior to the lifting of the embargo on the sale of commercial aircraft by American companies, Julean Arnold,

American commercial attaché in Nanjing, submitted to Dr. Julius Klein, director of the Bureau of Foreign and Domestic Commerce, Department of Commerce, a report that asserted that "The embargo is far from effective and, in reality, shuts out only the arms and munitions which might otherwise have been imported into China from the United States and Great Britain, and directly from certain European countries from whence they are now imported indirectly." Klein concluded that the United States should either abolish the agreement entirely or secure the acquiescence of the powers concerned in making it tight.[58]

Commerce officials were extremely impressed by Arnold's report and discussed the problems among themselves. A few months later, C. K. Moser, an official of the Commerce Department, wrote in a memorandum that on several occasions American businessmen had suggested to him that they were unfairly handicapped in China by a rather stubborn commitment to an ineffectual embargo. For example, a personal letter from a San Francisco business house suggested that the Commerce Department persuade the State Department to issue a permit for export of arms and munitions. Moser finally asked: "Don't you think this is a question in which the Department of Commerce, over the Secretary's signature, might make some observations to the State Department?" An officer of the Commerce Department, in a letter to Julius Klein, added the suggestion that the secretary's attention be called to that matter. "A portion of this profitable Chinese trade in munitions of war would mean much to our dying ordnance industry."[59] From those discussions, some officials of the Department of Commerce were clearly inclined to lift the Chinese arms embargo.

Arnold's report was so important that in March 1928 Klein called it to the attention of the secretary of commerce, then the highly esteemed Herbert Hoover. The secretary discussed the question informally with both the State and the War Departments, and later the subject surfaced in a cabinet meeting in connection with the resolution proposed in Congress, which would require the United States to prohibit the exportation of arms and allied materials to belligerents under certain circumstances.[60]

Not only in the halls of Washington bureaucracy but also in state chambers abroad, the embargo received considerable attention. The French minister in China, in late November of 1928, called a meeting of representatives of the embargo signatories to consider, in the light of the embargo's ineffectuality, whether to cancel or to maintain the agreement. At that meeting in Peking, Britain's Sir Miles W. Lampson, minister in Peking, argued that it would be premature to give up the agreement, and the Italian representative shared this view. The Japanese representative opined that lifting the arms embargo would encourage mass importation of arms into China at levels far higher than were being smuggled in at that time. John MacMurray for the United States suggested that the Chinese government should first request the abolition of the agreement. In the end, unable to reach an alternate agreement, the ministers decided to recommend to their governments preservation of the status quo.[61]

Early in the next year, however, the attitude of Great Britain changed. In January, the London government authorized Lampson to discuss the question of the termination of the embargo with his colleagues in Peking.[62] Lampson informed his American counterpart that in view of the general recognition extended to the

Nanjing administration as the legitimate government of China, London was inclined to suggest cancellation of the arms embargo. MacMurray, it turned out, also favored taking the opportunity to end the embargo, a sentiment favored by Secretary Kellogg.[63]

MacMurray then presided over an informal preliminary discussion with the British, Dutch, and Italian ministers, and the French and Japanese chargés d'affaires regarding cancellation of the China arms embargo. Lampson presented for consideration the other countries' drafts of notes addressed to the Nationalist government with regard to both the Arms Embargo Agreement and the Naval Agreement (1922).[64]

All of the representatives, with exception of the Japanese, now favored cancellation of the embargo. Japan's representative cautiously maintained that it would be premature to take such action until the outcome of the Disarmament Conference in Nanjing demonstrated that a stable government could and would be maintained in China. To reduce their armies, the Nanjing government was holding the Disarmament Conferences with the local warlords at the beginning of 1929.[65] The Japanese apprehension seemed to have been that were the embargo lifted, the Americans and the British would quickly extend their military influence in China by delivering large quantities of arms, including aircraft, to the Nanjing government.[66]

Having considered the opinions expressed in the Peking meeting, Secretary Kellogg repeated his views in a telegram to MacMurray on February 27, that "Department favors immediate cancellation of agreement concerning withholding of naval assistance as well as arms embargo agreement of 1919."[67] A month later, the Chinese foreign minister stated that the Nationalist government had unofficially asked Great Britain and the United States to lift the embargo, but Japan's opposition delayed the action.[68]

The Japanese chargé d'affaires informed MacMurray on April 2 that the question must be discussed with the other interested ministers and approved by the Japanese cabinet, perhaps requiring some weeks. London was no longer willing to wait and decided to take action unilaterally. Lampson received instructions from London to notify his colleagues that Great Britain desired to terminate the arms embargo and the corollary agreement regarding naval equipment, and the British government would withdraw from the arms embargo by itself if the other powers failed to agree unanimously with the British proposal. Under that pressure, Japan caved in and consented to agree to the termination of the embargo.[69] At a meeting of the diplomatic corps 12 days later, the representatives of the powers agreed that a note declaring the cancellation of the arms embargo should be addressed to the foreign minister of China. The representatives of the powers that had participated in the naval agreement also adopted a resolution terminating it on the same day, April 26, 1929.[70]

The China arms embargo, enforced only fitfully for a decade, was finally abandoned. The major reason for Great Britain to lift the arms embargo was economic in nature. At the beginning of 1929, Commander W. F. Barr, an officer of the Chinese Navy, visited London on a secret mission on behalf of the Nationalist government. Barr's visit was to ask Great Britain to lend China the services of two or three naval officers and to seek the aid of Great Britain in reconstructing its navy.[71] The existence of the China arms embargo and the naval assistance agreements, however, prohibited the signatory powers from supplying ships to the Chinese Navy. Since Germany was not a signatory party to those agreements, the Chinese government would probably turn to Germany for assistance; that would have been a great loss to British industry. As Austen Chamberlain, secretary of state for foreign affairs, stated on March 20, in replying to questions in the House of the Commons, as long as the China Arms Embargo Agreement was in force, the British government could not provide assistance in helping the reconstruction of the Chinese Navy. The British government would be very pleased to extend aid as soon as the restriction had been removed.[72] There is no doubt that helping China build a modern navy constituted the major reason for the British government to lift the arms embargo.

The American motive for lifting that agreement was also obvious. MacMurray telegraphed Secretary Kellogg in early 1929: "I am myself in favor of taking the opportunity to get rid of the embargo which has never worked enough to accomplish anything of its purpose but has given unfair advantages to those capitalizing their disregard of it."[73] Stanley K. Hornbeck, chief of the Division of Far Eastern Affairs, Department of State, repeated those sentiments more explicitly in a memorandum nearly three months later. He held that in the opinion of this division, the cancellation of the 1919 Agreement was a development distinctly advantageous. The agreement was one which would not be and which had not been lived up to by certain governments. The agreement "has in fact simply penalized the arms trade of certain countries (especially the U.S.) without measurably restricting the flow of arms into China."[74] It is clear that to compete with other powers, mainly Britain, in the China arms market, the United States had to abandon the embargo.

The abandonment of the China arms embargo meant that effective cooperation among the powers to restrict or control the China arms trade was virtually impossible because each power was certain to pursue its own economic interests in China. Every country was reluctant to lose profits from that potentially large market. For the sake of its economic interest in China, Great Britain took the lead in lifting the China arms embargo. The United States also was no longer interested in undertaking a cooperative policy with other countries, mainly Britain, in regard to the arms sales to China because the U.S. government did not want American businessmen to be put at a disadvantage as compared with other nations' exporters. Therefore, the United States supported Great Britain.

The cancellation of the decade-old China arms embargo had some immediate consequences. First, the Nanjing government was able to purchase foreign aircraft openly, which would consolidate its military power and help to extend the GMD's influence throughout the country. Second, the regional warlords which were independent from the control of Nanjing also tried to purchase foreign aircraft, openly or secretly, to strengthen their forces to resist the Central government's unification movement. Finally, lifting of the China arms embargo would contribute to an intense competition among the major powers for the China aircraft market during the 1930s.

NOTES

1. For a study of the British effort to introduce commercial air service to China after the First World War, see Noel H. Pugach, "Anglo-American Aircraft Competition and the China Arms Embargo, 1919-1921."

2. The Commission on Aviation Affairs of the Nationalist Government, "Hongjun Yangeshi Chugao" [The first draft of the history of the development of the air force], Record Group 787, File Number 581, the Second Historical Archives of China, Nanjing. Hereafter cited as SHAC.

3. "Use of Airplanes in Chinese Warfare," February 13, 1930, Report Number 7659, File Number 2078-102, Records of the War Department General and Special Staffs, Record Group 165, National Archives, Washington, D.C. Hereafter cited as RG 165, NA.

4. "The Progress of Aviation in China during 1929," by J. C. Huston, American consul in Shanghai, March 10, 1930, File Number 893.796/78, Department of State Records, Record Group 59, National Archives, Washington, D.C. Hereafter cited as RG 59, NA.

5. Lawrence M. Chen, "Aviation in China," *Information Bulletin* (Nanking), 3 (May 1, 1937): 255–71.

6. Lincoln C. Reynolds, "Aviation in China," 1935, File Number 893.796/193, RG 59, NA.

7. Anthony Chan, *Arming the Chinese: The Western Armaments Trade in Warlord China, 1920–1928*, pp. 118–19.

8. Ibid.

9. For the background of the China arms trade after World War I, see Zunkong Chen, *Lieqiang tui zhongguo de junhuo jinyun, 1919–1929.*

10. The dean of the diplomatic corps in China (Jordan) to the Acting Minister of Foreign Affairs (Ch'en Lu), May 5, 1919, Department of State, *Foreign Relations of the United States, 1919*, 1: 670. Hereafter cited as *FRUS.*

11. For a scholarly study of the China arms embargo, see Elton Atwater, *American Regulation of* Arms *Exports*, pp. 122–36.

12. For a scholarly study of the Vickers contract, see Pugach, "Anglo-American Aircraft Competition and the Chinese Arms Embargo, 1919–1921," 351–72.

13. Department of State to the American Legation in China, September 17, 1920, File Number 693.119/346, RG 59, NA.

14. For the text of the Joint Resolution, see Department of State, *FRUS, 1922*, 1: 726–727. See also Jacob Schurman, American Minister in China, to the Secretary of State, June 28, 1924, File Number 893.113/711, RG 59, NA.

15. For the text of the proclamation, see ibid.

16. The Secretary of State to the secretary of treasury (Mellon), May 19, 1922, Department of State, *FRUS, 1922*, 1: 728–29.

17. "History of Aviation in China," by Parker G. Tenney, Assistant Military Attaché, Report Number 7557, File Number 2078, RG 165, NA.

18. The Division of Far Eastern Affairs, "Curtiss Aeroplane Contract," June 29, 1922, File Number 893.113/377, RG 59, NA.

19. The minister in China (Schurman) to the secretary of state, October 4, 1922, Department of State, *FRUS, 1922*, 1: 742-3.

20. Schurman to Hughes, April 13, 1923, and Hughes to Schurman, April 17, 1923, File Number 893.113/463, RG 59, NA.

21. *Zhongguo jindai junshishi lunwenji* [A collection of papers on modern Chinese military history] (Beijing: Junshi kexue chubanshe, 1987), pp. 551–52.

22. Ibid.

23. The minister in China (MacMurray) to the secretary of state, July 30, 1926, Department of State, *FRUS, 1926*, 1: 735.

24. MacMurray to the secretary of state, August 5, 1926, Department of State, *FRUS, 1926*, 1: 736.

25. *New York Times*, May 5, 1929, p. 23.

26. "British Aircraft for China," *Times* (London) January 16, 1929, p. 12. See also *North China Herald*, January 26, 1929, p. 145.

27. MacMurray to Department of State, February 15, 1918, File Number 893.113/Airplanes/6, RG 59, NA.

28. Consul General Douglas Jenkins to Department of State, May 7, 1928, File Number 893.113/Airplanes/7, Department of State Records, RG 59, NA. The French Consul was M. Denjou.

29. Aeronautical Chamber of Commerce of America, *Aircraft Year Book, 1928*, pp. 156–7.

30. The minister in China (MacMurray) to the secretary of state, June 20, 1928, Department of State, *FRUS, 1928*, 2: 303.

31. The secretary of state to the minister in China (MacMurray), June 23, 1928, ibid.

32. On September 12, 1928, in a letter to the secretary of state, Senator Hiram Bingham informed the Department of State that "I believe that there will be a good market for airplanes in China and I am anxious to let some of my friends know whether the embargo has been raised," Bingham to the Secretary of State, September 12, 1928, File Number 893.113 Airplanes/10, RG 59, NA.

33. The secretary of state to Senator Bingham, September 24, 1928, Department of State, *FRUS, 1928*, 2: 307–8.

34. *China Weekly Review*, January 26, 1929, pp. 354, 360.

35. Jenkins to the secretary of state, December 12, 1928, File Number 893.113/Airplanes/18, RG 59.

36. *China Weekly Review*, February 23, 1929, p. 526.

37. Walter A. Adams, American consul in Hankow, to the secretary of state, December 18, 1928, File Number 893.113 Mahoney Aircraft Corp./26, RG 59, NA.

38. "American Planes Tested at Hankow; Mrs. Gale Flies," *China Press* (Shanghai), February 20, 1929, p. 1.

39. Adams to the secretary of state, File Number 893.113 Mahoney Aircraft Corp./26, R G 59, NA.

40. Lincoln C. Reynolds, "Aviation in China."

41. Douglas Jenkins, "Aviation Situation in Canton," December 14, 1928, File Number 893.113 Advance Aircraft Co./5, RG 59, NA.

42. Ibid.

43. Jenkins to the secretary of state, December 28, 1928, File Number 893.113/Advance Aircraft Co./14, ibid.

44. Jenkins to the secretary of state, April 6, 1929, File Number 893.113 Airplanes/26, ibid.

45. For an account of the American aircraft businessmen's activities in China in 1929, see Ernest B. Price, "Possibilities for the Operation of Air Transport Lines in China," May 12, 1930, Price Papers, Hoover Institution on War, Revolution, and Peace, Stanford University, California. Price was the former president of the China Airways, Federal Inc.

46. "Major Robertson and Party Studying China as Market for American Airplanes," *China Weekly Review*, February 23, 1929, p. 526.

47. "The Sino-American Air Partnership," April 27, 1929, ibid., pp. 356–8; "China's New Aviation Contracts," *The Far Eastern Review*, May 1929, pp. 197–200.

48. *New York Times*, April 20, 1929, p. 1.

49. Jenkins to the secretary of state, October 23, 1928, File Number 893.113 Airplanes/15 Department of State Records, RG 59, NA.

50. Jenkins to the secretary of state, April 10, 1929, File Number 893.113 Airplanes/27, ibid.

51. For the case of the Curtiss Robin and Falcon airplanes, see "Memorandum—Curtiss Robin in the Hands of the Chinese Military," March 31, 1930, Price Papers.

52. The secretary of state to the Minister in China (MacMurray), April 6, 1929, Department of State, *FRUS, 1929*, 2: 536.

53. "Memorandum—Curtiss Robin in the Hands of Chinese Military," March 31, 1930, Price Papers.

54. Memorandum by the Division of Far Eastern Affairs, Department of State, January 24, 1929, File Number 893.113 Airplanes/23, RG 59, NA.

55. For the scholarly study of the China arms embargo from 1919 to 1929, see James Valone Stephen's dissertation, "A Policy of Such Common Interest: The United States and the Diplomacy of the China Arms Embargo," University of Rochester, New York, 1989.

56. List drawn from "Importation of Arms and Munitions into China," June 10, 1931, File Number F3446/52/10, General Correspondence, Foreign Office Archives, Record Class 371, Public Record Office, London. Hereafter cited as PRO. These figures cover only arms and munitions imported through the Customs at the Treaty Ports and do not include the consignments entering into China overland from Russia. The equivalent of a Haiguan Tael during the year 1928 was 2 11s 6d.

57. Ibid. Since Norway was not a signatory country of the China arms embargo, some countries exported arms and munitions to it and then delivered them to China. As a result, Norway became the major exporter of arms to China during that time.

58. Julean Arnold, "Embargo Against Importation of Arms into China," December 29, 1927, Box Number 968, File Number 222, Records of the Bureau of Foreign and Domestic Commerce, Department of Commerce, Record Group 151, National Archives. Hereafter cited as RG 151, NA.

59. Moser to Domeratzky, February 14, 1928, Box Number 968, File Number 222, ibid.

60. Klein to Arnold, March 30, 1928, ibid.

61. MacMurray to the secretary of state, December 11, 1928, File Number 893.113/1157, Department of State Records, RG 59, NA.

62. See Lampson to Austen Chamberlain, April 30, 1929, File Number, F3463/120/10, China Confidential Print, Record Class FO 405, PRO.

63. MacMurray to the secretary of state, January 30, 1929 and Kellogg to MacMurray, January 31, 1929, Department of State, *FRUS, 1929*, 2: 523.

64. The American minister in China proposed the Naval Agreement to the Diplomatic Body in a memorandum on January 25, 1923, and subsequently other governments concerned accepted it on various dates.

65. See George F. Botjer, *A Short History of Nationalist China, 1919–1949* (New York: G. P. Putnam's Sons, 1979), pp. 101–2.

66. The Minister in China (MacMurray) to the Secretary of State, February 8, 1929,

Department of State, *FRUS, 1929*, 2: 524–5.

67. Kellogg to MacMurray, February 27, 1929, Department of State, *FRUS, 1929*, 2: 525.

68. MacMurray to the secretary of state, April 8, 1929, ibid., 2: 526. The Japanese government opposed lifting the China arms embargo at the time because Japan had not recognized the Nanjing government while the Washington and London governments had. Also, Japan feared that if the China arms embargo were lifted, the United States and Britain would extend their military influences in China against Japan by exporting quantities of arms and munitions because the Nanjing authorities were a pro-America and pro-Britain government. See Lampson to Austen Chamberlain, April 30, 1929, File Number F3463/120/10, China Confidential Print, Record Class FO 405, PRO.

69. Thc minister in China (MacMurray) to the Secretary of State, April 16, 1929, *FRUS*, 2: 527–8.

70. The official note to the Chinese foreign minister read as follows: "The Governments of the Netherlands, Belgium, Brazil, Denmark, Spain, the United States, France, the United Kingdom of Great Britain and Northern Ireland, Italy, Japan, and Portugal, having reviewed the Arms Embargo Agreement of May 5th, 1919 in the light of the changed situation resulting from the establishment of the National Government, consider that there is no longer any reason for the continuance of that Agreement and have decided to regard it as canceled from to-day's date." See the Senior Minister in China (Oudendijk) to Chinese Minister for Foreign Affairs (C. T. Wang), ibid., 2: 529–30.

71. "The Arms Embargo and China," *North China Herald*, April 6, 1929, p. 41.

72. "Question Up in Commons about Raising Embargo on Export of Arms to China," *China Press* (Shanghai), March 20, 1929, p. 7. Subsequent to the abandonment of the arms embargo, Foreign Minister Wang and Sir Miles Lampson, the British Minister in China, signed an agreement on July 1, 1929. The official text of the communique read: "An agreement has been signed by the Chinese and British Governments providing for the training of Chinese naval cadets in England and for engagement by the Chinese Government of a British naval mission to assist in the development of the Chinese Navy. It is understood that the Chinese Government had a comprehensive program for the building of a substantial navy and that it is its intention to construct certain vessels in Great Britain." See *New York Times*, July 2, 1929, p. 1. That announcement came as no surprise because some months earlier President Chiang Kai-shek declared that China intended to proceed with a large naval building program and to train officers with foreign assistance, although he did not indicate what form the program would take.

73. MacMurry to the secretary of state, January 30, 1929, Department of State, *FRUS, 1929*, 2: 523.

74. Stanley Hornbeck, "Embargo on Arms to China," April 25, 1929, File Number 893.113/1178, RG 59, NA.

Chapter 2

The Early Development of the Nationalist Air Force, 1929–1931

From 1929 to 1931, China was going through a difficult period, while the local warlords and Communists were fighting against the Nanjing government. The completion of the Northern Expedition in 1928 contributed to nominal unification of China under the Nationalists led by Chiang Kai-shek. The Nanjing government was not in effective control of the provinces beyond the Lower Yangzi region. The GMD, indeed, efficiently controlled only two provinces and partially controlled eight provinces. National politics had not penetrated into the local areas where landlords and gentry continued to dominate life. Warlords still ruled much of the huge country of China (18 provinces) in 1929: Feng Yuxiang in the northwest, Zhang Xueliang in Manchuria, Yan Xishan in Shanxi province, Li Zongren in Guangxi, and Lung Yun in Yunnan. Yet other warlords or powerful local leaders controlled many other smaller areas. These warlords maintained their own bases and armies and established their own taxation systems. The military strength of the three army groups commanded by Feng, Yen, and Li far exceeded that of Chiang's army. The power of the regional militarists was one of the major threats to Guomindang.

The Communists also posed a threat to the Nanjing government. In the early 1920s, agents of the Russian-dominated Communist International helped to reorganize Sun Yat-sen's Guomindang Party while sponsoring the establishment of the Chinese Communist Party, which demanded fundamental social and economic changes within China. Under the pressure of the Russians, the CCP and GMD cooperated in 1924 and launched the Northern Expedition against warlords in 1926 in order to unify China. After the GMD armies led by Chiang Kai-shek occupied Shanghai in April 1927, Chiang issued orders to execute thousands of Communists in the GMD government and armies. Following the collapse of the First United Front, the Communists began to incite peasants to rebel and launched a series of unsuccessful armed uprisings against the Nanjing government. By 1929 they had established Red Army bases in Hunan, Jiangxi, and Anhui provinces. The bases, only a few hundred miles from Nanjing, continued to expand rapidly and attracted many peasant converts.

In the face of that threat, Nanjing, which was eager to bring the country under the GMD flag, determined to suppress the rebellions by sending in the Nationalist Army. The prospects for the Nationalist government to dispatch military forces to destroy the warlords seemed dim, in part because railways and roads were scarce and China lacked other modern means of transportation. It was also difficult for the Nationalist Army to eliminate the Communist bases because most of the bases were situated in inaccessible and easily defended mountainous areas.

The advantages of having an extensive and efficient air force became increasingly apparent. A modern air force might help the Nationalist Army to destroy both the local warlords and the Communists and to extend the Central government's influence to more distant provinces.

It is not surprising that Nanjing formulated a comprehensive scheme to that end. In February 1929, the Nanjing government announced that in view of the air force's importance for the country's defense, the Aviation Department of the Ministry of War established in November 1928 would be reorganized and expanded. At the time, the Aviation Department had only five squadrons of five to nine aircraft each, but the pressure mounted to secure more aircraft. In another two months, the department, renamed the National Aviation Administration, secured independence from the Ministry of War, under the direct control of the Nationalist government council in time of peace; it would be transferred to the commander-in-chief of the Army, Navy, and Air Force in time of hostilities. The designation "Air Force" was formally used by the Chinese for the first time in April 1929. By that time, the Aviation Administration had grown from six sections to three departments and nine sections. These administrative changes marked a necessary step in the reorganization of the Nationalist Air Force.[1]

In 1930, the Aviation Administration of the Nationalist government had only four squadrons, with a total of about 70 obsolete airplanes, and the Nationalist Air Force was of no military importance. A U.S. Army Air Corps captain commented in late 1930, "The Air Force of the Chinese is small. The equipment is not uniform and aircraft are old types with the exception of a few recently acquired. The military efficiency is poor. These forces could be quickly destroyed by one pursuit squadron."[2]

To speed the process of the development of aviation in China, a five-day National Aviation Conference was held in Nanjing beginning on April 20, 1931. More than 70 officers represented the Central Headquarters of the Nationalist Party, the government itself, various committees and ministries, and a number of provincial and urban political organs. In the opening ceremonies, President Chiang emphasized the importance of aviation in communications and pointed out that airplanes not only would be effective in war but also would play an important part in the development of the country's industries.[3]

General Zhang Weizhang, the overall director of the Aviation Department, set the conference agenda in a lengthy memorandum containing a number of suggestions. After extensive deliberations, the conference adopted five major recommendations:

1. Expansion of the Air Force
2. Setting of a definite policy for aviation administration
3. Development of an airplane industry
4. Floating of a national aviation loan of 50 million U.S. dollars
5. Construction of airfields and landing strips throughout the country [4]

The conference served as a milestone in the history of military aviation in China. The Aviation Administration lost no time in implementing the broad recommendations in several different ways:

1. Reorganizing the existing squadrons
2. Stepping up the training of aviation students
3. Establishing military airplane factories
4. Creating aviation administrations in important provinces[5]

In the years that followed, the Aviation Administration of the Nationalist government built military airdromes in all directions from Nanjing, purchased airplanes abroad, and negotiated with foreign businessmen to establish aviation schools and airplane factories in China. The United States, naturally, became one of the main sources for the development of the Chinese military.

AMERICAN AIR ASSISTANCE TO THE NATIONALIST AIR FORCE

In the post-embargo years, the U.S. Departments of Commerce, War, and State differed in their attitudes regarding sales of military airplanes to China. Since the U.S. government endeavored to develop an aeronautical industry for military purposes in the late 1920s and early 1930s, the American aeronautical industry had to maintain a surplus productive capacity of nearly 50 percent beyond the peacetime needs of the U.S. Army and Navy because if a war erupted, the plants had to begin a construction program for the reinforcement and expansion of the air corps. To solve the problem of surplus military airplanes, the U.S. government pursued a policy of encouraging American businessmen to sell airplanes to other countries, believing any foreign business obtained would alleviate their overhead.[6]

After the outbreak of the Great Depression in 1929, there was not a great demand for aircraft. The domestic demand for military airplanes was not sufficient to maintain the aircraft factories as an economically viable industry. To assist American aircraft manufacturers in aggressively seeking export business, the Department of Commerce established the Aeronautics Trade Division in the Bureau of Foreign and Domestic Commerce in 1929. In April 1930, the U.S. Congress appropriated funds to send aeronautics trade commissioners to South America and to East Asia.[7]

The Commerce Department became interested in the China market. In a report of the spring of 1930 to Leighton W. Rogers, chief of the Aeronautical Trade Division, Edward P. Howard, aeronautics trade commissioner, enunciated a comprehensive view of the military aviation situation in China: "The market for aircraft in China appears at the present writing to be largely military. . . . Any American advantage in the sale of military aircraft should, therefore, be immediately followed up and our position secured rather than relinquished to other nationals."[8]

By the end of 1930, Howard became assistant trade commissioner in Shanghai, and later trade commissioner. The Commerce Department seemed to have considered the China market as especially promising. Howard, a well-known American aviator and aviation technician, was experienced in the organization of aviation schools. He had been an aircraft and engineer inspector with the Bureau of Aeronautics in 1929–1930.[9] Howard not only assisted American businessmen in selling airplanes but also persuaded Chinese authorities to employ American instructors and import American aircraft factories.

Like Commerce, the Department of War approved of selling military airplanes to China. With its encouragement, the aeronautical industry struggled to maintain a surplus productive capacity of nearly 50 percent during peacetime so that the aircraft factories could be equipped quickly for reinforcement and expansion of the U.S. Army Air Force if war came. To keep their factories working at optimal capacity, aircraft factories must sell airplanes elsewhere during peacetime.[10]

The Department of State was a different tale. After the end of the embargo, the Department of State announced a new China arms policy:

1. The Presidential Proclamation of March 4, 1922, regarding shipment of arms to China will be left undisturbed.
2. Exportation of arms and munitions of war for use of the Chinese Government will be authorized upon request by the Chinese Government through its representative here, prospective exporters being required to submit application for export licenses to be approved by this Department.[11]

The department clarified its policy with respect to airplanes in a press release of September 3, 1929: export licenses issued by the department would be required for the shipment of arms and munitions, including "aircraft, when fitted with armor, guns, machine guns, bomb-dropping or other military devices, or mounting for such guns or devices."[12] In practice, this policy required U.S. citizens who wished to export arms or munitions, including military airplanes, to apply for permits first from the Nanjing government. Only after receiving such documents might they apply for export licenses from the State. When the Chinese Embassy in Washington confirmed with the Department of State that the Nationalist government had issued such permits, the Department of State would grant export licenses to the American businessmen. This licensing system was enforced from the time the China arms embargo was lifted by the great powers. Great Britain and other countries also used similar procedures. The Department of State maintained this licensing system because it regarded the political situation in China unstable, and civil fighting continued. Consequently, the department was unwilling to see American arms and munitions, including military airplanes, passing into the hands of local governments without the approval of the recognized central government.

The officials of the Department of State held that granting export licenses to American businessmen without the permission of the Chinese Central government would have jeopardized relations between Nanjing and Washington.[13] As Stanley Hornbeck wrote in a staff memorandum: "The political (military) situation in China is chaotic. The present moment is not opportune for making the exportation of arms from this country to China a question entirely removed from the control of the government and for making announcements that arms may freely and without restriction be exported from this country to China."[14] In addition, it was the policy of the Department of State to discourage the sale of arms and munitions, including military airplanes, to countries engaged in civil wars because, as Hornbeck said, "The Department did not want this country subjected to the possible criticism that Americans were concerned only with the money that might be made out of the arms traffic."[15]

In fulfillment of these policies intended to strengthen the Nationalist government vis-à-vis local authorities, the Department instructed its officials and agents on the scene in China to refrain from active promotion of sales of arms and munitions. This course of action, or inaction, as it were, was expressed in a telegram from the acting secretary of state W. R. Castle to the American consul in Nanjing: the Department did not wish to take an initiative or any active part in the promotion of purchase by the Chinese or sale to the Chinese of arms or munitions, including aircraft.[16] To be sure, U.S. officials should lend appropriate assistance to American businessmen, but they were to avoid becoming involved in serving as an agency in connection with the purchase and sale of arms.[17] Under department instructions, diplomatic and consular officials in China were required, as far as possible, to disassociate themselves completely from involvement in the sales of military airplanes.

No doubt, the varying and even conflicting policies of various departments regarding arms sales to China reflected differences in American arms sale policy. That could be seen clearly when American military airplanes were first sold to China in 1929. On August 24, 1929, General Zhang Weizhang, director of the Aviation Administration, who was the key figure in aeronautics policy in Nanjing, announced that the Nanjing government recently had received $500,000 contributed mainly by overseas Chinese for the purpose of improving the aviation equipment of the Nationalist Air Force. As a result, Nanjing had a total of $1,250,000 which would be spent immediately on aircraft, including the initial purchase of 62 airplanes, the first step in a two-year program designed to reorganize the Nationalist Air Force.[18] On September 9, General Zhang called at the American commercial attaché's office in Shanghai to outline plans formulated by his aviation administration, including purchases of American airplanes.[19]

At that time, a number of representatives of American and European airplane manufacturers were competing keenly for aircraft orders from the Chinese. Gale, the sole agent in China for Corsair planes manufactured by the Chance-Vought Corporation of Long Island, succeeded in drawing the attention of the Chinese military aviation authorities to Vought Corsair military planes and quoted prices that were favorably received.[20] But Gale soon learned that he faced sharp competition.

Bert Hall, an American veteran of the French Air Service during World War I, arrived in China in May 1929. He served as a representative of an airplane manufacturer, but later undertook to serve as aviation adviser to the Nationalist government.[21] Hall, like Gale, submitted quotations for the Vought Corsair planes. Hall's quotations were considerably lower than Gale's, which naturally led Chinese purchasing authorities to suggest that the prices quoted by Gale might be unreasonably high. Therefore, the Chinese authorities decided to deal directly with the Chance-Vought Corporation rather than through intermediaries, the legitimate authority of whom they could not ascertain. They asked the American consul in Nanjing Walter A. Adams to obtain a direct quotation from the company, but before complying with the Chinese request, Adams obtained Gale's consent. On October 5, Adams asked the Department of State to transmit the request of the Nanjing government for a price quotation directly to the Chance-Vought Corporation.[22]

But then, curiously, Earl Baskey, a representative of Gale's company, managed to conclude a contract in October with the Aviation Administration under which the Chance-Vought Corporation would sell 12 Corsairs for about $1,000,000 without waiting for communication directly from the company. The Corsairs, armed with machine guns in the wings and equipped for bombing, were modern planes with a top speed of 158 miles per hour. The first two planes were to be shipped by the third week of November 1929.[23]

After the contract was initiated, the Chinese Legation in Washington asked the Department of State to grant permission to the United Aircraft Exports, Incorporated (the sales subsidiaries of Chance-Vought Corporation) to export the 12 planes. Several days later, the Chinese Legation additionally requested that permission be granted to purchase and install machine guns and bomb racks in the planes and to supply the necessary ordnance. But to mixed feelings on the Chinese side, the department issued a license to ship to China 12 military airplanes without armament equipment.[24]

When Nanjing received the news, the Ministry of Foreign Affairs announced this outcome to local American businessmen: "Export permit for Vought Corsair airplanes ready; but armament equipment has to be purchased from American government who will not sell. State Department suggests purchasing planes without equipment."[25] U.S. consul Edwin Cunningham in Nanjing lost no time in telegraphing the secretary of state, emphasizing, "I venture to express that no difficulties will be placed in the way of this contract, I cannot but view with great concern the very serious effect the cancellation of this contract may have upon the future trade in American airplanes."[26]

Two days after receipt of the consul's telegram, the Department of State felt the need to explain its reason for disapproving of the arming of the craft, and informed the Chinese Legation that "the grant of permission to Chance-Vought Corporation (United Aircraft Exports, Incorporated) to purchase this material is not within the province of the State Department. In this matter, the Corporation should approach directly the branch of the Government concerned." In a near-classic case of buck-passing, the department forwarded a copy of the Chinese Legation's letter of November 5 to the War Department.[27]

The granting of permission to American airplane companies to sell military equipment, arguably, was not within the province of the State Department, at least not the sole province, because guns, bomb racks, bomb-release mechanisms, and the likes were manufactured exclusively for the U.S. military. Accordingly, to sell military aircraft to foreign governments, U.S. manufacturers were required to obtain the permission of War Department, a definition of bureaucratic turf which one of his assistant secretaries reminded the secretary of state of in early December 1929.[28]

Informed that the Department of State authorized only the sale of planes without equipment, Nanjing bluntly threatened to cancel the order, which would amount to 700,000 gold dollars, and to turn to the British, whose government raised no objections to the shipment of such armaments to China. Having already established letters of credit for that purpose, the Nationalist government was in a hurry to obtain military airplanes capable of bombing in what it saw as an impending renewal of the civil war.[29]

As it turned out, though, the War Department also lacked the independent authority to authorize the sale of military equipment to China. In 1923, President Warren G. Harding forbade War Department sales of surplus military equipment to foreign governments or even to American citizens, except under guarantees that the equipment would not be used in civil wars, a policy that the War Department had been urged to implement since its announcement.[30] As in 1919, Harding's administration adopted the view that the civil wars should not be supported or countenanced, partly on practical political grounds of preference for dealing with stable, orderly governments favorable to U.S. interests, and partly on pacifistic, humanitarian grounds.

The War Department favored export of military aircraft by supplying foreign governments with surplus military aircraft and equipment, provided no military secrets were revealed. If the United States were to shrink from such sales, the aircraft industry would be weakened, and rival industries abroad would certainly take up the slack. This judgment was articulated in a memorandum to the secretary of war in early December 1929.[31]

In the case of China, it was obvious that Nanjing would use the Corsairs in the civil war. Were Harding's 1923 order rigidly followed, these planes should not have been exported to China. But the lifting of the China arms embargo left the issue somewhat ambiguous, requiring the State and War Departments to wrestle with it.[32]

Nelson T. Johnson, assistant secretary of state; Trubee Davidson, assistant secretary of war; and Colonel Foy of the War Department struggled to resolve the uncertainty and at the same time define departmental spheres of authority. After lengthy discussions, they failed at their task and decided to ask President Herbert Hoover to allow an exception to Harding's policy.[33]

In an effort to sell surplus military airplane armaments to China, Secretary of War Patrick J. Hurley submitted such request to President Hoover, stating:

American aircraft manufacturers cannot compete with foreign manufacturers in the sale of military aircraft to foreign governments unless the former can deliver their products fully equipped with military equipment. . . . American aircraft manufacturers' needs must, therefore, depend upon the government for such equipment or arrange for its development and manufacture by commercial concerns.[34]

A similar letter was addressed to the secretary of the navy. Chance-Vought Corporation had already shipped three two-seat Navy-type Vought Corsair fighters to the Nationalist government by December 3, and worked overtime to complete the rush order.[35] On January 10, 1930, Hoover formally authorized the War Department to amend President Harding's policy to sell military aircraft equipment to China, replying:

In view of the special situation referred to, and to the end that the American aircraft industry may not be adversely affected in the development of American trade relations with foreign countries, authority is hereby granted to make exception to the policy established by the President's letter of April 23, 1923, in so far as pertains to the selling of surplus military aircraft equipment, that is not an article of common commercial production, to foreign governments or to their agents when the sales of such aircraft equipment have been approved by the Department of State and when such sales will not disclose any military secret.[36]

With the president's authorization, on January 13, 1930, the War Department issued permission to ship military equipment to China and notified the Department of State of that permission.[37] Although the 12 aircraft finally did arrive in Shanghai, the Chinese government was less than happy with the American experience. The delay in delivery caused considerable consternation among the Chinese generals, who were convinced they needed these fighting craft immediately in the war between Nationalists and the warlords in North China. The Nanjing government was displeased with that manner of doing business and threatened to have no further dealings with the Gale Company. The sale of these 12 Corsair airplanes represented the first time that American airplanes equipped for war entered China. The consequences and significance of the sale were not negligible. A precedent had been established, which paved the way for the export of ever-increasing numbers of military airplanes to China. The episode of the Corsairs also graphically illustrates differences in the making of crucial policy on armament sales abroad within governing circles.

After the purchase of 12 Corsairs in 1930, the Nanjing government had been taking steps to expand its air force by importing more American military planes, while at the same time ordering a number of military planes from other countries, notably Germany and Great Britain. After the Corsair purchase, the next early order was 20 Junker pursuit planes from Germany in the next year. When delivered, they proved so inferior to the American planes that most of the Chinese pilots were reluctant to fly them. The American assistant military attaché in China recorded, perhaps a bit self-serving, that all the Chinese pilots had become much afraid of the German planes. And upon the arrival of six more Junker planes, it was unlikely that the pilots could be persuaded to fly them. All pilots and the Aviation Bureau officials anticipated the arrival of 20

American Douglas planes ordered in May, which they had found easy and safe to handle. The Chinese government declined to order any more German-made Junker planes.[38]As for the British airplanes, American consul in Shanghai J. C. Huston commented, "The British have sold a few planes, but some of these, which were of light construction, proved unsatisfactory."[39] Recognizing the superior quality of American airplanes, Chinese pilots preferred to fly them, and American airplanes continued to enter the Chinese market in large quantities. The United Aircraft Exports was one of the major corporations to sell American warplanes to China.

The United Aircraft Exports was licensed in late April of 1930 to export 20 military airplanes, each equipped with two bomb-racks, two machine guns, and all other necessary equipment.[40] These 20 Vought Corsair planes arrived in Nanjing in September, bringing to 32 the number of Vought planes acquired within six months.

American aircraft had become so important in Nationalist calculations that Chiang Kai-shek approved the dispatch to the United States of four high-ranking officers led by the long-powerful Zhang Weizhang on a general mission to visit aircraft factories and to interview U.S. officials. That air mission was charged with placing orders for yet additional airplanes and investigating recent developments in aviation in the United States with a view to improving Chinese military aviation. During the mission, the distinguished group visited Boeing in Seattle, Chance-Vought in New Jersey, and Curtiss factories, and interviewed officers of the Departments of War, Navy, and Commerce.[41]

In June 1931, the Central government placed an emergency order for 20 Douglas bombing-observation planes, each of which could carry two fixed machine guns (one Browning and one Lewis) and two bomb-racks. These 20 bombers were delivered to Shanghai on August 9. The Nationalist government purchased, in fact, a total of 30 Douglas bombers in a single year from 1930 to 1931.[42] In 1931, the total value of American aircraft, engineers, supplies, and accessories imported into China was $1,101,170 in gold. Forty-three airplanes, most of which were military airplanes, were exported to China from the United States in 1931.[43]

With the acquisition of the modern American airplanes, the Nationalist Air Force moved to a higher plateau of effectiveness. The first 12 Corsairs formed the nucleus of the Air Force, augmented by Douglas and additional Corsair aircraft. The total number of serviceable planes attached to the Nationalist Air Forces in mid-1931 was estimated by the U.S. consul general as follows:

Douglas	30
Vought Corsair	20
Junker	15
Others	15
Total	80[44]

The squadron, the largest unit, consisted of nine airplanes each divided into three groups of three planes. Twenty flying officers were assigned to a squadron, commanded by a colonel with a lieutenant-colonel as an assistant.

Three flight commanders bore the rank of major. The commanding officer had no particular plane or flying position; he could fly whichever he chose. By 1931, the Air Force had grown into eight squadrons, seven composed of light bombing or observation planes. The other one consisted of two-seater pursuit planes.[45]

In the local areas yet under the jurisdiction of the Nanjing government, air units were also being developed. For example, in Hunan province, after acquiring a number of American airplanes, General He Jian, chairman of the Hunan provincial government, organized an air contingent force, composed of two Fairchild, two Corsair, two Waco, and two Moth planes, all purchased in 1931. A number of other planes were assigned by the Nanjing government to this province.[46]

Americans not only sold aircraft to the GMD, but also helped to organize the Nationalist Air Force and participated in the air fighting. Several Americans had played important parts in the reorganization of the Nationalist Air Force from 1929 to 1932. Not the least among them was Floyd Newman Shumaker. Shumaker, born in Fullerton, Nebraska, in 1889, was a skilled telegrapher by the age of 18, and was employed in various locations in the West for railroads, newspapers, and stockbrokers. Later, he boarded a ship bound for the Philippines and became an operator at a government relay transmitter in Davao, Mindanao.[47]

The beginning of World War I found Shumaker in Shanghai, working as an electrical contractor, converting buildings from gas lighting to electricity. When the war broke out in Europe, he enlisted in the British Army in Hong Kong, but transferred to the new U.S. Army Air Service in 1917, where he attained the rank of major. He remained in service until 1929, when he returned to Shanghai as a vice-president of the Aviation Corporation of America. He planned to spend a year surveying the China aircraft market for possible aeronautical development, including the establishment of transport lines, the building of airdromes and hangars, and the erection of radio stations. At the same time, he also served as a representative of the Douglas Aircraft Corporation of Santa Monica, California.[48]

After completing his study, Shumaker first suggested to the Chinese officials in February 1930 that civil aviation be brought under the direct jurisdiction of the Ministry of Aviation, rather than the Ministry of Communication. Later, Minister of Communication Wang Penchun worked out a plan that followed Shumaker's advice.[49] Next, Shumaker made strenuous efforts to reorganize Nanjing's air force. He believed that a strong military air force would guarantee the stability of the Nanjing government, and China would be ready for immense growth in commercial aviation.[50] To detail his ranging and far-sighted ideas, Shumaker presented a comprehensive scheme running to 133 typewritten sheets to the Nationalist government, embracing equipment, instruction, and strategic advice. As part of his scheme, Shumaker planned to sell the Nanjing government 72 Douglas observation planes that were also effective for bombing and air-to-ground attack, and capable of flying for eight hours on a full load of fuel. An ellipse of airdromes 100 miles by 400 miles centered on Nanjing was recommended. From these airdromes, a military attack could be made to almost any important part of China proper and the aircraft still could return to base. Later, with the assistance of Shumaker, the Nanjing government hired two Americans who had recently graduated from the U.S. Army Air Corps at Kelly Field, Texas, as a flying instructor and as an engineer officer.[51]

To implement these grand plans, Shumaker, as president of the China Aviation Equipment Company, representing the Aviation Corporation of America, concluded a contract in the spring of 1930 with the Nanjing government for 10 Douglas airplanes. The Department of State swiftly issued to the Douglas Aircraft Corporation a license to export those warplanes, each equipped with two machine guns and bombing equipment.[52] Shumaker, bringing with him success, was appointed aviation adviser to the Nanjing government in May 1930.[53] The U.S. minister then in China carried on conversations with Shumaker at Nanjing concerning the opportunities for sales of American aircraft to China. Among other things Shumaker was of the opinion that the only way China could make a success of air mail would be to put it into the hands of the military authority as a military service.[54]

Shumaker not only tried his best to sell American military airplanes to China and to help the Nanjing government reorganize its air force but also participated in airplane bombing operations by the Nationalist military forces against local warlords. During the civil war, both Shumaker and Bert Hall personally directed operations from the Nationalists' airbases, Shumaker on the front and Hall in Hangzhou. They informed the U.S. military attaché that they were involved in military operations because the Nanjing government was experiencing great difficulty in obtaining sufficient pilots to carry on operation in an effective manner.[55]

On July 14, 1930, the U.S. consul general in Tianjin Mahlon F. Perkins reported to Minister Johnson that "an American named Shumaker is understood to have taken part in the airplane bombing operations of the Nationalist Government military forces in the present civil war, having attempted to damage the Yellow River Bridge on the Beiping-Hankou Railway near Chengzhou. Shumaker is stated to have admitted such participation in bombing operation to persons at Hankou some weeks ago. Shumaker is an aviator."[56] The evidence is incontrovertible that Shumaker participated actively in bombing operations of the Nanjing government's military forces because the Nationalist Air Force did not have enough pilots to carry out operations. He was certainly the first American pilot to serve the Nanjing government in a civil war.

Robert M. Short, a reserve lieutenant in the U.S. Army, was another American who played an important role in the early development of the Nationalist Air Force, albeit not as broadly and dramatically as Shumaker. Before coming to China, Short spent six months at March Field, a primary flying school, but was discharged from the school in 1928 without completing the course. After 1928, he accumulated flying experience privately and became a second lieutenant in the Air Corps Reserves in February 1931. He came to Shanghai as a salesman in the same year to show the Boeing airplanes.[57] The Chinese engaged him in June of 1931 as an expert in the Military Aviation Bureau at Nanjing.[58] On August 17, 1931, the Aviation Administration of the Ministry of War announced that a pursuit squadron would be organized for purposes of bandit suppression (a euphemism for fighting local warlords or Communists) and that an American aviator, Lieutenant Robert M. Short, an ex-American army flier employed by the Military Aviation Bureau for the preceding two months, would be engaged as an instructor.[59] The next six months found him not only instructing Chinese pilots but battling the Japanese who had begun military operations in Manchuria and elsewhere in September 1931. Soong personally asked Short to pilot a recently acquired modern U.S. pursuit plane, and on February 20, 1932, Short engaged three Japanese pursuit airplanes. In 21 minutes of combat, he damaged two of them and escaped unharmed. While he was flying a Boeing pursuit plane in an air battle several days later, six Japanese planes attacked him. This time, he did not escape and was shot down at Langfang, a small village near Suzhou.[60] The Chinese proclaimed Short's supreme sacrifice an act of heroism symbolic of American sympathy with the cause of China. His body was interred with honors in a special plot at the entry to Hongjiao airfield. The State Department informed his wife, Elizabeth, that Short had lost his life near Suzhou while in the service of the Chinese Army.[61] Short went down as the first foreign pilot to die in the Sino-Japanese conflict while fighting for the Chinese cause.

THE ROLE OF THE NATIONALIST AIR FORCE

To bring China under the Nationalist flag, Chiang tried to reduce the power enjoyed by regional rulers. From 1929 to 1931, in the name of national unification, Chiang clashed with several powerful military leaders who revoked their allegiance to him and plotted against the centralization of authority. He fought against his former ally, the Guangxi group led by Li Zongren, and faced a succession of rebellions led by Tang Shengzhi, Feng Yuxiang, and Yan Xishan. The Guangxi clique's removal of Chiang's appointee from the Hunan governor in March 1929 provoked the arrest of the Guangdong governor at the Third Nationalist Party Congress and an anti-Guangxi coup in Beiping, the official name for Peking from 1928 to 1949. War immediately followed but Guangxi forces failed to fight effectively in the Wuhan area, giving Nanjing an easy victory. Realizing that he and the Guangxi clique shared a common enmity toward Nanjing under Chiang, Feng Yuxiang maneuvered his troops in North China to prevent the completion of any Nanjing campaign against the Guangxi remnants. As a result, all the major enemies of Chiang allied together, linking their military effort with the political call of Wang Jingwei for a reorganization of the Nationalist Party. The intention of the alliance was to strike southward from North China, using the forces of Feng Yuxiang and Yan Xishan and to strike northward from Guangxi under Li Zongren, thereby forcing Chiang into a two-front war. With the assistance of the North-Eastern Army led by Zhang Xueliang, Nanjing suppressed all rebellious opposition to the Central government and established its authority in North China.

Although the Nationalist Air Force, in its early stages, was small in comparison with those of many other countries, it did have a great impact on the Chinese scene and did alter the political situation in China. When the feisty warlord Feng Yuxiang rebelled against the Central government in October 1929, a Nationalist military force was dispatched to put down the rebellion, and the Nationalist Air Force helped by bombing Feng's army. The Second Squadron of the Guangdong Air Force was also ordered to help the Nationalist Air Corps.[62] The Nationalist Air Force's bombing proved effective; the command headquarters of Feng's army at Luoyang was destroyed, and approximately 80 army officers of high rank, including the heads of divisions and brigades, were killed. Feng's army was forced to retreat to Shanxi province.[63]

About this same time, in November 1929, the army of warlord Li Zongren of Guangxi province in the south and the Fourth Route Army of Zhang Fakui were deployed to Guangdong province for the purpose of attacking Nationalist-held Canton, while the Nationalist forces concentrated on Henan province in the north. The Nationalist Air Force immediately sent airplanes to Canton to help the Cantonese defend the city. In cooperation with the ground forces, the Nanjing Air Corps and the Guangdong Air Force rained large quantities of bombs on its enemy in the suburbs of Canton. Lacking the support of an air force of their own, Li's and Zhang's armies were defeated and escaped to mountainous territory of Guangxi province, pursued by the Nationalist Army.[64]

During the summer and fall of 1930, the military resources of the Nationalist government were used extensively in suppressing Feng Yuxiang's and Yan Xishan's armies in Henan and Shandong provinces, respectively, and the Air Corps played a prominent part in those campaigns, although at the time they consisted of only 10 or 12 recently designed, American-made light bombing planes and a few obsolete German Junker transport airplanes. The rebels were even more limited, about 30 obsolete primary trainers and a small number of transport planes.[65]

Wang Binghang commanded the Nationalist government's Air Corps. The topography of Henan and Shandong provinces, a great plateau essentially, was exceptionally suitable for the operation of aircraft. The Nationalist planes took off from Guide, Honan province, and bombarded Kaifeng, Chengzhou, and Jinan every day, all cities of great military and strategic importance.[66]

In addition to causing much damage and distress through bombing, aerial support greatly helped in ground battles. On June 6, the Nationalist Air Corps heavily bombed the town of Xuchang. Fan Zhongxiu, the highest-ranking army officer in town, was killed, and his soldiers fled, allowing the Third Regiment of the Nationalist Army to take over the city on June 7 almost without resistance.[67] Clearly, the bombing there led directly to a major land victory. On August 27, the Nationalist government's airplanes flew over Tianjin city and fired their machine guns on the East Station, in what appeared to be a means of demoralizing the enemy troops. Three more planes flew over Beiping and dropped six or seven bombs for the same purpose.[68] After preliminary heavy aerial bombing by the Nationalist Air Corps, the Nationalist Army found no difficulty in occupying the key large cities of Kaifeng and Jinan. Lincoln Reynolds, American vice-consul in Tianjin, commented, "The fall of Kaifeng and Tsinan, resulting in the liquidation of the revolt, was greatly facilitated by aerial bombardments."[69] The revolt of Feng Yuxiang and Yan Xishan was suppressed. The campaign was over in October 1930. The success of that campaign extended Chiang Kai-shek's influence to North China and saved the Central government from a dangerous crisis.[70]

During the campaigns from 1929 to 1930, the Nationalist Air Force had about 40 to 50 airplanes, most of them Corsair and Douglas airplanes recently purchased from the United States. The Nationalist Air Corps dropped 15,209 bombs, photographed enemy defense works, directed the artillery fire, and signaled the infantry by pistol signals to attack. It was the first time in Chinese history that modern military aircraft were used in war.[71] The advantage offered by modern American airplanes enabled the Nationalist Air Corps to control the air and to assist the ground forces effectively. The rebels had only obsolete airplanes that could not defend against attack by American airplanes. That was one of the major reasons for the defeat of the rebels in such a short time. Without the assistance of the modern American planes, the Nationalist government would not have won the victory over the local warlords quickly, and the civil wars would have lasted longer.

After suppressing the rebellions in North China, the Nationalist government lost no time in opening campaigns against the Communist Army in South China, which posed no less a serious threat to the Nanjing government. In October 1930, Chiang Kai-shek undertook the first of what became known as the annihilation campaign to suppress the Communists. Officially designated as "Bandit-Suppression Campaigns," these offensives were initially carried out by local Hunan troops who came increasingly under the command of Chiang's Whampoa forces, which still relied on the militarists in Guangdong and Hunan for assistance.

To help the Nationalist Army, the Aviation Administration sent the First, Third, and Fifth Squadrons to Jiangxi province to support the Nationalist Army. The First and Fifth Squadrons were stationed in Nanchang, the Third Squadron in Changshu, each equipped with nine Corsair or Douglas airplanes.[72]

At the beginning of the campaign, each squadron sent up only one or two planes daily to observe the Communists' operations. Later, the field commanders of the Army, Navy, and Air Force at Nanchang ordered the Air Corps not only to observe the enemy but also to assist the ground forces in attacking the Communists. From then on, each squadron sent airplanes with bombs over the Communist districts for both observation and bombardment. The rest of the airplanes concentrated on bombing the Communists. With three bombers in a group, one group after another continued the bombing for several days. It proved difficult to observe and bomb the Communists because the weather was often overcast and poor, and the Communist Army troops hid in the forests. Communication equipment, such as telegraphic apparatus, was virtually nonexistent, and communication between the aircraft and the ground commanders relied entirely on orders dropped in cloth bags for coordinating operations between the Air Corps and the infantry. All in all, the Air Corps played only a limited role in that campaign, other than helping the infantry to transport arms and munitions, giving signals to the ground force for starting attacks, and observing the operations of the Communist Army. The first campaign was not successful, partly because of the ineffectiveness of the Nationalist Air Force.[73]

A second annihilation campaign against the Communists started in the early spring season of 1931. Three squadrons were sent to the front, each equipped with nine Douglas or Corsair airplanes. As in the first campaign, the Air Corps flew both observation and bombing missions in support of the ground forces. Again, poor weather and the lack of telegraphic equipment frustrated the assault. The second campaign also failed to achieve its objective.[74] And then came a third campaign. Mao Bangchu, now the deputy director of the Aviation Administration, ordered the First Squadron stationed in Nanchang, the Third Squadron in Chanshu, and the Fifth in Ji'an. In July, several bombing runs killed many Communist soldiers, which probably contributed to the flight of Communist units southward. On July 20, the Air Corps bombed the 5,000 retreating soldiers intensively and reportedly killed more than half of them.[75]

The first two campaigns failed. The Nationalist troops suffered heavy casualties in the face of Mao's guerrilla warfare. Air power was not successful in these campaigns. The topography was unsuitable for the operation of aircraft because most of the Communists' bases were in the mountains. Also, the Nationalist Air Corps lacked effective communications equipment, critical in effective air operations, particularly in such difficult terrain. In the campaigns that followed, after Nanjing had acquired more American airplanes, the Nationalist Air Corps attacked Communist Army heavily.

The Chinese Air Force harassed not only internal enemies but also the Japanese. The Great Depression started in New York in 1929 and spread all over the world in the following years. When world trade collapsed, Japanese exports decreased in Western markets. The decline of Japan's overseas trade forced the Japanese to shift their interests to China because it had been an important source of raw materials and a principal market for Japanese manufacturers. To the Japanese, Chiang would threaten their economic interests in Manchuria and China. For this reason, together with the deteriorated economic situation in the world, the Japanese initiated a military offensive against the Chinese authorities in Manchuria on September 18, 1931, and attacked Shanghai in January 1932. When the Japanese military forces were attacking Shanghai, the Nationalist Air Force flew against the Japanese Air Force, perhaps its first such use against foreign aggression. On January 12, 1932, following the outbreak of the Sino-Japanese hostilities at Shanghai, several Japanese planes were brought into action. They bombed Chapel, the North Station, and several other points in and around Shanghai without interference from Chinese aircraft. The Nationalist Air Corps did not take part in the battle because the Chinese officials feared that if Chinese planes were permitted to participate in the fighting in Shanghai, Japanese airplanes would bomb Nanjing, where the Nationalist Air Force was based.

Then, on the first of February, Japanese gunboats shelled Nanjing. As a result, the Chinese Air Force was permitted to move into Hangzhou, and finally to Shanghai. Several days later, Chiang Kai-shek instructed Defense Minister He to send airplanes to attack Japan's Air Force in Shanghai. Chiang instructed, "When our air corps takes part in the battles, our air corps should adopt some strategies to attack Japanese bases in China, in addition to helping our infantry. On the first day, it should attack Japan's Air Force first and destroy its ability.

Later, every day or every two days, it should attack the important military control centers."[76] On the same day, Chiang also instructed the Nineteenth Route Army of the Cantonese, "When the national air corps takes part in the combat, infantry should keep in touch with the air corps closely in order to operate in the same line."[77]

After receiving instructions from Chiang, the Nationalist Air Force dispatched a number of American observation and pursuit airplanes, two small British Lincock pursuit planes, and a few Junker K-47 pursuit airplanes, to the Hongjiao airfield on the outskirts of Shanghai. To increase the strength of the Nationalist Air Force in the Shanghai area, the Guangdong Air Force was ordered on February 10 to send the Second Squadron to Shanghai. The Chinese Air Force thus had about 30 planes in the front; Japan had 32 airplanes.[78]

On February 5, an aerial battle was fought over Shanghai—four Chinese planes against four Japanese planes. One Chinese airplane was damaged. On February 26, six Japanese bombers and nine pursuit planes appeared above Jianchiao Airport, Hangzhou. Before the Chinese airplanes were fueled and ready, the superior force of Japanese planes began bombing the airport, destroying three Chinese trainer airplanes and wounding two Chinese mechanics. Ten Chinese airplanes took off immediately. A genuine aerial battle began and lasted for half an hour. After the combat, the Bureau of Aviation of the Chinese government claimed no loss of airplanes and lives. Two Japanese airplanes were shot down.[79] On the same day, another aerial battle broke out in the air above Shanghai. The Chinese pilots flew one Corsair and one Junker against 15 Japanese airplanes. One Japanese airplane was shot down; the two Chinese airplanes were badly damaged. After that battle, the Chinese Air Force was ordered to leave Shanghai because of its heavy losses. The Japanese Air Force controlled Shanghai.[80]

In summary, after the Chinese arms embargo was lifted the National government began to develop its air force in an effort to strengthen its military power. The Nationalists sought Americans' help; American aircraft manufacturers were very interested in the aircraft market in China, selling many war planes to the Chinese. In Washington, the U.S. Departments of Commerce and State differed in their attitude regarding sale of military aircraft to China. The Commerce Department encouraged American businessmen to sell aircraft to the Chinese because its task was to promote the sale of American products to other countries, especially during the Great Depression. But the State Department discouraged the sale of warplanes to China because China was engaged in civil wars. As a result, there was a clash between the State and the Commerce Departments over the sale of aircraft to China in the midst of the Great Depression. The disagreement continued in the following years.

Between 1929 and 1931 Americans began to help the Nationalists develop military aviation by selling warplanes, offering advice, and participating in the civil strife against the warlords. Despite limited American assistance, the Nationalist Air Force played a part in helping the Nationalist Army in suppressing the local warlords' rebellions in North China, but it had only a limited role in the military campaigns against the Communists. Since the aircraft carried limited bomb loads and delivered them with marginal accuracy, the main role of aviation at this point during that time was psychological. After the Shanghai Incident, the Nationalist government analyzed the reasons for its defeat and further realized the crucial role of airplanes in defending China. For this reason, the GMD government decided to accelerate the buildup of its modern air force. With the help of American businessmen and advisers, that air force reached its peak in size before the Sino-Japanese war.

NOTES

1. See the Commission on Aviation Affairs of the Nationalist government, "Hongjun yangeshi chugao," SHAC.
2. Dache M. Reeves to Commanding General, Philippine Department, Manila, October 27, 1930, Box Number 481, Office of the Adjutant General Central Files, 1926–1939, Records of the Adjutant General's Office, National Archives, Washington, D.C. Hereafter cited as RG 94, NA.
3. "Nanking Aviation Conference," *North China Herald*, April 21, 1931, p. 75.
4. "Aviation Conference-Proposed for United Air Ministry," ibid., April 28, 1931, p. 112.
5. *Junshi zhazhi* [Military Magazine], 16: 187 (1929).
6. For the account of the history of the aeronautic industry in the United States, see Barry Bluestone, *Aircraft Industry Dynamic: An Analysis of Competition, Capital, and Labor* and William Glenn Cunningham, *The Aircraft Industry: A Study in Industrial Location* (Los Angeles: Lorrin L. Morrison, 1951).
7. "Congress Providing Funds for U.S. Aviation Trade Commissioners," *China Weekly Review,* April 12, 1930, p. 259.
8. Howard to Rogers, May 29, 1930, File 560, Box Number 2503, RG 151, NA.
9. Nelson E. Margetts, Military Attaché, to the War Department, January 15, 1931, Report Number 7923, File Number 2078, RG 165, NA.
10. Aeronautical Chamber of Commerce of America, Inc., "Report of Sale of Aeronautical Equipment to China," File Number 893.113 Airplanes/117, RG 59, NA.
11. Secretary of state to MacMurray, April 27, 1929, Department of State, *FRUS, 1929*, 2: 530.
12. "American Policy with Respect to the Shipment of Arms and Munitions of War to China," June 10, 1930, File Number 893.113/1265, RG 59, NA.
13. Memorandum of conversations among Samuel Stone, president of the Colt Patent Fire Arms Manufacture Company, Stanley Hornbeck, and others, May 12, 1930, File Number 893.113/1245, RG 59, NA.
14. Memorandum by Stanley Hornbeck, April 25, 1929, File Number 893.113/1178, ibid.
15. Conversation between Colonel Simone of the Du Pont Company and Stanley Hornbeck, August 6, 1929, File Number 893.113 Du Pont Co./9, ibid.
16. The secretary of state to the consul at Nanking (Adams), October 12, 1929, Department of State, *FRUS, 1929*, 2: 537.
17. Castle to American consul in Nanking, July, 24, 1931, File Number 893.796 Boeing Airplane Co./13, RG 59, NA.
18. "China Would Purchase 62 War Planes Here," *New York Times*, August 25, 1929, p. 24.

19. Granville Woodard to Leighton Rogers, September 19, 1929, File Number 2078, RG 165, NA.

20. Walter Adams to the secretary of state, October 8, 1929, File Number Chance-Vought Corp/13, RG 59, NA.

21. "Report on Civil and Military Aviation in Nanking," by C. M. Alport, cipher officer to the British Consul General in Nanking, March 31, 1930, File Number 4248, Embassy and Consular Archives, China: Correspondence: Series 1875 to 1956, Record Class 228, PRO. See also *North China Herald*, June 8, 1929, p. 381.

22. Adams to the secretary of state, October 8, 1929, File Number 893.113 Chance Vought Corp/13, RG 59, NA.

23. *Hangkong zhazhi* [Aviation Magazine] (Guangdong), 4 (1929): 5.

24. The license was issued on November 9, 1929 to the United Aircraft Exports Corporation. The Division of the Far Eastern Affairs, "American Policy with Respect to the Shipment of Arms and Munitions of War to China," June 10, 1930, File Number 893.113/1265, RG 59, NA.

25. Cunningham to the secretary of state, November 11, 1929, File Number 893.113 Chance-Vought Corp/14, RG 59, NA.

26. Cunningham to secretary of state, November 11, 1929, File Number 893.113 Chance-Vought Corp/14, RG 59, NA.

27. "Sale of Military Equipment by the United States Government," June 11, 1930, File Number 893.113/1265 and 893.113/1267, RG 59, NA.

28. Nelson T. Johnson, assistant secretary, to the secretary of state, December 6, 1929, File Number 893.113 Chance-Vought Corp./37 1/2, RG 59, NA.

29. Cunningham to the secretary of state, November 11, 1929, File Number 893.113 Chance-Vought Corp/14, RG 59, NA.

30. Ibid.

31. Memorandum for the secretary of war by F. Trubee Davidsion, assistant secretary of war, December 2, 1929, Box Number 482, Office of the Adjutant General Central Files, 1926–1939, RG 94, NA.

32. Memoranda for the chief of staff by C. F. Humphrey, Jr., Acting Assistant Chief of Staff, January 16, 1930, ibid.

33. Cunningham to the secretary of state, November 11, 1929, File Number 893.113 Chance-Vought Corp/14, RG 59, NA.

34. Hurley to Hoover, December 26, 1930, File Number 893.248/30, RG 59, NA.

35. *New York Times*, December 4, 1929, p. 2.

36. Hoover to Hurley, January 10, 1930, File Number 893.248/30, RG 59, NA.

37. Edwin Cunningham, "Complaint Against United Aircraft Exports Inc. of New York for Non-Fulfillment of Contract," April 1, 1930, File Number 893.113 Chance-Vought Corp./45, RG 59, NA.

38. Captain Parker G. Tenney to the War Department, September 2, 1931, Report Number 8094, File Number 2078, RG 165, NA; *North China Herald*, August 25, 1931, p. 262.

39. J. C. Huston, "The Progress of Aviation in China during 1929," March 10, 1930, File Number 893.796/78, RG 59, NA.

40. License requested by Chinese Legation on April 22, 1930, issued by State Department on May 5. The Division of Far Eastern Affairs, "American Policy with Respect to the Shipment of Arms and Munitions of War to China," June 10, 1930, File Number 893.113/1265, RG 59, NA.

41. The mission left Shanghai on February 2, 1931 for the United States. The officers were: General Zhang Weizhang, director of the Bureau of Aeronautics of the Ministry of War; General Wang Pinghang, vice director of the Bureau of Aviation of the Chinese Minister of War; Major Wu Jiashi, and Captain Zhang Zhaojiang. See Meyer, American Consul at Nanking, to Secretary of State, February 9, 1931, File Number 893.248/33, RG 59, NA. See also *New York Times*, April 1, 1931, p. 1.

42. Military attaché in China to the War Department, August 19, 1931, Report Number 8083, File Number 2078, RG 165, NA.

43. "Aircraft Import-1931," Box Number 503, File Number 560, RG 151, NA.

44. "Estimated Strength of Chinese National Air Force," by Consul General Meyer, August 14, 1931, File Number 893.20, RG 59, NA.

45. Tenney to the War Department, January 16, 1931, Report Number 8182, File Number 2078-95, RG 165, NA.

46. E. P. Howard to L. W. Rogers, October 27, 1931, Shanghai, File Number 560, Box Number 503, RG 151, NA.

47. For the background of Floyd N. Shumaker, see "Shumaker, Floyd Newman" in *The National Cyclopedia of American Biography* (Clifton, N.J.: James T. White, 1980), 59: 347–48.

48. John Magruder to the War Department, January 20, 1930, Report Number 7653, File Number 2078, RG 165, NA.

49. Shumaker to Charles N. Ferrier, February 13, 1930, File Number 893.796 Curtiss/61, RG 59, NA.

50. Magruder to the War Department, February 26, 1930, Report Number 7665, File Number 2078-95, RG 165, NA.

51. Ibid.

52. The Chinese Legation requested a license on April 10, 1930, and the State Department issued one two days later. The Division of the Far Eastern Affairs, "American Policy with Respect to the Shipment of Arms and Munitions of War to China," June 10, 1930, File Number 893.113/1265, RG 59, NA.

53. Mahlon F. Perkins to Department of State, May 2, 1930, File Number 893.796/79, RG 59, NA.

54. Memorandum of a conversation between Minister Johnson and Shumaker, May 2, 1930, File Number 893.796/79, Department of State Records, RG 59, NA; Nelson Johnson's Memorandum, May 12, 1930, Box Number 14, Nelson T. Johnson Papers, Library of Congress.

55. Lieutenant-Colonel Nelson E. Margetts, military attaché in China, to the War Department, July 4, 1930, Report Number 7737, File Number 2078, RG 165, NA.

56. See Johnson to the Secretary of State, July 14, 1930, Department of State, *FRUS, 1930*, 2: 23–24. In a dispatch to the Department of War, Margetts reported that Major Shumaker spoke of having taken actual part in China's civil warfare. He claimed to have directed the bombing of the arsenal at Gongxian and the Yellow River Bridge north of Zhengzhou. He described in detail the bombing of a bandit-infested village at Linxian, Henan, in which nine planes took part. See Margetts to the War Department, November 12, 1930, Report Number 7863, File Number 2078, RG 165, NA.

57. Cunningham to the Secretary of State, February 24, 1932, File Number 893.20/ Short, Robert M./4, RG 59, NA.

58. Dr. Arthur N. Young, a financial adviser to the Nationalist government, submitted to the highly influential financial minister T. V. Soong, a statement that "Prospective development of civil and military aviation in China justifies the

appointment of a well-qualified person for this position," and Young recommended that Major Edward Howard, who was an assistant to the American commercial attaché in China at that time, was qualified for the position of aviation adviser to the Nationalist government. Young also emphasized that Howard would give primary attention to military aviation. At the time, however, the Nationalist government was considering appointing Short to the Military Aviation Bureau of Nanjing as an instructor. On June 24, 1931, at the request of Soong, Young asked Nelson Johnson to make an inquiry to the Washington government concerning Short's army records and qualifications. After receiving the records, the Nanjing government appointed Short as adviser. See Memorandum for Minister of Finance, by Arthur Young, June 23, 1931, Arthur Young Papers, Hoover Institution on War, Revolution, and Peace, Stanford University, California. Hereafter cited as Young Papers; Memorandum for the Minister of Finance—Appointment of an Aviation Adviser, by Arthur Young, June 23, 1931, ibid.

59. Parker Tenney, Assistant Military Attaché, to the War Department, September 2, 1932, Report Number 8094, File Number 2078, RG 165, NA.

60. Minister Johnson to the secretary of state, February 26, 1932, File Number 893.2221/2, RG 59, NA.

61. Before Short's death, the Department of State apparently did not know that Short had been serving actively in combat in the Chinese Air Force. See Department of State to Mrs. Elizabeth M. Short, February 25, 1932, File Number 893.20 Short, Robert M./5, RG 59, NA.

62. "Hanghongshu gongzou zongbaogao" [General report of the Aviation Administration], 1930, File Number 16971, Record Group 787, SHAC.

63. As a newspaper commented, "In December 1929, when the Nationalist forces were in pursuit of Kuominchun [Feng's army], aeroplane did a great deal to bring about the defeat of the Kuominchun, and it was mainly through those bombings that Loyang, Honan province, was taken." *North China Herald*, December 14, 1929, p. 418.

64. "Junzhenbu hanghong disidui zhouchan jingguo" [The combat history of the Fourth Squadron of the Ministry of Military], 1930, File Number 1202, Record Group 787, SHAC.

65. Ibid.

66. Tenney to the War Department, May 19, 1930, Report Number 7730, File Number 2078-95, RG 165, NA.

67. *Jiangzongtong milu* [The private history of President Chiang] (Taipei: Central Daily, 1976), 7: 156.

68. Johnson, American minister in Peking, who observed that bombing, reported to the secretary of state, "Two Nationalist planes, apparently Douglas, recently delivered, flew over and dropped three bombs which fell in the southern part of the city this morning at ten o'clock." See Johnson to the Secretary of State, August 27, 1930, File Number 893.113 Douglas Aircraft Co./13, RG 59, NA.

69. Lincoln C. Reynolds, "Aviation in China."

70. *Jiangzongtong milu,* 7: 156

71. See "Junzhenbu hanghong disidui zhouchan jingguo."

72. *Jiaofei zhanshi* [The history of suppressing the bandits] (Taipei: Guofangbu shizhengju, 1967), 1: 190, 220.

73. Tuonian Wang, *Fangong kanluan* [To suppress the rebellion of the communists] (Taipei: Liming wenhua chubangongsi, 1982), 1: 105.

74. *Jiaofei zhanshi,* 1: 243; Wang Tuonian, *Fangong kanluan*, 1: 137–38.

75. *Jiaofei zhanshi,* 3: 350; Wang Tuonian, *Fangong kanluan*, 1: 161–163.

76. Chiang to He, February 5, 1932, *Zhonghua minguo zhongyao shiliao chupian: Duiri kangzhan shiqi* [The first edition of the important historical sources of the Republic of China] (Taipei: Zhongguo guomindang dangshi weiyuanhui, 1976), 1: 446–47.

77. Ibid.

78. For a detailed account of the air battles in the Shanghai Incident, see *Shijie hongzhong zouzhan pashinian* [The air combats of the world for 80 years] (Shanghai: Shanghai kexue chubanshe, 1988), pp. 58–60.

79. *Zhongri waijiao shiliao congbian* [A series of records of the Sino-Japan diplomatic history] (Taipei: Zhonghua minguo waijiaoshi xuehui, 1965), 3: 84.

80. *Shijie hongzhong zhouzhan pashinian*, p. 59.

Chapter 3

The Expansion of Chinese Military Aviation, 1932–1936

After a growing but solid beginning, Nationalist military aviation developed tremendously between 1932 and 1936. With American help, the Nanjing government rapidly in quick succession established the Central Aviation School, erected the Central Aircraft Factory, and reorganized the entire Nationalist Air Force. Americans completely dominated the Nationalist military aviation which reached a new stage.

The Nationalist government determined no later than 1932 to make China an air power of bright, if not first magnitude. The battles between the Chinese and Japanese Air Forces over Shanghai in February 1932 brought the government to appreciate, as it had not earlier, the importance of military aviation, not only in unifying China, but also in defending the country against external aggressors. During the Shanghai confrontation, the Japanese Air Force maintained superiority over the Chinese and attacked heavily. Afterward, the Nationalist government concluded that China's military failure at Shanghai and in northern China was largely due to the Japanese aerial attacks and to the lack of a Chinese Air Corps and air defense. The Nationalists also understood that many years and much money would be necessary to refurbish both the army and navy to catch up to the major powers. The Nationalist Air Force, although embryonic, could more quickly and less expensively match up to the Japanese Air Force. As a result, the Nationalist government grew increasingly "air minded," and decided first to concentrate on the development of the air force. In the meantime, it made wide use of the slogan "Hangkong Qiukuo" [National salvation by the use of aircraft] to arouse the interests of the Chinese people in developing the Nationalist military aviation.[1]

In 1932, Nanjing diverted 11 million dollars, originally intended for development of the Navy, to the development of the Nationalist Air Force.[2] By midyear, plans for improvement and growth were drafted. A Five-Year Program extending through 1936 laid out the main avenues of future development.[3] Schools must be established: two primary training schools, one advanced flying school, one mechanics school, and one consulting school, schools with the capacity to train 700 pilots and 200 mechanics. Twenty-seven squadrons with 432 airplanes were envisioned. Three aircraft factories and six repair shops must be erected from the ground up.

American businessmen were to play an active role in encouraging the Nationalist government to arrange for American assistance in the development of all of the elements of modern military aviation. George C. Westervelt, president of Intercontinent Aviation, is an excellent example. Captain Westervelt, a graduate of the U.S. Naval Academy, accumulated thereafter extensive experience in the aircraft industry. After three years in a postgraduate naval architecture course at the Massachusetts Institute of Technology in Cambridge, he took up aeronautics in 1914 and supervised the construction of approximately 1,500 airplanes for the U.S. Navy during World War I. Westervelt made two trips to Europe and submitted a report to the Navy Department that became fundamental for American aviation activity in the European area. Later, he spent more than seven years commanding the Naval Aircraft Factory in Philadelphia. In 1927, at his own request, Westervelt retired from the Navy with 30 years of good service, turning to commercial aeronautics in the design and building of factories and in the general administrative work of air transportation. That long and comprehensive aeronautic experience prompted him to offer advice to China while there in the interests of the Curtiss-Wright Corporation.[4]

Westervelt arrived in Shanghai in December 1930 and spent about five months in China. During that time, he extended the Yangzi line from Hankou to Chongqing, renovated the airline's fleet, and established good relations with the Chinese. When returning to China again in November 1931, he tried to encourage the Nationalists to develop military aviation in China, promising Nanjing that his company would supply the Chinese with American military planes. Japan's attack on Shanghai in January 1932 gave him an opportunity to convince the Chinese of the importance of air power in modern wars.

Westervelt observed the Japanese bombing of civil and military targets, and afterward, wrote a letter to T. V. Soong, Chinese minister of finance, that stressed the importance of aircraft in the Shanghai battles. Disclosing the vulnerability of Chinese cities to air bombings and the poor performances of the Nationalist Air Force, he argued that if China had possessed an air force one-half as great in striking power as that concentrated in China by the Japanese, and of an equal efficiency, the landing of the Japanese troops in Shanghai could have been prevented. More important, he pointed out the advantages of U.S. assistance in the development of Chinese military aviation:

> The United States is the proper aviation contact and source of supply for China, until China is able to supply its own requirements. This belief is due to my knowledge of the fact that in practically every class of military airplanes and in most classes of commercial airplanes, more efficient and up-to-date planes are manufactured in the United States than in any other country. In addition, due to a similarity in geographical characteristics, and in territorial extent, these planes are more suited for Chinese conditions than planes manufactured elsewhere would be.[5]

Westervelt went on to suggest to the Chinese government that China secure from the United States a high-ranking army officer of wide experience as overall adviser to the Chinese government in matters aeronautical. In addition, the Chinese government should arrange for a representative in the United States of wide experience who would take charge, in China, of the training and instruction of aviation personnel.

No less than military aviation, Westervelt encouraged the Nanjing government to develop commercial aviation, stating that commercial experience would contribute directly to military aviation. Indeed, Westervelt published an article in Shanghai, entitled "Effect on Military Aviation of Commercial Aviation,"[6] in which he argued that commercial aviation would furnish a considerable opportunity for the use of military aviators as pilots in time of peace, and even more, provide them with some business experience. Westervelt cited the example of the United States, pointing out that there was a direct relation between the growth in efficiency of commercial aviation and that of military aviation. Without doubt, Westervelt's ideas influenced the Chinese government in seeking Americans' assistance for the expansion of the Nationalist military aviation. Responding enthusiastically to Westervelt's proposal, T. V. Soong appointed Dr. Arthur N. Young, economist and financial adviser to the Nanjing government, to initiate this project. Young later decided to work with Edward P. Howard, trade commissioner in Shanghai for the U.S. Department of Commerce, to arrange an American air mission to China.

THE JOUETT MISSION

After the end of the China arms embargo, scores of military planes reached China monthly. There were, however, not enough pilots to fly them. Training pilots became an urgent task for the Nanjing government. As early as March 1929, Nanjing opened an aviation class for the training of competent aviators and mechanics at the Central Military Academy in Nanjing. The average student needed approximately 150 hours of flying before qualifying as a full-fledged pilot. The class used 12 British DH Moths as training planes, but they were far too light for satisfactory performance. Eighty students graduated from the class in the spring of 1930 and 83 students graduated in March 1931.[7] To speed up the process, the Ministry of War established a separate aviation school in Nanjing on July 1, 1931, with General Mao Banchu as its first president. The school initially admitted 100 students, 40 for the course in mechanics, and 30 each for flying and scouting. In December 1931, the school moved to Hangzhou.[8] In this manner, the Chinese government tried to train its pilots independently but could not keep up with its estimate of its needs. Moreover, the graduates found it difficult to fly American airplanes because they were much heavier than the British DH Moth airplanes. Only two out of 60 Chinese pilots were qualified to fly the newly purchased Vought Corsairs.[9]

Hence, at the beginning of 1931, Minister of Finance Soong attempted to send Chinese aviators to the United States for training. Soong confidentially told the American commercial attaché in Shanghai that the Chinese government wished to send 12 flying officers, all with actual fighting experience, to an Air Corps school in the United States for advanced instruction in organization and maintenance tactics.[10]

The Air Corps liaison office suggested to the Aeronautic Trade Division that a formal request be made through the Chinese Legation and that, if approved, U.S. policy mandated acceptance of no more than two foreign students in the military flying school at one time. That policy proved unacceptable to Nanjing because of the great sense of need at that time.

In addition to the abortive attempt to train Chinese pilots in the United States, the Nationalist government asked Washington to dispatch an American aviation mission to China to train the Chinese students at home. In March 1932, Soong inquired of Lt. Col. Walter Drysdale, the American military attaché in China, to forward a request to the War Department for a group of qualified Americans to organize and operate an air school. When the U.S. minister, Nelson Johnson, learned of the request, he strongly urged the Department of State support, noting, "The presence of such a mission in China will be invaluable in increasing the use of American planes and equipment."[11] Johnson linked a U.S.-staffed air school with sales opportunities for U.S. aircraft manufacturers.

When Johnson's message was shared with the War Department, the department responded that it was not interested in sending such a mission to China, and that the request might be forwarded to the Department of Commerce because of its commitment to civil aviation interests. The Department of State was correct in its presumption; Leighton Rogers for the Aeronautics Trade Division responded in a letter to State Department's Far Eastern Division, promising to extend "the kind of assistance which we can render, as well as the means of rendering it best adapted to the involved situation."[12] To send a mission was not a Commerce function, and the State Department should have dispatched such a mission. But the State Department was also not interested in it, telegraphing the American consul general at Shanghai, explaining, "The Department is of the opinion that it would be inadvisable for this Government to take any steps in connection with the sending to China of such a mission while the present Sino-Japanese situation continues as it is."[13]

The Department of State had many reasons for its refusal. Since the Manchurian Incident of September 18, 1931, diplomatic relations between the United States and Japan had decidedly deteriorated when Washington condemned Japan's aggression in Manchuria. Then, the death of Robert Short, who was piloting a Chinese plane against Japanese fighters, exacerbated matters. At that time, the Japanese were preparing a strong note of protest at the U.S. breach of neutrality. Were the United States to authorize boldly a military air mission, Tokyo may well have taken it as grounds for strong action.[14]

Not surprisingly, Nanjing turned elsewhere. The director of the Chinese Aviation Bureau informed the U.S. commercial attaché that the French air attaché had arrived in Nanjing on April 6, carrying an offer to send a French military air mission immediately. In addition, London offered to train Chinese aviators in British schools. And the Italian government offered to send an official air service mission. General Wang tried to let Americans know that China would ask other nations to help it build up a strong air force if the United States failed to provide effective assistance.[15] There can be no doubt, however, that the Chinese government still preferred American assistance, and Nanjing, therefore, asked

Washington whether it might be better to send a civilian mission than a military mission.

When the Commerce Department's aeronautic trade commissioner Edward Howard was discussing this latest proposal with the Chinese government, Rogers informed Thomas H. McConnell of United Aircraft Exports that such a mission "will be of tremendous assistance to American manufacturers in selling in the Chinese market."[16] With the Department of Commerce's approval, Howard, in the second week of April, held several conversations in Shanghai and Nanjing with all of key figures: General Wang Binghung, Minister Soong, Minister Johnson, Dr. Arthur Young, and Colonel Drysdale. Soong then decided to accept Howard's proposal for a civilian mission composed of ex-army officers with state-of-the-art military experience.[17] The State Department insisted it had "no objection to the trade commissioner's discussing civilian aviation with Soong," but cautioned that "it would be inadvisable for this government or any of its officers to be associated with military training."[18] Howard accepted this as authorizing his meetings with Soong, Dr. Young, and Bert Hall, then aviation adviser to the Nationalist government.[19]

The first choice of Rogers of Commerce to head this mission was John Jouett. Jouett, earlier in command of the U.S. Air Corps training and of the First Attack Group, resigned his command in March 1930 to accept the position as the head of the Aviation Department of the Standard Oil Company of Louisiana.[20] By May, the Department of Commerce had completed its plan and forwarded it to Howard. The goal of the plan was to "provide an adequate refresher course for sixty pilots who are already trained," and to schedule school operations "to graduate every six months a class of fifty pilots with 60 hours of primary instruction, 40 basic training and 15 hours of military training of an advanced type to include bombing, gunnery, observation, photography, radio and ground school work, all to be in accordance with the system and standard employed at Randolph Field." If the Chinese government requested, "the director will be prepared to reorganize the Chinese Air Force and prepare a budget after arrival. He will also be prepared to assume the duties of aeronautical military adviser."[21] There can be no doubt but that the civil air mission would very likely assume military duties.

Once the plan was completed, Jouett was asked whether he would accept the appointment. At first he was reluctant to undertake the tasks, but the assistant of secretary of commerce Clarence Young was insistent on playing on his duty to the U.S. government. Jouett finally accepted, flew to Washington, and sent a telegram to Nanjing.[22] Jouett's first task was to survey the duty records of 200 graduates of the U.S. military aviation schools, from which he identified a group of young reserve officers, the senior officer of whom was H. T. Rowland.[23] All of these officers were graduates of the U.S. Army flying schools, the youngest of whom had at least two and one-half years of service. Included in the group was a mechanical staff of five. The leader of the group was G. B. Clark, a Navy Reserve officer, with the rest mostly Army enlisted reservists. After the Americans signed contracts, the Chinese government advanced them money. All these arrangements with China were made by cable, at the expense of the Department of Commerce.[24]

Meanwhile, the Department of State developed some qualms about a civilian air mission training Chinese cadets in military aviation. Consequently, Hornbeck telephoned the Department of Commerce and on several occasions talked at length with Young and Rogers about the matter. Rogers insisted that as long as the mission was entirely civilian, there could be no objection, but Hornbeck, ever the prudent diplomat, held that, in view of current situation in the Far East, it would be inadvisable for the American government or any of its officials to associate in any way with a project involving the Chinese military. Rogers admitted that the mission was going to organize an aviation school for the Chinese military. Later, Acting Secretary of State Castle remarked, "If we had known about the present project at an earlier date, we should have raised objection, but since the present project had gone so far, there did not appear to be anything that the Department could do about it."[25] A convenient and face-saving explanation if anything untoward were to happen.

The Commerce Department officials then took a leaf from the State Department book. Before the departure of the Jouett Mission, Rogers wrote a confidential letter to Howard, asking him to make "every possible effort" to "keep the organization and departure of this outfit secret," because the Canton government might protest the United States' involvement in the civil struggle between the Canton faction and the Nanjing government. In addition, "Japan might object strenuously to the American mission teaching Chinese pilots how to fly."[26]

Colonel Jouett and his mission of 14 men sailed under the Golden Gate Bridge on the S.S. *President Hoover* from San Francisco on a brilliantly beautiful June day, arriving in Hangzhou on July 8.[27] This arrival was destined to usher in a new phase in Nationalist military aviation. The Jouett Mission undertook the staggering task of laying the foundations for a modern and effective air force through the establishment of a modern aviation school, the reorganization of the Nationalist Air Force, and the encouragement of the importation of American military airplanes.

The mission began work with an aviation school. At the beginning of September, Chiang Kai-shek renamed the Aviation School in Hangzhou the Central Aviation School and served as the president of the school.[28] The school was drastically reorganized. The first class of 50 Chinese pilots, astonishingly, started on September 15. The school used the teachings and policies of Randolph Field and Kelly Field, applied the U.S. Army training system, previously unknown to the Chinese, and adopted many of Jouett's voluminous recommendations concerning the training of squadrons. The courses at the school called for four months of primary training, four months of basic training, and four months of advanced training, much as in the United States.[29]

The school was distinctly under the immediate direction of Jouett. During his tenure, 335 qualified cadets graduated from the school (the class of December 1933: 55; the class of June 1934: 65; the class of December 1934: 64; and that of June 1935: 56). These graduates, including pilots who received refresher classes, shortly constituted the bulk of the flying personnel of the Nationalist Air Force.[30] "The graduates of the 1932 class," recalled John Jouett in 1937, "almost without exception, are now captains and squadron leaders."[31] The cost per pilot for training at the school totaled 21,000 Chinese yuan, only about one-third of the cost of training a pilot in the U.S. Army Air Corps. The school was so successful that Generalissimo Chiang Kai-shek expressed his great satisfaction with the result of the American mission's effort.[32]

Were that not enough, Nanjing called upon Jouett to lay out the reorganization of the entire National Air Corps. Before departing for China, Jouett understood that his mission included "assisting China to build its military aviation to a point on a par with other progressive nations of that day."[33] Nanjing conferred on Jouett full authority and power, at least on paper, to reorganize the Nationalist Air Corps, in Soong's words, "to give advice and assistance with respect to all branches of aviation development in China, both civil and military . . . to submit plans for and supervise the reorganization of the existing air force."[34] In addition, "All foreign advisers and experts and other foreign personnel connected with or to be connected with Chinese military aviation shall be subordinate and responsible to you." Left somewhat ambiguous was the extent to which Jouett became a line officer with the Chinese Air Force, empowered to order and capable of enforcing his orders to bring about change.

After signing a contract with the Nanjing government and receiving appointment as a principal aviation adviser, Jouett formulated a five-year program for the Nationalist Air Force.[35] Jouett's first step was to recommend that the Commission on Aviation Affairs be detached from the Ministry of Military, and constitute a separate ministry, equal to the Ministries of Military and Navy, all three under the direct control of the National Military Council. The Nationalist government adopted Jouett's suggestion in August 1932.[36]

The second step was to adopt a pay and promotion scheme for aviation officers comparable to that of foreign air forces. Before the arrival of the American mission, no uniform system of promotion or scale of salaries for Chinese aviation personnel existed. That meant that young officers were often required to serve without promotion for a number of years. The air force formally adopted his promotion scheme in February 1933, which paid second lieutenants $100 monthly, and with higher ranks increasing in pay roughly corresponding to that in the U.S. Army. Now, flying officers achieved a status shared roughly with other military officers, and the corps consequently attracted a higher class of personnel.[37]

John Jouett also made significant changes in the field location of units and their tactical use. He consolidated seven squadrons into three, and located them in strategically important places, such as Hangzhou, Nanjing, Nanchang, and Hankou. In addition, he directed the establishment of auxiliary schools, that is, motor maintenance, aviation medicine, and aviation radio. With the critical help of the Jouett Mission, Nanjing may be said to have established its modern field air force by the end of 1934. The following outline suggests the augmented strength and reorganization by that time:

A. Bombardment squadrons

First Bombardment Squadron

Station:	Nanchang
Planes:	11 Fiats and Vought Corsairs
Pilots:	17, all graduates of refresher class of Hangzhou

Second Bombardment Squadron

Station:	Nanjing
Planes:	7 Fiats
Pilots:	12, all graduates of Basic and Primary School at Hangzhou

B. Observation Squadrons

First Squadron	
Station:	Lanzhou (Gansu)
Planes:	2 Douglases and 3 Corsairs
Pilots:	14
Armament:	Rear 30 cal. machine guns and 38.5 pound bombs
Second Squadron	
Station:	Nanchang
Planes:	8 Douglases
Pilots:	16
Armament:	Rear 30 cal. machine guns

C. Pursuit Squadrons

First Squadron	
Station:	Hangzhou
Planes:	15 Curtiss Hawks
Pilots:	17
Armament:	Two forward guns, one 50 cal., one 30 cal.[38]

The Nationalist government promoted the status of the Commission on Aviation Affairs in mid-spring 1934. Chiang Kai-shek served formally as director of the commission.[39] The Generalissimo in a Nanchang speech intended both to inform and to argue that in order to save the country from extreme dangers, China must develop its own air force immediately. Lending his name and authority, Chiang assumed personal supervision of the Air Corps because he was determined to develop it into an efficient and powerful fighting unit.[40]

In addition to its organizational and training efforts, the Jouett Mission contributed greatly to increasing sales of American military airplanes to China. During the 1930s, airplanes were still very expensive for most national budgets, and especially for a fiscally weak country like China. Although the minister of finance allotted substantial funds for the development of military aviation, those funds were not enough. The five-year program of Nationalist military aviation called for an expenditure of $32,260,000, and for the fiscal year 1933–1934, the Central government appropriated $2,333,000 for military aircraft. A monthly allowance of $133,000 for the Central Aviation School, however, absorbed almost the entire allotted amount.[41]

Thus, the Nationalist government was forced to scavenge funds from other sources to purchase foreign airplanes. In doing so, the government began to organize a propaganda campaign in late 1932 through newspapers, motion pictures, and airplane demonstrations. First, a portion of monthly salaries of practically every government employee was held back. Beginning in July 1933, the Ministry of Finance inaugurated a state lottery as a medium through which to raise funds for the development of military aviation. That summer, the Ministry netted $750,000 for the Nationalist Air Force.[42] Simultaneously, the Nationalist government encouraged its citizenry to raise funds for aviation development. Numerous patriotic bodies and private Chinese commercial enterprises, including Chinese overseas, conducted campaigns to raise funds to purchase aircraft. In 1933, the

Chinese donated $200,000 in Chinese currency to the aviation program.[43] On Chiang Kai-shek's fiftieth birthday (October 31, 1936), for example, the Chinese people subscribed $3,500,000 in Chinese currency to buy 69 airplanes as a gift to the Nationalist government, although certainly that was a highly unusual day.[44] With these funds, the Chinese government placed orders for many foreign airplanes. Although the United States began to export modern airplanes to China in 1928, the total sales were small before 1933. Beginning in 1933, however, American salesmen increased their sales immensely and virtually came to corner the Chinese market. Some factors contributed to this increase. Not the most consequential but undoubtedly the most spectacular was the marketing by Major James H. Doolittle.

Doolittle, a noted American speed and stunt ace, was well known in China. In 1930, he worked as manager of the Aviation Department of the Shell Oil Company. With the support of the large American aircraft manufacturer, he demonstrated the Curtiss Hawk for China during a trip around the world. He arrived in Shanghai in April 1933, and put on a number of demonstration flights in the following two months. When demonstrating a high-powered American combat plane, the Curtiss Hawk, Doolittle startled Chinese observers with his exploits above the rooftops.[45] The Curtiss Hawk was a magnificent craft to showcase Doolittle's skills and vice versa. It was equipped with a 700-horsepower Wright cyclone engine, similar to the one that powered Lindbergh's survey plane, with a top speed of more than 200 miles an hour. It was the fastest pursuit airplane in Asia. Doolittle's exhibition resulted in the largest single order placed by the Nationalist government up to that time, 50 modern Curtiss Hawk single-seat pursuit biplanes.[46] Nanjing even purchased the plane used by Doolittle himself. Before that plane was incorporated into the air corps of the Chinese Air Force, the Nanjing government held an elaborate public dedication ceremony in Shanghai. At the end of 1933, 50 Curtiss Hawks were delivered to China, which gave the Chinese the nucleus of one of the best equipped air forces in the world.[47]

Although Doolittle contributed to the sale of American airplanes to China, U.S. airplanes were able to dominate the China market in the following years mainly due to the success of the Jouett Mission. As a member of the board in charge of the purchase of aircraft and accessories for the Central government, Jouett was "in a position to advance the sales of American aircraft." Although the officials of the mission, responsible for preparing purchase recommendations, endeavored publicly to adhere to a policy of strict impartiality among the several manufacturers, their preference for a particular plane or type of equipment with which they were most familiar most certainly influenced their selections.[48]

Jouett's five-year program for acquisition, which specified the number, type, and make of airplane, illustrates his projection for the first year's purchases:

First Year:
15 Fleet primary trainers
10 Douglases with Pratt and Whitney engines
15 Boeing P12 pursuits with Pratt and Whitney engines
20 Italian Fiat B. R. 3 bombers

Second Year:
30 Boeing P12 pursuits with Pratt and Whitney engines
15 Curtiss Falcons[49]

Eighty percent of Jouett's suggestions required U.S. aircraft.

Furthermore, through their acquaintance with Jouett, American businessmen were able to make contacts with high Chinese officials and to obtain useful information about Chinese military aircraft requirements that put them in an advantageous position. As a representative of the United Exports, Carl A. Nahmmacher, concluded, "The foreign nations have the aid of their diplomatic organizations—an advantage which I regret we are not afforded by Washington under its present policies. However, to date we enjoy a prestige with the high Chinese officials concerned that can be turned into a distinct advantage."[50]

Ultimately, China imported 120 airplanes in 1933, of which 90 percent were from the United States. In all, the United States and Europe shipped $8,102,000 worth of aircraft to China in that year, over 80 percent of which were military planes. In money terms, the U.S. accounted for 70 percent of the China market in 1933:

United States:	$5,634,000
Germany:	$ 364,000
Britain:	$ 353,000
France:	$ 72,000
All other countries:	$1,679,000[51]

The following year, China purchased 215 American planes, including 25 Northrop bombers, 12 Douglas basic trainers, 22 Douglas observation planes, 38 Corsair observation light bombers, 81 Fleet trainers, 20 Northrop bombers, 10 Boeing pursuit planes, two Douglas Dolphin flying boats, and one Douglas transport DC-2.[52] By mid-1935, the Nationalist Air Force already had about 500 airplanes; "the majority have been bought from the United States," the *China Weekly Review* commented.[53]

In the light of these striking successes, it is truly surprising that the Chinese government did not renew Jouett's contract when it expired in June 1935. Nearly two years earlier, A. N. Young telephoned Jouett that Italian instructors would replace the American air mission.[54] Jouett fought back. Although Soong was not a military figure, Soong's original interest in Jouett's work carried a good deal of influence, or so Jouett thought. In October 1933, Jouett went to Shanghai and Nanchang to see both Chiang Kai-shek and Soong, but in vain.[55]

After a conversation with Chiang Kai-shek on October 9, 1934, W. S. Drysdale, American military attaché in China, admitted that the American group was being severely criticized by key Chinese officials, and the great personal popularity of Mussolini insured a prominent position for an Italian aviation mission. Colonel Jouett and his mission would probably not be reemployed, Drysdale learned, upon the termination of their current contract.[56]

Even at the end of 1934, however, there were indications that some within the Chinese government still wanted to retain an American mission, but only if Washington would send army officers on the active list to substitute for Jouett, a development they must have understood was unlikely. After receiving such request, on January 26, Secretary Cordell Hull telegraphed C. E. Gauss, American chargé in China, informing him as follows: "an invitation by the Chinese Government to any officer of the United States Army on the active list to accept appointment in the service of the Chinese Government, intended toward or leading to the 'lifting' of such officer from service of the American Government, would be regarded with definite displeasure by this Government," because the shift of American Army officers to the Chinese government would "give the impression that American Government is inciting the Chinese Government to military preparedness," and such activities "would have a net effect disadvantageous to each and to both countries."[57]

But why Italy? Benito Mussolini was anxious to spread Italian influence not only to African and neighboring lands but in Asia as well. When Minister H. H. Kung visited Rome in 1933, for example, Mussolini told him that the Italian government would make 12,000,000 taels of the remitted Boxer Indemnity fund available to China if Nanjing accepted an Italian aviation mission. The funds were to be used for the purchase of a first installment of Italian planes, and for the payment of the salaries of the Italian mission (the salaries of the mission were to be less than one half that of the Americans). In addition, Italians would open a flying school and build a factory for the Nationalist government. In May 1933, the Italians sent to China a mission of 12 reserve officers.[58] For his part, Jouett did not play his cards as adroitly as he might have; he did not prove to be an adept politician. He did not succeed in allying himself closely with General Chiang Kai-shek while the Italian adviser definitely did. In the Chinese anti-Communist campaigns of 1933, Jouett refused to allow American pilots to take part in combat, while the Italians did.

A second element in the decision to abandon the Jouett Mission was rooted in Japanese protests to the Chinese government regarding the activities of the American mission. The Japanese made no secret of their dislike of American participation in the development of Chinese air power, a sentiment known well to the Chinese and one which they soon made public. In January 1933, a Japanese foreign office spokesman asserted that some 40 American officers had been sent to China to take part in hostilities against Japan, apparently referring to the Jouett Mission. The Japanese, it should be noted, did not protest the later Italian aviation mission.[59]

A third reason for the replacement of the Jouett Mission was that Kung, who favored the Italian fliers and airplanes, did not wish to retain the American mission. In the winter and spring of 1932–1933, as a special envoy to investigate the economic and industrial conditions in Europe and the United States, Kung took the opportunity to study the development of military aviation in various European countries. On his return, he lost no opportunity to impress upon the Nationalist government the continuing necessity of building an adequate air force. In spring of 1933, the National Military Council and the Commission on Aviation Affairs jointly entrusted Kung with the task of negotiating the purchase of aircraft. That autumn, Kung replaced Soong who, as finance minister, had supported Jouett. Thus, Kung was responsible for finding the ways and means to finance the purchase of airplanes, which jeopardized Jouett's situation.[60] By the end of 1934,

Kung completed an agreement with the Italian aviation mission, providing that Italy would establish an aircraft factory in China. Kung wrote Chiang Kai-shek soliciting a location for the new factory; Chiang suggested that the new factory should be in Nanchang.[61]

The final element of the explanation for the decline of Jouett's fortunes in China was Chiang Kai-shek's suspicions of Soong's attempt to control the Nationalist Air Force. He first instructed Soong in March 1933 that the Nationalist Air Corps be trained by several powers, not the United States alone. If Germans, French, or Italians expressed an interest in training pilots, they should not be discouraged.[62] Chiang's instruction, indeed, paved the way for the employment of the Italian aviation mission to weaken Soong's influence in the air force. Chiang then moved the Hangzhou aviation school to Luoyang in Central China in March 1933 in order to control it directly by himself. This location proved a disaster because dust storms affected operations in Luoyang; the school had to return to Hangzhou in several weeks. Finally, Chiang replaced Soong with General Ke Qingan, chief of the Aviation Bureau, who was to be in charge of the aviation affairs. Since Soong who had supported Jouett lost his authority in the school, Jouett's situation deteriorated.[63]

The consequence of all of these considerations was that, in June 1935, the Nationalist government did not renew Colonel Jouett's contract. He returned to private business and sailed from Shanghai on June 7. Prior to his departure, the Nationalist government awarded the failed contractor a medal, the Order of the Commander of the Jade, and the Central Aviation School presented to the corps of American instructors a magnificent incense burner, and the latter gifted the school with a silver loving cup. Colonel Jouett, on the eve of his departure, was the guest of honor at a dinner given by new Finance Minister Kung, at which wine flowed freely and testimonials of good will were unrestrained on all sides. The next morning the cold light of dawn reminded the Americans that, in spite of their contributions, they had been discharged.[64]

THE FIRST MODERN AIRCRAFT FACTORY

At the beginning of Nationalist rule, China had only three small aircraft factories located in Shanghai, Canton, and Fujian province. In the entire year of 1929, they succeeded in producing only six airplanes with a top speed of only 70 miles per hour and a ceiling of only 15,000 feet.[65] Soong, who desired to establish China's own aviation manufacturing industry and to reduce the cost of airplanes and equipment to the Nationalist government, had long cherished the idea of establishing an aviation manufacturing industry, and in early 1931, he approached several foreign companies, inviting them to submit proposals. The Boeing Company won out and the U.S. commercial attaché in Shanghai asked the Commerce Department to instruct Boeing to send a representative who possessed the utmost patience, tact, and authority to negotiate a contract with the Nationalist government. The request was marked urgent, otherwise the business might be awarded to others.

While negotiations were under way, the Shanghai Incident heightened the urgency felt by Soong and now much of the government.[66] As soon as he learned of the Chinese sense of urgency, George C. Westervelt, president of the Intercontinent Aviation and a representative of the Curtiss-Wright interest, lost no time in submitting to Minister Soong a proposal for building and operating an airplane factory. This proposal would commit the company to furnish the entire capital for building and equipping a factory and to design new constituent corporations to support the factory. Soong was favorable in principle but passed the proposal on to Arthur Young for some suggestions.[67]

In response to Soong's letter, Young "strongly urged accepting the Curtiss-Wright proposal" because "this company is one of the biggest and strongest in the world and is successfully manufacturing planes in Chile."[68] He reminded Soong that China was then arranging for Americans to assist in developing military aviation, and if the nationals of another country were to handle the factory project, there might be conflict as to methods and policies, and certainly conflict of interest. A German proposal might be superficially attractive, but Germany could not openly build military planes because of the Versailles prohibitions. Soong was still not entirely satisfied with the Curtiss-Wright proposal; he was apprehensive about China being limited to Curtiss products. Soong held that any acceptable proposal should include products of the other leading manufacturers, such as the United and the Consolidated companies.[69] As a result, in August 1932, Soong asked E. P. Howard[70] to request the U.S. Department of Commerce to inquire whether the principal American companies could produce a single joint proposal. But Captain Westervelt demurred: "Combining of competitors is entirely unnecessary and unworkable. . . . In view of our long interest and large investment in China, we [Curtiss] should have preference."[71] To try to resolve these differences, the president of the United Aircraft Exports, Francis H. Love, flew to Nanjing to confer with Soong. The United Aircraft Exports, a trading association, desired to cooperate with Soong in every possible way and hoped there would be a sound basis for getting together on the whole project.[72] Later, Love went to Shanghai and submitted a proposal to H. H. Kung.[73]

Anxious to secure the important contract as quickly as possible, the Curtiss Corporation determined to send William D. Pawley,[74] a top company executive in whom the company reposed great confidence. As Westervelt said, "Our choice of Mr. Pawley for this position, and our taking him away from other important duties and activities indicates the importance we ascribe to our proper relations in China."[75]

When Pawley reached China in February 1933, Kung arranged an interview with Chiang Kai-shek, in the course of which Pawley proposed that the China National Aviation Corporation's civilian service from Shanghai southward to Canton and northward to Beiping be extended and that China purchase Douglas DC-2 planes for these overland flights. Chiang approved Pawley's plan and asked Pawley to build an aircraft factory for the Chinese government.[76] In April, Pawley sent Kung a brief sketch of a factory for Kung's approval, a proposal almost the same as Westervelt's proposal to Soong a year earlier.[77] In response to the requirement of the Chinese government, at the end of April 1933, the Curtiss-Wright Corporation, the Douglas Aircraft Corporation, and the Intercontinent Aviation Corporation organized the Central Aircraft Manufacturing Corporation, with an initial capital of $250,000. Those companies believed that by assisting China in that way, they would help their American manufactured products and materials receive preference.[78]

After a long negotiation, Pawley, on behalf of the Curtiss-Wright Corporation, the Douglas Aircraft Corporation, and the Intercontinent Aviation Corporation, signed a contract with Kung, on behalf of the Military Council, on December 8, 1933. This contract provided that these American interests would erect a modern factory at Hangzhou, Zhejiang province, with a production capacity of 60 pursuit planes annually. At the expense of the Nationalist government, 12 American engineers would work as department heads in charge of 500–600 Chinese workers. The factory was promised by May 31, 1934. The cost of this factory was to be $250,200, repaid by the Chinese government in five equal annual installments, after which the plant would become the property of the Nationalist government Military Council. The American interests also agreed to give one year free instruction in the Curtiss-Wright and the Douglas factories in the United States to 30 Chinese over a period of five years.[79]

The factory did not quite meet the construction deadline but did open in August 1934. The first project consisted of rebuilding 10 crashed Hawks. As each airplane was somewhat differently equipped, the reconstruction of those planes proved far more difficult than simply manufacturing them. The 10 Hawks, however, were eventually returned to service. Subsequently, 22 Douglas O-38s and Northrup attack airplanes were built from scratch. The factory then turned to build 30 Douglas planes of three different designs. Engines and propellers were purchased from the United States, but all other plane parts were manufactured locally. The factory was under American management. Eight Americans served as instructors in the factory at monthly wages varying from $300 to $700.[80] One hundred Chinese workers initially were employed, but the number increased to approximately 300 in 1936. Their daily wages varied from 40 to 85 cents in Chinese currency. Labor costs, of course, were far below those in the United States, roughly 50 percent. American inspectors and foremen proved extremely capable and were successful in utilizing and training the unskilled Chinese labor.[81] Prior to the end of 1936, 127 airplanes were manufactured in the factory at a total calculated savings of $600,210, when compared with the purchase prices of those planes in the United States. From the savings, the Chinese government was able to recover the entire cost of the factory and still post a net profit of $271,133. Nanjing, therefore, became owner of the factory within the first five years.[82]

The factory also benefited American businessmen. The establishment of the factory placed American aviation interests in an extremely favorable position because American interests operating the factory acquired from the contract the sale of aircraft engines and equipment to China. More important, the establishment of the Central Aviation Factory encouraged the expansion of the Nationalist military aviation by training the Chinese engineers and laborers, and by manufacturing modern military airplanes for the Nationalist Air Force.

THE CENTRAL AVIATION SCHOOL

American influence existed not only in the aircraft factory, but also in the Central Aviation School. After John Jouett left for the United States in June

1935, only one officer and two mechanics of the American personnel remained at the school. Because of its great success, the Chinese government still wanted to keep the Central Aviation School in operation in spite of the canceling of Jouett's contract. General Zhou Chirou, in command of the aviation school, had received authorization to renew the contracts of these three Americans as he thought it was in the best interest of the school to employ someone directly from the United States to fill the vacancies in the school.[83] Although Roy L. Holbrook's contract was not renewed by the Central Aviation School in June 1935, he accepted a new two-year contract with the Chinese government in the capacity of a technical adviser to the Central Trust of China, the newly created purchasing department of the Central government. In August 1935, the Central Trust commissioned Holbrook to visit the United States to examine certain aviation equipment and make recommendations for purchase. He was also to endeavor to obtain for the Central Aviation School the services of five American flying officers to replace other American officers whose contracts were not renewed. The new flying officers were not to be military officers of the U.S. government and would be employed only as instructors. Holbrook called at Joseph C. Green's office in the State Department's Division of Western European Affairs to inquire whether the U.S. government would have any objection to his assignment and was told that there was nothing whatever in his mission, as he had explained it, that contravened American law. As Green expressed it, "Certainly the Department of State would have no desire to put any obstacle in the way of his carrying out the commission with which he had been entrusted."[84]

Several months later, Holbrook informed Stanley Hornbeck that he had selected four instructors for the air mission to China, three of them reserve officers, but Hornbeck made no comments on Holbrook's mission.[85] Holbrook needed a dozen pilots and engineers for the Chinese flying school at Hangzhou and solicited recommendations from Claire Chennault.[86] Three pilots and three mechanics, far fewer than he needed, under the supervision of L. R. Holbrook, arrived at the Central Aviation School in August 1936.[87]

In October, Lieutenant Colonel George E. Reinburg, a retired officer from the United States Air Corps many years earlier, arrived in Shanghai. He met various aviation officials and renewed his old acquaintanceship with Colonel F. D. Price, commanding officer of the Fourth U.S. Marines in Shanghai. After Reinburg met with Chiang Kai-shek, the Chinese government assigned him as a special aviation adviser to Chiang, and placed him in charge of the Central Aviation School at Hangzhou, with greater authority than previously enjoyed by Colonel Jouett. He then went to Hangzhou to take up his new duties.[88] By late 1936, 11 Americans had been attached to the school led by Reinburg.[89] The school gave a flight training course and conducted primary and advanced training in the Douglas observation and Curtiss Hawk airplanes. About 160 students underwent flight training. The school also trained technical nonflying officers. These students studied aircraft maintenance for a year and supervised work in the assembly and repair department for another year. The assembly and repair department was under the direction of Americans. After the departure of the Jouett Mission, the school apparently survived without unusual difficulties.[90]

AMERICAN–BRITISH AIRCRAFT COMPETITION IN SOUTH CHINA

In the 1930s, foreign aircraft businessmen were competing actively in China's aviation market when the weight and influence of the air forces became increasingly important in settling political and military differences in China. The Italians established a 20-member mission in Nanchang, organized a primary training school at Luoyang, and erected an aircraft factory in Nanchang in October 1936. No Italian aircraft, however, had been completed before the Sino-Japanese war broke out in July 1937. The Italian mission offered a less stringent flying course than that provided by the Americans, and the Italian-trained Chinese pilots were found to be ill-prepared to combat the Japanese. At Nanchang Aviation School, the headquarters of the Italian air mission, Italians carried out primary training. At the Luoyang Primary Training School, the Chinese students, who reportedly preferred American training systems, did not like the Italian system. "The quality of the instruction put out by the Italians was distinctly inferior to that" given by the American instructors, and there was a "growing dissatisfaction with the efforts of the Italians in attempting to build up China's air force," observed an American intelligence officer.[91] The Italian air mission left China after the outbreak of the war because the Italian government did not want to help China fight against Japan. The Italian presence in the air of China was limited. As Claire Chennault later recalled, "The Italian flying school at Luoyang was unique. It graduated every cadet who survived the training course as full-fledged pilots regardless of his ability. . . . The Italian assembly plan at Nanchang was also a fraud. It turned out large quantities of a Fiat fighter that proved to be a firetrap in combat."[92]

The Germans also were active in China. The Eurasia Corporation was a success of German industrial effort in China. In 1931, the Chinese government signed a contract with Lufthansa, the German national airline, establishing the Eurasia Aviation Corporation, capitalized at $1 million. Eurasia inaugurated the Shanghai-Berlin service on May 30, 1931.[93] By 1936, Eurasia had developed two main services. One route was from Shanghai to Lanzhou and the other from Beiping to Canton. Eurasia aircraft of Junker manufacture were among the most modern in the world at that time.

In the field of military aviation, the German government was determined to gain for German enterprise the right to construct China's first airplane factory. The Junker firm negotiated with the Chinese to construct a plant and train Chinese personnel under an arrangement by which the Nanjing government, the Chinese banks, and the German firms would each put up one-third of the capital. Chiang agreed to this proposal in March 1934, and a preliminary contract was signed on September 29, providing that Junker supply all aircraft materials, including engines, and train the Chinese in manufacture of the planes. In the first year, 54 bombers and 24 other aircraft were to be produced and 200 to 250 airplanes were to be built in the next three years. The site of the assembly plant was set for Hangzhou. The Nanjing government approved the final agreement with Junker on October 1, 1936, and construction began thereafter. No German airplane, however, had been produced before the outbreak of the Sino-Japanese war.[94]

The aircraft competition was also severe in South China in the 1930s. The provincial governments in South China strove to develop air power because it was helpful not only in maintaining local control, but also in protecting against Nanjing's attempts to dominate the South. Like the Central government, the local authorities sought foreign assistance in developing military aviation programs. Foreign aircraft businessmen took this opportunity to extend their influence in South China by selling aircraft, sending air missions, building aircraft factories, and establishing aviation schools, which led Americans to dominate military aviation in Guangdong province, the British in Guangxi, and the French in Yunnan.

American Influence in South China

Founded in 1924, the Canton Aviation School was reorganized in 1928 with General Zhang Weizhang as commandant. Students were all volunteers with at least a high school education. Primary instruction consisted of a short ground course, followed by 15 hours of elementary flying, after which the student was expected to solo. Among other things, students were required to take off, land, loop, split S, and Immelmann before being qualified for their licenses. Thirty-five students received their pilot licenses in 1928, and about 100 pilots and 50 mechanics were trained in the next year.[95] British Avro-Avion training planes were used for primary training because they were lighter and easier to handle than American planes. Advanced students had to receive some instruction in American Ryan Brougham and Waco planes.[96]

The Cantonese became more interested in American aircraft. In 1932, several high-ranking military officers of the Guangdong Aviation Bureau went to the United States to study American aircraft and to raise funds among Chinese-Americans for the purchase of airplanes. After coming home, they decided to ask Americans to send an air mission to Canton. Following up his role in the Jouett Mission, Edward Howard, American commercial attaché at Shanghai, turned to promoting an air mission for the Canton Aviation School. He proposed to send to Canton either three flying instructors (one specialized in bombardment, one in pursuit, and one in observation) and one chief mechanic, with monthly salaries of $200 each, or one adviser and one mechanic, with salaries of $600 and $200, respectively. When he informed Leighton Rogers of his plan, the latter agreed with his project and recommended the second alternative. The first alternative, according to Rogers, offered salaries too low for capable pilots, but mechanics' salaries were the same in both plans: $200.[97]

With Rogers' approval, Howard traveled to Canton in December 1932 with his proposition of an unofficial American mission on a much less elaborate scale than Jouett's in Hangzhou. The Canton Aviation Bureau replied that they needed only two pilots of noncommissioned officer type who would receive $200 a month and two mechanics who would receive $150. In spite of the competitively low pay, the pilots had to be competent to instruct in all phases of combat flying, including artillery observation, bombardment, and pursuit, and the mechanics were required to instruct in gunnery and photography. Further, one of the ground men was to be qualified to set up a complete photographic laboratory. These instructors, in place of more highly qualified and richly paid advisers, were to work under the command of Cantonese officers.[98] At this time, Rogers selected Edward Deeds, a man with considerable experience in aviation training in United States, as head of the prospective mission, and contacted a number of airmen, presumably recommended by Jouett, as to whether they would be interested in such a mission.

Rogers sent the proposed list of names to such aircraft companies as Curtiss-Wright, Douglas, and United, and insisted that those companies "must necessarily handle this very confidentially." While negotiation with the Canton government was in process, Rogers repeatedly instructed Howard that "under no circumstance should there be any publicity on the matter."[99] Rogers must certainly have been aware that the Department of Commerce was anxious not to publicize this mission, in light of the Japanese already using the arrival of the Jouett Mission group for propaganda purposes, unfavorable to both the United States and China. If the Japanese knew of the negotiations between Howard and Canton, it might embarrass the State Department sufficiently to force it to ask the Department of Commerce to stop all activity. Rogers reaffirmed that "it is our function . . . to promote the sales of aeronautic equipment and this is an excellent way of doing it."[100]

The arrangement moved ahead. The Department of Commerce finally selected Deeds and two others for the Canton Mission. Before his departure for China, Deeds visited the Boeing plant, which contributed to another big order from the Canton authorities for Boeing airplanes. On February 8, 1933, the air technicians appointed by the Department of Commerce sailed from Seattle to take up duties at Canton.[101] Deeds was a cautious and loyal man at the advanced training school. When the Chinese pilots wanted to fly some new American planes, he judged those aircraft too fast and refused to let them take them off until he personally tested every one. Unfortunately, when he tried a new Curtiss Hawk, an extremely fast pursuit plane, the plane crashed and he was killed on the first of July in 1933.[102]

The Canton Aviation Bureau asked that Deeds be replaced with another American instructor with an equal knowledge of pursuit, army cooperation, gunnery, observation, and map making. The new instructor, whose salary would be $200 monthly plus traveling expenses, was to be preferably a captain or a major. Thus, on July 6, Julean Arnold, commercial attaché at Shanghai, suggested that Rogers cooperate with William Pawley to choose a flight instructor for the Canton Aviation Bureau. Several months later, on September 7, Captain Claiborne, again appointed by the Department of Commerce, was en route from Los Angeles to Canton. In 1934, the Canton Aviation School employed four foreigners as instructors.[103]

American instructors contributed to the successful development of the Canton Air Force. From January 1933 to July 1936, the Canton Aviation School graduated approximately 150 pilots trained by the American instructors. The graduates of the school formed the bulk of the flying personnel of the Canton Air Corps.[104] More important, American instructors extended American influence on military aviation in South China by introducing the U.S. training system. After suppressing the Canton rebellion in July 1936, the Nanjing government took over the Canton Aviation School, renaming it the Central Aviation School at Canton.[105]

In addition to an aviation school, Americans were interested in an aircraft factory in Canton. Aircraft assembly and fabrication began early in South China, although initially at a very modest level. As early as 1922, Yang Senyi, who studied aviation in the United States, returned to Canton, accompanied by two

Americans, hundreds of cases of airplane equipment and accessories as well as three assembled JN 9 and one JN 4 American airplane. With the support and advocacy of Dr. Sun Yat-sen, president of the Canton Revolutionary Government, an airplane repair shop was built in Daxiatou, Canton. A training plane designed by Americans and equipped with an American engine was successfully manufactured at that "repair shop" in July 1923, the first plane manufactured in China. Not until a Chinese engineer, Mei Lunkan, who studied aviation in the United States, returned to Canton in 1927 was further progress made. Mei Lunkan enlarged the repair shop, enrolled more talented workers, and fashioned the shop into an airplane factory at Dongshan, Canton. The next winter, Chinese mechanics manufactured a monoplane, constructed entirely with Chinese materials except for the engine. It had modest capabilities, a top speed of 100 miles an hour and a ceiling of 15,000 feet, but it made a successful test flight on December 29, 1928, and proved serviceable thereafter. From 1928 to 1933, the aircraft factory at Dongshan manufactured a number of small airplanes called "Yangcheng."[106]

With these small achievements behind them and with nearly two years' experience with the training mission of Deeds and others, the Canton military authorities decided in 1933 to appeal for American assistance in erecting a modern airplane factory in Guangdong province. The advantages of U.S. supply of manufactured equipment and aircraft parts were several, as Canton saw it. First, sea passage took only 22 days on the average from American west coast ports, even with one intermediate calling port, while European passage took about twice the time, with many calling ports. Second, American businessmen were constructing an aircraft factory in Hangzhou for the Nanjing government, so they could start another one without difficulties.

Events moved quickly. Responding to an official request from Canton, William Pawley, president of the Intercontinent Aviation Incorporated, forwarded the news that "The Curtiss-Wright Corp. agreed to purchase for the account of the Canton Air Force machinery, tools, and equipment necessary for a factory capable of producing sixty airplanes per year."[107] On February 22, 1934, the Canton Aviation Bureau wrote a letter to Francis H. Love, president of the United Aircraft Exports, asking that company to establish an airplane factory on Canton's account. Several days later, Love replied, "We could take no financial part in the proposed factory but would render all assistance feasible in a technical way."[108]

In June, Pawley concluded a contract with the Guangdong province for the sale of $140,000 worth of machinery and special tools to be used in the projected factory.[109] The factory was to be modern in every respect with a monthly production capability of three bombers or six two-seater training planes. Machinery and tools for the factory was to be shipped from the United States on or before July 9, 1934, and the factory should be in production before the end of 1935. Unlike the factory being constructed at Hangzhou for the Nationalist government, the Guangdong factory was financed entirely with Chinese capital, Intercontinent Aviation refusing to float a loan.

To cement this arrangement, General Wang Guangrui, commander of the Canton Air Force, came to the United States in June 1934 with other officers of the aviation headquarters to study aircraft manufacturing and recruit aircraft engineers. Charles Healy Day, formerly associated with the New Standard Aircraft Corporation of

Paterson, New Jersey, was hired to assist in directing the operations of the new airplane factory being constructed at Shaoguan, 200 miles from north of Canton. Day, a noted flyer and designer, served as the vice president and chief engineer of the New Standard Aircraft Corporation, Paterson, New Jersey from 1927 to 1930. With his recommendation, two mechanics were hired for the Canton Air Force.[110]

One of these, Constantine L. Zakhartchenko, a Russian-American in his mid-forties, was educated at the Russian Imperial Naval Academy and received a master's degree at Massachusetts Institute of Technology. From 1923 to mid-1934, he was employed as a design engineer by various aircraft designers and manufacturers. On August 1, 1934, he contracted with the First Group Army Corps of Canton as assistant engineer-in-chief at Shaoguan Aircraft Works at a salary of $350.[111]

In January 1935, over half a million dollars' worth of airplane materials, accessories, and engines was delivered in Canton. In July, Arthur D. Engle, vice president of the Austin Company of New York, who was at that time supervising the construction of the airplane factory in Hangzhou, was assigned to the Guangdong factory. Under Engle's supervision, the projected airplane factory began construction on the south side of the city of Shaoguan on the fork of the Jin and North Rivers.[112]

Chief Engineer Charles Healy Day did not reach Canton until mid-autumn of 1935, in time for the assembly line to begin production on December 1, 1935. On that day, the Shaoguan Aircraft Factory came into existence, only a year after negotiations began. Without building any prototypes, the assembly of 20 planes began without designs being fully detailed. Such was the sense of need for great speed.[113]

The factory and its personnel, independent of the Canton Air Force, were directly subordinate to Marshal Chen Chidang. Zhou Baohang, superintendent of Canton Customs, served as general supervisor of the airplane factory and its approximately 450 employees. The factory employed five American engineers.[114] On February 3 and February 10, 1936, the Intercontinent Aviation Corporation signed two contracts with the First Group Army Headquarters of Guangdong. Under the first contract, the Intercontinent Aviation Corporation would supply the Guangdong military with one Vultee and one Hawk airplane as models, as well as with the necessary materials to manufacture 29 planes of each type. Most of the materials were due to arrive at Canton in June 1936. The Shaoguan Aircraft Factory was to be turned over to the Intercontinent Aviation Corporation before July 15.[115]

The first craft called Fuxing was test flown on May 20, 1936, and it demonstrated outstanding flying characteristics and performance. Ten days later, Charles Day proudly flew the first Fuxing to Canton for a preliminary demonstration for Marshal Chen Chidang. With such success, the design department immediately started to work on a more advanced design while production continued on the basis of the original design program that called for 10 advanced trainers and 10 all-purpose planes based on the original design.[116] Because of the American engineers' success at Shaoguan, the Canton military authorities asked them to enlarge the small airplane factory at Dongshan so as to manufacture Fuxing airplanes. The Shaoguan Aircraft Factory would produce only Vultee and Hawk planes.[117] Five American engineers replaced Day and his associates at the Shaoguan factory. Thus, the Shaoguan Aircraft Factory would concentrate in building Hawk III and Vultee planes.[118]

Formerly under control of Colonel Mei Lunkan, the factory at Dongshan was in need of refurbishing and modernization, and Day submitted a plan to the Canton military authorities for roughly $30,000 in equipment to be ordered from the United States, a very small investment indeed. The Shaoguan Factory was to transfer all unfinished airplanes and materials to Dongshan, and Zakhartchenko and other engineers were to be transferred. Zakhartchenko grandly claimed that "the engineer department will be in position to produce designs at least three or four years in advance of the rest of the world."[119]

As the factory transfer continued, General Chen of Guangdong province and General Li of Guangxi declared independence from the Nanjing government on June 1, 1936. In only two months, the Nationalist government suppressed the rebellion, and Chen was forced to flee to Hong Kong in August. As a result, the National Aviation Commission of Nanjing simply assumed ownership and control of the Shaoguan Aircraft Factory. Although it underwent only slight modifications, the old personnel were retained, including Americans on substantially the same terms and in the same capacities they formerly enjoyed.[120] The Shaoguan Aircraft Factory, together with the Hangzhou, proved an essential contribution to the development of Chinese air power in South China.

In dealing with the rival factions in China, the Commerce Department's project conflicted with the State Department's China policy, which was to discourage civil wars in China and not to provoke Japan. The publicity of American air assistance to Canton would contribute to Japan's and Nanjing's protests. The Commerce Department, however, could not be criticized for sending the air mission to Canton and building an aircraft factory for the Cantonese, because Washington did not have a clear-cut political policy toward China. In addition, promoting the sale of American aircraft to other countries was very important to U.S. aircraft industry hit by the Great Depression. C. L. Egtvedt, an officer of the Boeing Airplane Company, described the worsening situation in a letter of February 21, 1934, to the chief of Air Corps in the U.S. War Department. Egtvedt wrote that at the present time his company had no prospects of any commercial business and had only an insignificant order of Navy parts. The company's plant was running at about 60 percent capacity and its primary shops were already running out of work, necessitating the layoff of men in these departments. By summer, a great number of employees would be out of work, and in the fall, all of the company's current orders were to be completed. Because of these conditions, they were very anxious to obtain some immediate business which would help to keep a portion of their employees at work.[121] The situation of Boeing suggests that American aircraft manufacturers were struggling for their survival during the Great Depression.

The efforts of the American manufacturers in cooperation with the Aeronautic Trade Division proved highly successful in China from 1929 to 1934. A statistical list giving a complete picture of the U.S. export trade in aeronautical products from 1929 to 1934 is as follows:

Year	Value of Aeronautic Products	Exports to All Countries	Exports to China
1929	$70,334,107	$9,202,385	$ 783,790
1930	$60,846,177	$8,806,396	$1,210,329
1931	$54,520,627	$4,809,440	$1,006,656
1932	$34,861,185	$7,633,070	$ 367,925
1933	$33,485,576	$9,155,882	$2,246,362
1934	$42,733,865	$17,548,181	$3,842,510[122]

From this list, one may see how important the Chinese aircraft market was for the American aeronautic industry during that period. The China aircraft trade encouraged the development of the American aircraft industry and had significance for employment in the U.S. aircraft industry. Employment in the U.S. aircraft industry was 18,600 in 1929, and only 9,600 in 1933. The average weekly salary was $30 in 1931,but only $25 in 1925. In the 1930s, 75 percent of the cost of constructing aircraft went into labor. As analyzed by the Aeronautical Chamber of Commerce of America, it was estimated that $15,000,000 of export business a year meant the employment for the year of at least 4,000 or 5,000 American skilled workmen whose experience would be invaluable in wartime and who otherwise might have to be maintained at public expense. Promoted by the Commerce Department, the China aircraft trade impacted in some small measure on the survival and development of the U.S. aeronautical industry in the 1930s.

British Influence in South China

British aircraft businessmen, like Americans, were very active in South China. Four leading firms in Hong Kong represented British aircraft manufacturers: the Far Eastern Aviation Company; Messrs. Arnold and Company; Reiss, Massey and Company; and the Jardine Engineering Corp. In 1929, the Arnold firm established an aviation department for the purpose of selling aircraft to China,[123] but the Far Eastern Aviation Company, established in Hong Kong in 1930 with business offices in Guangxi, Guangdong, and Fujian, was the most active company in South China. Guangxi province was of special interest to the Far East Company, and subsequently it became a major market. After the Guangxi Air Corps became independent of the Guangdong military clique in 1931, it underwent steady development under the guidance and advice of R. Vaughan-Fowlcr, managing dircctor of thc Far East Aviation Company.[124]

With the help of a British air mission, the Guangxi government established an aviation school in Liuzhou in 1932 with two classes: one flying and one mechanics, each of which had 30 students initially. The instructors were all Royal Air Force reservists who were closely affiliated with the Far Eastern Aviation Company. In addition to sending instructors, the Far Eastern Aviation Company furnished all equipment, including 30 or so Armstrong Whitford pursuit planes and Avro planes, all shipped between 1933 and 1935.[125] The British, therefore, dominated the aircraft market in Guangxi.

The Guangxi authorities were not content to import aircraft; in July of 1932, they began to build a factory at Liuzhou that would manufacture and repair aircraft. Smaller than the other factories, this one employed 120 mechanics and

100 apprentices and was capable of producing two airplanes per month. Twelve Avro Avians and two Avro Cadets had already been assembled by April 1935, and six Avro Cadets were in the course of construction.[126] The result was a substantial order from the Guangxi regime for British aircraft.

To assist the Far East Aviation Company in its marketing in South China, John Higgins, chairman of the board of Armstrong Whitworth Aircraft, traveled to Hong Kong and concluded that a flying school in Hong Kong would contribute to the increasing British influence in military aviation in South China. On March 1, 1934, the Far East Flying Training School under the control of the Far East Aviation Company began operations at the enlarged airdrome in Hong Kong with four instructors. The expansion of the airfield cost HK$100,000 and created a facility large enough to accommodate 70 planes. It housed not only the planes of the Royal Air Force, but those of the school as well. The establishment of this school heightened British influence on aviation immediately because students from the Guangxi Aviation School were compelled to attend the Far East Flying School for advanced training.[127]

To be sure, with British cooperation, the Guangxi Air Force underwent rapid growth. In 1935, Guangxi could put a total of 46 airplanes in the air, all of which were of British manufacture.[128] The entire force was located within the province of Guangxi with its principal base and school at Liuzhou. Other airfields within the province were known to be at Nanning and Wuzhou. The Air Corps was directly under the control of the provincial military leader, Li Zongren, who was opposed to Chiang Kai-shek and allied with Chen Chidang of neighboring Guangdong.

British diplomatic officers played an important role in promoting British aircraft sale to China. Only about two months after the China arms embargo was lifted, British minister to China Miles W. Lampson suggested:

> Although the Americans may be relied upon to make effort to exploit this promising field of enterprise, it is still too easy to assume that Americans are on the road to creating a monopoly, or even that they have yet secured a firm foothold. . . . British manufacturers should lose no time in interesting the Chinese authorities in their respective types of aircraft.[129]

In order to help its nationals to compete with America on the airplane market in South China, the British government decided to send a representative to China. The London government tried to provide assistance to the British businessmen as much as did the Commerce Department to U.S. business interests. In 1932, E. M. B. Ingram, British ambassador in Beiping, first urged London to dispatch an air attaché to China: "As to the air side of the picture, I have long been thinking that if we had an Air Attaché out here we might have been able to do more than we have and that the Air Mission would not have come to nothing." Later, Ingram again stressed the significance of an air attaché and encouraged the Foreign Office to send one when possible. He held that London was not taking proper advantage of its opportunity and was letting other countries, especially the United States, surpass the British. Americans, for example, had established their new flying school at Hangzhou and had been able to attach aviation experts to their legation.[130] Ingram tried to draw the attention of the British government to the Chinese aircraft market by waving the red flag of success. As a result, Anthony Eden, under-secretary of Foreign Affairs, announced on July 3, 1933, in the House of Commons that steps were being taken to appoint an air attaché to the British Legation in Beiping. Several months later, Wing Commander R. P. Willock was nominated as the first occupant of the post. This would "prove of value in connection with the growth of aviation in the Far East," commented the *Times*.[131]

Willock arrived in Shanghai early in 1934 and assumed his duties with his headquarters in Shanghai rather than Beiping. Shortly, thereafter, he met with the British aircraft representatives, including C. H. Arnold (Messrs. Arnold and Company), F. A. E. Smith (Far East Aviation Company), and Brian Lewis (Reiss, Massey and Company), in order to coordinate their efforts in the whole of China. Hope rose among the local representatives of British airplane manufacturers that the air attaché could stimulate the interest of the Chinese government in British machines. To increase the sales of British aircraft to Canton, Willock went to Canton in April to visit the headquarters of the Guangdong Air Force, accompanied by Herbert Phillips, the local British consul general. While touring, Willock invited General Lin Fuyuan of the Guangdong Air Force to visit Hong Kong for the purpose of seeing the British Flying Boats. After visiting the Far East Flying School, Willock confidently commented, "At the present time, the majority of Chinese pilots and mechanics are American trained. The Far East Flying School, by means of bringing Chinese in touch with British aircraft and methods of training, should go a long way in assisting British aviation interests in China in the future."[132]

The seed Willock planted grew and bore fruit. In October, the Far East Aviation Company concluded a contract with the Guangdong Air Force, which provided that the company would sell one Vickers Bygrave bombing airplane to Canton. That was the first contract obtained by British aircraft businessmen with the Canton Air Force since 1932. Although the order was small, it was important. As Herbert Phillips commented, "It marks the beginning of the attempt by a British organization to break the strong American monopoly in air matters at Canton."[133] In the following years, the British increased their aircraft sales to South China and came to dominate the military aviation in Guangxi province.

The Problems of Aircraft Sales

Military airplanes purchased by South China provinces abroad angered the Nanjing government, which was attempting to unify China by military force. Facing the resistance of the local warlords, Nanjing relied on Germany for military advice and arms[134] and on American aircraft and aviation advisers to strengthen its military power. At the same time, after the China arms embargo was lifted, Nanjing tried to control tightly the importation of arms, including aircraft, in an effort to ban the local warlords from gaining Western arms and airplanes. In 1929, Nanjing promulgated and disseminated orders that no military planes were to be purchased by individual military units or provincial governments, and that import permits would be only granted for training and commercial planes. In May 1930, Nanjing decreed that war materiel imported into China would be considered illegal unless it approved each particular consignment through its diplomatic representative in the exporting country.[135] Since Nanjing could not effectively control the whole country, it needed the cooperation of the major powers to prohibit exportation of arms and aircraft to the provincial warlords.

The United States, Great Britain, and other powers continued to regulate their arms export to China after the China arms embargo was terminated in 1929 by requiring licenses for these exports and issuing them only for shipments designed for and approved by Nanjing. The goal of the policy was to promote the stability of the recognized Nanjing government and discourage the continuation of civil strife. To strengthen the National government and to reduce the military power of the local governments, the State Department, complying with the Nanjing government's decree of May 1930, issued in July a statement mandating that firms wishing to make shipments of arms, including military planes, must have both the import permits of Nanjing and export licenses issued by the Department of State. Export licenses were not required, however, for commercial aircraft except for planes that could easily be converted for military use.[136]

However, the arms sales to South China were not so simple. The Nationalist government was recognized by the major powers as the government of China, but Guangdong and Guangxi provinces as well as some provinces in South China remained functionally independent of Nanjing. Relations between Nanjing and Canton were sometimes good but at other times strained. At the times when the Canton government was not openly hostile to Nanjing, import permits were sometimes issued to the Canton authorities for the importation of arms and airplanes from other countries. Even so, the Canton government declined to approach Nanjing with what it saw as a demeaning request for a Hujiao [import permit] covering airplanes to be imported; the Canton government regarded that as "losing face." The Canton authority could purchase directly arms and airplanes from the Western powers without difficulty. In addition, under the terms of the Barcelona Convention of 1921, direct military supplies were permitted to pass through Hong Kong to Canton if the latter's permits were obtained. Military shipments originating in Hong Kong also were allowed to reach Canton. The Hong Kong escape hatch provided Americans with a channel to sell military airplanes to Canton in spite of Nanjing's dismay. In addition to military craft, they sold "commercial aircraft" that could be converted easily into military planes.[137]

In 1931 the Canton government rebelled against the Nanjing government. On February 28, 1931, Chiang Kai-shek, chairman of the Nanjing government, placed his colleague Hu Hanmin under house arrest. Hu, a senior leader of the GMD and head of the Legislative Yuan at that time, had long been dissatisfied with Chiang's monopoly of power in the party and government. He enjoyed a significant following within and outside of the Nationalist government. Enraged by the house arrest, his followers in the government soon left Nanjing in protest. They found an ally in their opposition to Chiang in the Canton militarist Chen

Chidang. In May, they gathered in Canton and, with Chen's support, organized an "Extraordinary Session of the GMD Central Committee" to change Chiang's authority. The "Extraordinary Session" immediately established its own "National government" in China. A full-scale civil war seemed imminent. The Canton authorities, therefore, desperately craved fighting planes during their rebellion. And in July, 24 American fighter planes ordered by the Guangdong Air Force were delivered to Hong Kong to be transported to Canton.[138] At this time, the Canton Aviation Bureau informed the American consul general at Canton that an order had been placed with an American company for 12 additional military airplanes to be exported as commercial aircraft to Hong Kong, where they were to be armed with Browning machine guns and then delivered to Canton.

Of course, these purchases drew the attention of the Nanjing government. Minister T. V. Soong informed Willys R. Peck, American consul general in Nanjing, that the Canton government "had purchased eight Lockheed Vega planes manufactured in Burbank for immediate delivery consigned to Canton by way of Hong Kong" and that "sellers had furnished Canton authorities blueprints showing how these commercial planes could be easily converted into prohibited planes by mounting machine guns."[139]

In July 1931, the State Department permitted 20 Douglas planes, each of which was equipped with two aircraft guns and two bomb racks, to leave the United States for Macao as commercial airplanes. There could have been no doubt that the final destination of those planes was Canton and that additional armaments would be added to those planes when they were delivered to Canton. American minister Nelson Johnson expressed his bewilderment as to how these "twenty Douglas planes were permitted to leave the country as commercial planes."[140]

To stanch the influx of American airplanes into Canton, Nanjing asked Washington at the end of July 1931 to "take all steps possible to prevent the issuance of export licenses for planes consigned to Canton, or to destinations which would make it evident that Canton was to be the ultimate destination." Although those planes were civilian models, they could easily be converted to military planes or to be used for military purposes. Several days later, Johnson informed the Ministry of Foreign Affairs that his government was taking active steps to prevent the shipment of the eight Lockheed Vega planes to Canton.[141]

The shipment of military airplanes from the United States to Canton by way of Hong Kong or Macao soon was treated at length in State Department offices. Joseph Green, chief of the Division of Western Affairs, argued that if the export of arms to China via Hong Kong became a serious problem, the U.S. government should insist upon export licenses. That, he reminded the Department, had been done in the case of arms shipped to Honduras with a probable destination of Nicaragua. Stanley K. Hornbeck, chief of the Division of Far Eastern Affairs, firmly supported this view.[142] After some discussion, the Department of State decided to restrict the export of arms from the United States to South China.

William Castle, acting secretary of state, instructed Minister Johnson that under the current political conditions in China, the State Department viewed the exportation from the United States to Hong Kong and to Macao of arms and munitions of war, including military aircraft, as destined presumably for China, Nanjing, or elsewhere. He further pointed out that unless that presumption could be verified by the exporter in the United States, any such exportation should be accompanied by an export license duly issued by the department.

Why did the State Department enforce these restrictions on exporting military planes to China? The answer may be found in a memorandum of a conversation between Hornbeck and Leighton Rogers of the Bureau of Aeronautics of Commerce Department, wherein the latter took note that some of the aircraft companies were inclined to attempt to "get around" the U.S. regulation and policy wherever they saw a possible sale in sight. Hornbeck replied that "this would indicate shortsightedness on their part . . . while it may mean increase of small sales temporarily, must mean in the long run a retarding of large purchases which might and probably would be made by the *Central Government*."[143] Hornbeck later amplified in persuasive words:

> [T]he less the number and the severity of the civil wars in China, the more the chance of the Central Government to build up an effective military force, and, therefore, the greater the number of the purchases of first class equipment which it may make . . . the policy of the American Government is based on the principle of discouraging civil wars and contributing to the maintenance of peace in China.[144]

Considering commercial airplanes as tantamount to the export of unassembled war materiel to China, the State Department began in August to require licenses for their exports within the terms of the embargo. For the first time, the Department of State regulated arms and commercial planes trade not only to China but also to Hong Kong, the British colony, and Macao, the territory under Portuguese authorities. Obviously, to prevent the possibility of the civil war and to maintain stability in China, the Department of State attempted to enforce restrictions on exporting aircraft to South China.

When the U.S. government began to control tightly the sale of aircraft to South China, the Cantonese turned to the British. In 1931, the Guangdong regime purchased six Armstrong Whitford airplanes fitted with military equipment from the Far Eastern Aviation Company. It was apparent that they were strictly pursuit airplanes built to carry no cargo other than a pilot, guns, and ammunition. Nanjing protested to the British Legation in Peking, requesting that "transport of the consignment to Guangdong be prevented."[145]

The sale of these British planes to Canton also drew the attention of American businessmen. The local agent of the L. E. Gale Company complained to the American consul general in Hong Kong that shipments of military supplies of various kinds were being made freely from several European countries to Canton, while the U.S. government refused to permit the export of such articles to Canton. Thomas H. McConnell, vice president of the United Aircraft Exports Corporation, complained to Willys Peck that the British authorities at Hong Kong placed no restrictions in the way of sales of fighting planes by British merchants to the Chinese authorities at Canton. Moreover, he had an assurance that military planes consigned to Canton by way of Hong Kong would not be interfered with at Hong Kong.[146]

The complaints of American businessmen aroused the attention of the Department of State, which thereafter recognized the need to coordinate action with the Foreign Office in an effort to control effectively aircraft sales to South China from Hong Kong. At the end of July, Under Secretary of State William Castle telegraphed the U.S. ambassador in London, Charles G. Dawes, to inquire informally of the appropriate British authorities with regard to British policy and procedure respecting these shipments.[147] Within days, the Embassy reported back that London's policy was that no arms or munitions of war including military airplanes could leave Britain without export licenses issued by the Board of Trade. The Foreign Office gave their concurrence to shipments destined for China only with the sanction of the Nanjing government, and the Hong Kong government had been requested to conform to this policy. It seemed that the British government had an identical policy and position to that of the U.S. government. The Foreign Office, however, added that commercial airplanes could be exported to China or anywhere without reference to the Board of Trade or the Foreign Office and that the Hong Kong government could not interfere with transit shipments of arms through Hong Kong from other countries.[148]

This British attitude aroused the dissatisfaction of Nanjing authorities, and on September 7, the Ministry of Foreign Affairs lodged a protest with the British minister against the sale of Armstrong Whitford pursuit airplanes to Canton. The Ministry demanded that the British government take steps to prevent the delivery of those airplanes. Two days later, one of those planes flew from Hong Kong to Canton, although the delivery of the other Armstrong Whitfords was held up.[149]

The Far East Aviation Company delivered a second Armstrong Whitford pursuit plane in late October to the Canton Aviation Bureau, and the Canton Aviation Bureau confirmed that an order for three more British airplanes of the same kind had been signed and that additional purchases of British planes were contemplated. It was not unpredictable that representatives of L. E. Gale complained to Washington that "unless British competitors were prevented from selling military aircraft to Canton, American aircraft trade would be adversely affected."[150]

The delivery of these craft needled the State Department in part because, beginning in August 1931, the department imposed more rigid regulations on the export of American military airplanes to Canton. Ambassador Dawes, therefore, registered a protest with the Foreign Office, but no reply was received.[151] The State Department then instructed Dawes to draw discreetly and informally the attention of the Foreign Office to this reported delivery to the Cantonese authorities. Nearly a month later, an official of the Colonial Office told Dawes that no license had been issued from the London government for the export of airplanes destined for Canton, begging the question.[152]

Nanjing's and Washington's remonstrance to London notwithstanding, six more British fighting planes were delivered at Guangdong in December 1931, each costing $170,000 in Chinese currency. These aircraft boasted a top speed of 200 miles per hour and were equipped with machine guns, planes hardly intended to carry civilians from one city to another.[153] London took no steps to prevent the delivery of British airplanes to Cantonese officials and defended the practice. The Foreign Office used a loose and sliding definition of military airplanes, dividing aircraft into two categories only, armed and unarmed. So long as aircraft of a military type were not equipped with machine guns or machine gun mountings, they were not regarded by the British government as military aircraft within the meaning of the arms embargo. Hence, export licenses were not required. Under the regulation, British businessmen were able to deliver military airplanes to South China without export licenses. When British military planes were delivered to South China, armaments were sent separately to Canton. The planes that were transferred from Hong Kong to Canton in October contained no machine-gun mountings, and, therefore, the Hong Kong government had no power to prevent the clearance. Airplanes with machine-gun mountings in place, it was insisted, did not go to Canton through Hong Kong.[154]

The Department of State, on the other hand, defined unarmed military craft as military weapons. In short, export licenses were required for release of unarmed military airplanes. For that matter, even commercial planes, which could easily be converted to military use or were evidently intended for military use, required export licenses. Furthermore, the Department of State held that the exportation of military planes from the United States to Hong Kong or Macao as destined presumably for China should be accompanied with an export license.

The different definitions were only part of the tale. The Hong Kong government also permitted the shipment of military airplanes on a Hujiao issued by the local governments in South China. As American Trade Commissioner Thomas C. Barringer pointed out, "the shipment of aircraft to various South China provinces from Hong Kong is a comparatively simple matter because both Hong Kong authorities and the Chinese Maritime Customs recognized the Hujiao issued by the South West Political Council at Canton. A Nanjing Government's Hujiao is not necessary."[155]

The result of these policies and practices was clearly predictable. In 1932, for example, only 15 airplanes from the United States were exported or transported through Hong Kong to South China, while 40 British aircraft were sold in that area. At that time, the Cantonese authorities were "anxious to obtain American aircraft if possible," but they were not inclined to make these purchases through the Nanjing government.[156]

With the prospect of losing substantial sales to British firms, American diplomatic and commercial officers in China and Hong Kong complained vociferously about the China arms policy of the Department of State. After a meeting with the representatives of aircraft businessmen in July 1932, Frederick W. Hinke of the Far Eastern Division encouraged the State Department to consider "the advisability of liberation for its construction of the term 'military airplanes' to coincide with the British view." Trade Commissioner Barringer pointed out that American exporters were placed in a noncompetitive situation, an opinion which Consul Douglas Jenkins confirmed. In the best diplomatic fashion, Jenkins suggested that "the Department endeavor to ascertain just what the British attitude now is respecting the shipment of military planes to China especially to Canton via Hong Kong, and to afford American manufacturers to compete with British aircraft on more favorable terms."[157] The *China Trade Monthly Report* of the Commerce Department concurred, "due to a restriction on export to the southern provinces, the Chinese aircraft market was at least temporarily lost to American trade resulting in a considerable increase in British participation."[158]

Again, expressing irritation with the British relaxation of aircraft sales to Canton, the State Department was stirred to action. First, the State and Commerce Departments held a conference in November and reached an agreement. This agreement allowed exportation of commercial airplanes from the United States to Hong Kong for the Guangdong government without the specific approval of the Nanjing government, or the confirmation of it by the Chinese Legation in Washington. American aircraft manufacturers should mark the documents and the packages clearly as commercial aircraft and ship them to civil organizations rather than military institutions in Canton.[159]

Second, the State Department exerted pressure on the British government to the end that if Britain did not restrict the sale of military airplanes to Canton, the American government would relax its rigid control of the sale of military airplanes to Canton. On December 29, 1932, Under Secretary of State Castle had a conversation with Sir Ronald Lindsay, British ambassador in Washington, warning that if London did not honor restrictions as did Washington, it might prove necessary to open the field generally. However, no agreement came from that conversation.[160]

European competition on aircraft sales to China became severe in 1933. The British government appointed Wing Commander R. P. Willock as air attaché to China. The Italian government agreed to refund Boxer indemnity payments to China as credits for purchase of aircraft and decided to send an air mission to Nanjing, helping the Chinese establish an aviation school and a modern aircraft factory. The French government waived transit taxes and duties for its aircraft to be delivered in Yunnan province on the Indochina frontier, considered as France's sphere of influence.[161]

Under these circumstances, officials from the State and Commerce Departments opened discussions again on the problem in the spring of 1933. Hornbeck admitted that the State Department, in effect, had placed American commercial interests attempting to serve the markets of China under a severe handicap. The regulations governing the export of arms and munitions of war to China might be modified, he suggested, so as to eliminate the requirement that the Chinese Legation at Washington inform the State Department. Leighton Rogers of the Aeronautics Trade Division confirmed "the ineffectiveness of our policy in preventing the various Chinese provinces being supplied with aircraft."[162]

The pressure from the American aircraft industry, the efforts of the Commerce Department, and growing American suspicion of British arms policy toward China gradually effected a change in the State Department's position. The State Department announced on June 30, 1933, that the exportation of arms and munitions of war for the use of Chinese official authorities would be permitted in all cases when the firms in the United States, desiring to make a shipment, had submitted to the Department of State an application for license to export and the Chinese Embassy failed to notify the State Department that the shipment was prohibited by the Central government.

The previous regulations of the exportation of arms to China provided that licenses were granted when (a) an application for license to export had been submitted by the firm or firms in the United States, which desired to make shipments, and (b) the Chinese Legation at Washington had informed the department that it was the desire of the Chinese government that export of the shipment be authorized. In the new modified regulation, the provisions of section (b) were eliminated.[163] The acting secretary of state explained that section (b) of the pertinent regulation would "incur heavy cable charges" when American businessmen attempt to "cause the Chinese Government to issue the necessary instructions to its Legation at Washington."[164] The busy work had proved irksome to buyers in China, particularly in South China, and valuable orders had been lost to the American trade for no reason except that the Chinese government was negligent and other governments apparently did not insist on similar procedure.

Apparently, the purpose of the elimination of the provisions of section (b) was to remove the disadvantages under which American military airplane exporters to South China were placed in comparison with exporters of other countries because military airplanes were a very large portion of arms exported from the United States to China.[165]

Chinese customs figures suggest that the U.S. attempt to improve the conditions under which American manufacturers competed provided no quick successes. At the beginning of 1934, the value of airplanes to China from Great Britain was more than four times that of the United States, according to the following list:

Value of Aircraft and Accessory Imports into China (In customs gold units[166])

	1932	1933	Jan. 1934
Great Britain	604,372	217,553	388,514
United States	516,763	4,594,544	78,204[167]

Most British aircraft were shipped through Hong Kong into Canton, according to the customs report.[168]

The increasing consignment of British military airplanes to the South continued to disturb Nanjing. Through diplomatic channels, Nanking urged all foreign countries to conform to the Chinese request in regulating the export of arms and munitions, including airplanes, to China. Nanjing had occasion to address Britain directly in March 1934 when it informed the British minister in China that a recent consignment of British war materiel had not been authorized for import to Canton. Nanjing particularly drew London's attention to the Hong Kong channel. The minister, Sir A. Cadogan, passed that request on to London. The Foreign Office was not favorably inclined to the Chinese request, rationalizing that the United States, together with other countries, did not enforce effectively the Nanjing government's Decree of 1930. The foreign secretary complained to Cadogan that the Far Eastern Aviation Company recently had lost a contract with the Guangxi government for the supply of aircraft with armaments to the value of 115,000 to American and Italian competitors. Without Nanjing permits, the Far Eastern Aviation Company was unable to obtain a license from the Foreign Office. The foreign secretary further emphasized that unless all nations concerned enforced effectively Nanjing's Decree of 1930, the Foreign Office would not feel compelled to enforce it.[169]

After this cool reception from London, the Chinese government turned to the United States. Minister Johnson was informed that Guangdong and other Chinese provinces had secretly ordered and purchased a large number of airplanes from the United States, not authorized by Nanjing's permits. At the same time, the Chinese minister wrote Secretary Cordell Hull that shipment of war materiel into China must cease without Nanjing's permits issued through the proper legation. It was clear that were the United States to comply with the Chinese request, it would mean a restoration of the former procedures of the Department of State. A department staff held that compliance would place South China "in the hands of British merchants" and strengthen "British views to the effect that aircraft are in no sense military unless actually armed at the moment of importation."[170]

Because of the importance of London's attitude, in May 1934, the Chinese foreign minister turned to the British again. During a meeting with British minister Cadogan, the Chinese were assured that London was to comply with the Chinese request, but Cadogan also wanted to know whether other countries, such as the United States, would be willing to accede to China's petition. On the next day, Cadogan met with American minister Johnson, pointing out that American aircraft sales to South China seemed unfair to the Chinese government and informing Johnson of Britain's new stand on this issue.[171]

Assured that London would comply with the Chinese request, the State Department issued revised regulations on the export of arms and military equipment in late May of 1934. After so much diplomatic backing and forwarding, the State Department simply restored its former procedure. But at the end of its statement, the department instructed "the Consulates General at Hong Kong and Canton to study carefully British procedure at Hong Kong . . . whether such authorities continue in their past practice of considering all aircraft as commercial, and therefore not subject to restrictions unless actually armed at the moment of importation."[172] The department, it seems, doubted whether the British government would comply with the Chinese request. For its part, London was equally skeptical about Washington's promise to comply with the Chinese regulations. Secretary Simon instructed Lindsay that he must learn whether the American government had limited "their exports along the lines hitherto consistently followed by His Majesty's Government or not, and, if so, what categories of armaments are covered, whether they include aircraft in any or all forms."[173]

Although London and Washington decided to maintain effective control of arms entering into China, especially into South China, other powers might not comply with the Chinese request, which would impose an unfair handicap on both the British and Americans. International cooperation seemed necessary in regulating arms sales to China, and the British became a major advocate of such cooperation by addressing the inquires of the French, Italians, and others about their control of arms sales to China.[174]

Both London and Washington were to comply with the Chinese request, but the differences between them were still far from resolved definitively. First, the State Department continued to regard as military aircraft "(a) all types of aircraft actually fitted with armor, guns, machine guns, gun mounts, bomb dropping or other military devices, and (b) aircraft presumed to be destined for military use, whether actually fitted with armament or not." The Foreign Office, however, still persisted in dividing "aircraft into two categories only, namely, armed and unarmed" and did not "consider that under existing conditions the application of the stricter procedure in force in the United States is practical." The Department of State asked the Foreign Office to adopt the same strict definitions as the United States, but the Foreign Office refused unless all other countries also adopted them, causing the State Department to threaten that if "the continuance of this procedure should be found to result in placing American exporters at an unjustifiable disadvantage as compared with their principal foreign competitors," the Department would not "continue to apply this definition in respect to exports to China." Second, the State Department still resented the Foreign Office's policy with regard to the loopholes in Hong Kong. The secretary of state warned that if the Hong Kong government could not close the loopholes for restriction on arms import into China, the State Department would not obtain "the assent of the Chinese Government" before issuing a license for export of arms and airplanes.[175]

While working a system of international cooperation to prevent the export of arms to China without permission of the Nanjing government, London urgently needed Washington's cooperation. Considering the Americans' suggestions reasonable, the British decided to tighten Hong Kong's control of the arms traffic to China. British ambassador Lindsay informed the State Department in March 1935: "the Governor of Hong Kong was already considering how to deal effectively with the loophole pointed out in your note of January 10th; and would be very glad to cooperate with the United States Government with a view to closing it."[176]

Although Washington was glad of London's cooperation in closing the loophole in Hong Kong, Americans regarded that the disadvantage at which American exporters were placed still was not removed several months later. American businessmen claimed that "unless the regulations of the United States Department of State are amended immediately, many thousands of dollars worth of American business will be lost to foreign competitors." As Allen L. Paterson commented, Americans "find themselves in the unfortunate position with respect to the purchase of raw materials for the manufacture of these airplanes. A great deal of material is coming from Britain and Germany. The head of the Canton Air Force has expressed to me on numerous occasions that this is not their desire."[177]

After Paterson's report reached the State Department, the acting secretary of state briefly and succinctly informed Ambassador Lindsay on August 1, 1935, "the British procedure licenses are not required for engines, spare parts and equipment for use therewith." The secretary cited a recent example of a British airplane engine company that sold aircraft engines to Canton. Yet an American firm was refused an order from the Canton government because the latter declined to apply for a Hujiao from Nanjing. The Hong Kong authorities still permitted shipments of aircraft and aircraft engines to China on the basis of import permits issued by the local governments. Meanwhile, the British government was still unable to exercise control over transit shipments through Hong Kong, and the Department of State remained unsatisfied with what the British government had done in regard to the exportation of airplanes to China. In the end, the acting secretary suggested that the Foreign Office regulate aircraft sales to China along the line similar to those imposed by the Department of State.[178]

The British, however, refused this suggestion, suspecting that the restrictions adopted by the United States and Great Britain alone would not settle the matter of military aircraft sales to South China. Air Ministry Secretary Samuel Hoare commented that "bi-lateral action confined to the U.K. and U.S. to control the export to China of such aircraft and aircraft engines, etc. by license would not prove satisfactory."[179] At the beginning of 1936, when Lindsay informed the Department of State of the opinion of the London government, the Department of State firmly believed that the action would "do much to achieve their objectives."[180] Since the British as well as the French continued to export aircraft to South China without licenses from the Nanjing government, the United States became the only country that did not export aircraft to the Guangdong authorities without Nanjing's permits.

In fact, some American businessmen, with the occasional help of the U.S. commercial attaché, still circumvented the State Department's regulations. On December 9, 1932, for example, Lin Fuyuan, on behalf of the Canton Aviation Bureau, signed a contract with Max S. Polin of the China Airway Federal, Inc., which provided for the sale by China Airway to Canton of four Curtiss training planes, 10 Browning machine guns, and ammunition. Polin then informed George Westervelt of the China Airway that the airplanes should be consigned to China Airway, Hong Kong, rather than to the Canton Aviation Bureau, because the local U.S. commercial attaché's office had informed him that Washington would require no export permit if these planes were routed to the Crown Colony. The Canton Aviation Bureau, therefore, requested that all packing cases be marked "CHINA AIRWAY, HONG KONG."[181] In yet another instance, when Donald W. Douglas, president of the Douglas Aircraft Corporation, informed his staff how to sell six DC-2s to Canton, he wrote, "My suggestion to you in selling any airplanes to Canton would be that you allow yourself enough time to ship them in some other way than directly from here to Canton, in the event of any difficulty over this license."[182] These two examples illustrate methods by which Americans bypassed or defeated the precise spirit of State Department policy and regulation. Those cases also reveal continuing differences between the policies of the State Department and those of Commerce Department. It is fascinating to observe an American commercial attaché in the field assisting and advising American businessmen to evade the State Department's regulations.

Because the British continued to sell aircraft to Canton, as a result, the State Department in April 1936 asked Nanjing to grant permission for exportation of

American aircraft into Canton in order to compete with the British there.[183] The Anglo-American cooperation with the Nanjing government on regulation of aircraft sales to China seemed to have failed. Due to the increasing influx of Anglo-American airplanes to South China, Chen Chidang, governor of the Guangdong province, was able to build up his air force, spending over $1,400,000 for purchases of new equipment for the aviation school and air corps between 1933 and 1936. By June 1936, the Canton Air Force had expanded to nine squadrons with 150 airplanes, 300 pilots, 20 airfields, one airplane factory, one airplane repair shop, and one aviation school.[184] It became the second largest air force in China following the Nationalist Air Force. In neighboring Guangxi province, with the help of the British, Li Zhongren raised his air force to rank third highest in China. The combination of both the Guangdong and Guangxi air forces created the strongest and probably most effective air force in China. With their strength in the air, General Chen Chidang and Li Zongren rebelled against the Nanjing government again.[185]

Anglo-American cooperation in helping the Nationalist government control importation of aircraft throughout China failed. Anglo-American cooperation was difficult because both governments necessarily served their own commercial interest, especially critical defense interest, such as aircraft, endangered by the ravages of the Great Depression. Anglo-American aircraft competition in South China disclosed and underscored the distrustful and accumulated grievances that existed between the Washington and London governments. Economic competition in China's aircraft market increased suspicion between Americans and the British.

THE IMPACT OF THE NATIONALIST AIR FORCE

American influence was still strong in Nationalist military aviation before 1937. By the end of 1936, the Nationalist government had politically unified most of China. All of the provincial air forces were now under control of the Commission on Aviation Affairs of the Nationalist government, which provided the country with a substantial air corps really for the first time:

Numbers of Aircraft According to Chinese Sources

The Central government	383
Guangdong province	115
Guangxi province	53
Shanxi province	15
Sichuan province	20
Yunnan province	11
Navy	24
Miscellaneous	24
Total	645[186]

The Chinese Air Force, consisting of three bombardment squadrons, three pursuit squadrons, two observation squadrons, and four transportation squadrons, had become a modern, well-trained, and well-equipped air force. In addition, the government established a number of modern aircraft factories and modern aviation schools, as well as building 262 strategically placed airports.[187] Between 1932 and 1936, the Nationalist government stepped up activity on all fronts of military aviation development. The Nationalist Air Force, to a large extent organized, trained, and equipped by Americans, expanded and improved. American influence was the strongest foreign influence in China's military aviation, and Nationalist wings were almost completely American in equipment and training. The Nationalist military aviation had reached a turning point; China had come of age in military aviation.

With strength in the air, Chiang Kai-shek's military power gained a preponderance over his internal rivals. His strong air force not only assured his retention of power but significantly aided in extending his influence in uncontrolled or poorly held regions of China. The Nationalist Air Force helped the Nationalist Army in suppressing the Fujian uprising swiftly, quelling the Communists in South China, and putting down rebellion in Guangdong and Guangxi. For the first time, really, since 1912, most of China was effectively unified.

The Nationalist Air Force made itself a significant factor in Chiang Kai-shek's 1934 campaign against the Fujian revolt. After the Shanghai Incident of 1932, Nanjing ordered the Nineteenth Route Army, the famed defender of Shanghai, to leave Shanghai for Fujian province to participate in the anti-Communist campaign. After the move, the Army's leaders came to believe that Chiang had treated them unfairly, and nursed a grievance against him. The next year the Japanese invaded along the Great Wall, and the Nineteenth Route Army requested to be sent to North China to reinforce the Chinese resistance there. But Chiang rejected the request. Influenced by leftist political groups, Chen Minshu, leader of the Army, decided to revolt against Chiang in the name of anti-Japanese nationalism. In October 1933, the Army reached a secret "anti-Japan, anti-Chiang" agreement with the Chinese Communist Party in Jiangxi province and in November it established the People's Revolutionary Government at Fuzhou, the provincial capital, to challenge Chiang. This government declared an end to Chiang's policy of domestic pacification (against the Communists) prior to external resistance (against Japan), and implementation of land reform.

Chiang lost no time in launching a general campaign against these malcontents, dispatching 11 divisions of the Nationalist Army. In addition to sending observation missions over the Yenping city, General Mao Bangchu himself flew low and closely investigated the situation. Afterward, he conferred with Chiang, suggesting that an attack be made over the northern hill with the support of the Air Corps.[188] With Chiang's approval, the assault got underway on January 7, 1934. Six Fiats, carrying six 250-pound bombs, and 10 Corsairs and Douglas planes, each carrying two 120-pound bombs, took part in the attack. Bombing from 1,500 feet, they attempted to destroy the upper part of the wall of the city of Yenping, thereby demoralizing the rebels. Further, an observation plane directed army artillery fire by using pistol flare signals.[189] After the airplanes had partly demolished the wall and the defenders were in a panic, the observation plane signaled the infantry to attack; they captured the city in a day. The Generalissimo awarded all participating pilots and observers cash bonuses. The target of the next attack was Kutien. On

January 12, General Mao ordered the First Squadron to Shuikou. Nearly a week later, six Fiat bombers and 12 Douglas and Vought Corsair observation planes, loaded with a few 125-pound bombs and many 35-pound bombs, supported the Nanjing Army. The air attack was so successful that when the infantry advanced, there was almost no opposition. The defeated rebel leaders fled to Hong Kong and the Nanjing government brought Fujian under its control.[190] This campaign proved that the Nationalist Air Corps was capable of effective support of ground forces. As W. S. Drysdale, American military attaché, commented: "The recent rapid and successful suppression of the Fujian Revolt has been attributed largely to the prompt and effective deployment of Nanking's air corps."[191]

In these years, airplanes also played a prominent part in the Nationalist government's drive against the Communists in Jiangxi, Fujian, and Hunan provinces. When the Fourth Annihilation Campaign commenced in January 1933, Nanjing sent 18 observation planes and bombers assigned to the Third and the Fourth Squadrons to help.[192] As Mao Zedong's main Red Army force approached Linqun, the Nationalist commandant ordered the Third Squadron into action as ground support. Two pilots flew two Douglas airplanes from Nanchang to Linqun in Jiangxi to bomb and machine-gun the Red Army. More than 10 bombing missions forced the Communist soldiers to abandon Linqun.[193] Then, in late April, while the main forces of the Red Army fled to the South Village, Chiang Kai-shek ordered the Air Corps to observe their movements. Simultaneously, the Third Squadron bombed the South Village and succeeded in killing the commander of the Second Division of the Communist army, although most of the soldiers escaped. During this campaign, the Communist army reportedly suffered heavy personnel losses.[194] Although the Fourth Campaign failed to annihilate the Communists, the Nationalist Air Corps had dealt the Red Army a serious blow.

After the Fourth Campaign ended, the Nationalist Army immediately began preparations for the Fifth Campaign as it had done after earlier unfulfilled campaigns. Chiang moved his military headquarters to Nanchang and started the Fifth Campaign. Assisted by German adviser Hans Von Seeckt, Chiang adopted the block house-blockade strategy in surrounding and diminishing the Communist base. In the meantime, he used the traditional system of mutual responsibility to tighten government control over the Communist-dominated rural areas. Moreover, Chiang himself traveled to Jiangxi province to command this effort, enlarged the Nanchang airfield, and assigned the First, Second, and Fifth Squadrons to join the Third and Fourth Squadrons already at the scene. Each squadron was equipped with 12 Douglas planes. Fifty-one bombers flew in the campaign, and "more than a score of airplanes recently bought from the companies in the United States are expected to deliver the final blow to the Communists in the Southern Kiangsi province," the *New York Times* commented. Chiang requested General Chen Chidang of Guangdong province to provide airplanes, and the Cantonese authorities promised to undertake intensive bombing of the Red areas once the campaign started.[195]

When all of these preparations were completed, more extensive by far than earlier campaigns, the Fifth Campaign kicked off in May 1933. Airplanes played an important preliminary role in daily photographing the Communist areas in the southern Jiangxi, in which about 100,000 soldiers gathered.[196] Then, on May 2, intensive air bombing began and the photography paid off; many Red Army soldiers were killed. The Red Army subsequently withdrew in part at least to escape the air attacks. In the meantime, the Third and the Fourth Squadrons were bombing the core area of the Red Army. On May 13, they dropped 90 heavy bombs, which destroyed all the important military offices and a general hospital. Several days later, the Third and Fourth Squadrons bombarded Xingkuo in Jiangxi, demolishing all the defense works and military strongholds.[197] In June, 40,000 soldiers of the main force of the Red Army encircled Yiwang town and attempted to take it over. Because of the intensive daily bombing over the suburbs of Yiwang, the air raid killed hundreds of soldiers and the Red Army had to abort its mission.[198] In August, the Communist Army commanded by Peng Dehuai marched into the border between Guangdong and Jiangxi provinces. The Guangdong Air Force assigned the Third, Fourth, and Fifth Squadrons to the stations at the border between Guangdong and Jiangxi in order to prevent the Red Army from entering Guangdong.[199]

From August 10 on, the Guangdong Air Force began observing and bombing the Red Army in Jiangxi. On August 14, the Nationalist infantry captured Jiunshanwei in the southern Jiangxi after the second squadron of the Guangdong air corps completely demolished its solid defense works and strongholds in the bombardment. The Nationalist Army captured almost 200 soldiers of the Red Army in the battle. Primarily due to the air observation that provided detailed intelligence of the operations of the Red Army to the nationalist front commandant, the effort of the Red Army to flee to Guangdong province failed.[200]

In the northern Jiangxi province, the battle at Guangchang was one of the most important of the Fifth Campaign because the town was the gateway to Ruijin, the headquarters of the Communist bases in the province. The Communists mobilized their freshest and most experienced troops to meet the Nationalist North Route Army, which began its advance on April 9. The Nationalists found that the Communists had constructed numerous concrete blockhouses around the town and ordered the Third Squadron to destroy them. After the bombing, the Nationalist Army advanced on Guangchang and occupied it on April 28.[201] Under this situation, the Central Committee of the Communist Party called an emergency conference at Ruijin, the Communist capital, to assess the reasons for the loss of Guangchang and concluded that it was attributable to infantry attacks, artillery bombardment and bombing.[202] The Nationalists assaulted the last line of the Communist defenses in Jiangxi province at Yijian in early July, and finally destroyed it in August. The Nationalist Air Force played an important role in the final defeat of the Communists. As historian Keiji Furuya commented, "It was not until the latter part of August that the Government forces were able to destroy the [Yijian] defense by artillery fire and bombing from the air and the Communists were finally forced to withdraw."[203] He added that the Air Corps "carried out heavy bombing raids on Communist-held villages and on Mao's capital, Juichin."[204] The Fifth Campaign successfully broke into the central CCP base and wiped out most of its forces. In mid-October the Communist leaders decided to abandon their bases in South China. The remnants of the Red Army were obliged to escape to the western provinces of Hunan and Sichuan. The fleeing Communist forces started what was later called the "Long March," a retreat over almost 6,000 miles for a whole year. Of the 90,000 troops and 30,000 civilians who departed from Jiangxi, only 30,000 of the original force arrived in Shanxi province in October 1935.[205] During the Long March, the Nationalist Air Corps continued observing and bombing the Communists, inflicting heavy casualties on them.

What factors brought success after four intensive but failing campaigns? First, the Nationalist Army adopted a new strategy of building concrete strongholds to encircle the Red bases. Second, the Nationalist Army enjoyed the close support of airplanes and artillery while the Red Army had only poor military equipment. The Red Army was no longer able to resist the attack of the Nationalist Army and finally relinquished all of its bases in South China.[206] The effectiveness of the Nationalist Air Force assisted by the Guangdong Air Force contributed in major ways.

French historian Michel Georges writes of the efficiency of the GMD Air Force during the Fifth Campaign:

> The Nationalist Air Force played a significant role in the Fifth Campaign. It was responsible for annihilating the Communist troops who were besieged by the Nationalist Army. The Communists were forced to abandon their previous successful guerrilla tactics. The GMD Air Force mainly used Douglas and Vought Corsairs to attack and bomb the enemy army, and the Communist soldiers suffered heavy casualties. This tactic had great psychological impact on those survivors.[207]

Even the Chinese Communist historians admitted the effectiveness of the GMD air force during the Fifth Campaign. They wrote:

> During the Fifth Campaign from October 1933 to October 1934, Chiang mobilized about one million troops and sent 105 airplanes in 11 groups to attack the Red Army. Before the Nationalist Army attacked the township of Lichuan, the Nationalist Air Force dispatched airplanes to the Communist bases to take photos from the air of their strongholds, tunnels, and defense equipment. With these photos the Nationalist Army knew in detail the Red Army's defenses and occupied Lichuan in short time. After the fall of Lichuan, the Red Army sent a large number of reinforcements in attempt to retake Lichuan, but the Nationalist Air Force helped the Nationalist Army to resist the attack of the Communist Army by transporting arms and ammunitions and by bombing the Communists. The Red Army suffered heavy losses and failed to reoccupy Lichuan.
>
> During the battle at Guangchang in April 1934, the Red Army suffered heavy losses again. Before the campaign, the Nationalist Air Force enlarged both airports at Nanping and Longyan, which were close to the Communist bases. After the Nationalist air planes bombed severely the Red Army's defenses for several days, three missions each day, the Nationalist Army began to attack the township of Ningdu, the Red Army's military center. The Communists were forced to retreat and abandon Ningdu. After losing many battles, the Communists were forced to leave Jiangxi province and move to North China. When the Communist Army was on the march, the Nationalist Air Force continued to strafe the Communists in Hunan, Yunnan, and Sichuan provinces and provided material aid to the Nationalist Army in hot pursuit of the Red Army. The participation of the Nationalist Air Force in the Fifth Campaign inflicted heavy casualties on the Red Army and the people in the Chinese Communist bases.[208]

The effectiveness of the Nationalist Air Force also helped to suppress the Canton rebellion quickly in June 1936. Prior to 1936, Chiang Kai-shek had made no effort to bring Guangdong and Guangxi under the control of the Central government largely because the Communist bases in South China served as a dangerous buffer zone between Nanjing and Canton. After eliminating the Communists in South China, Chiang determined to bring the southern provinces completely into line.

Since 1931, Guangdong and Guangxi had been independent and in open defiance of Nanjing's authority. Chen Chidang and Li Zongren used the Southwestern Political Affairs Committee in Canton, supposedly a branch of the National government in Nanjing, to continue the practice of regional warlordism. Hu Hanming, a high-ranking and respected official of the Nationalist Party, had been their political patron. In May 1936, Hu died in Canton. Realizing that Chiang would not tolerate their independent status once their protector was gone, Chen and Li decide to make a move. In preempting Nanjing's strike on them, the southern provinces, seizing the issue of resistance against Japan, planned to revolt against Nanjing again. As a result, Chiang Kai-shek went to Canton to attend the funeral of Hu and to negotiate with both Chen and Li. Chiang requested that they give up their semi-independence and place their armies under the command of the Nationalist Army. These latter-day warlords rejected Chiang's overtures, and on June 1, they established a so-called Nationalist government in Canton and declared war on Nanjing. With this rupture, Chiang lost no time deploying units of the Nationalist Army to the South. A new civil war started. While the infantry of the Nationalist Army marched, three squadrons of airplanes of the Nationalist Air Corps defied the Guangdong air defense by flying over Canton from Nanchang and dropping leaflets on the Tianhe airfield in Canton. The Guangdong pilots did not rise to challenge them because the Nationalist Air Force had a great psychological impact on the Canton Air Force.[209]

In addition, Nanjing tried to induce and encourage Guangdong and Guangxi pilots to defect with their airplanes to Nanjing. Six airplanes did fly from Canton to Nanjing on June 30, and 12 Guangxi airplanes deserted to the Nanjing government in early July. To stanch the flow, Chen personally urged the aviators of the Guangdong headquarters of the air force and the aviation school to remain faithful to their duties; he promised to improve their living conditions. In spite of his rosy predictions, on July 7, nine airplanes of the Guangdong boldly flew to Nanjing in formation. Replacing the carrot with the stick, Chen detained Captain Ding Zexu because these nine planes were under his command. Both the Guangdong and Guangxi authorities took every precaution: all officers of their air forces were deprived of their Sunday leaves and of the right to receive friends and relatives in their barracks.[210]

This disciplining, however, did not prevent further desertions. On July 18, Wang Guangrui, commander general of the Guangdong Air Force, and 130 pilots flew in 73 airplanes from Canton to Shaoguan! The entire Canton Air Force had deserted to the Central government; and the Guangxi pilots were planning to defect to Nanjing. As the *China Weekly Review* stated, "Encouraged by the wholesale defection of the Guangdong air force, part of the Guangxi Air Force was contemplating to follow suit."[211] Because of the loss of most of his airplanes, Chen, who placed great reliance on his air force, lacked the capacity to resist the powerful Nationalist Air Force and Army. Therefore, he decided to resign and fled to Hong Kong on July 18. After the Nationalist Army took over Guangdong, Chiang Kai-shek tried to force Li to put his armies under the control of Nanjing. After weeks of negotiations and bargaining, Li accepted a new appointment by Nanjing and the crisis ended peacefully.[212] The Nationalist government succeeded in suppressing the uprising without a fight and its authority was truly effective in both Guangdong and Guangxi provinces for the first time since 1926. Chiang's successful solution of these challenges effectively enhanced the unification of China. Airplanes played an important role in the unification of China. As Madame Chiang Kai-shek said, "Of all the inventions that have helped to unify China perhaps the airplane is the most outstanding." [213]

JAPAN'S PROTESTS AGAINST AMERICAN AIR ASSISTANCE

During the late 1920s and early 1930s, both American and Japanese leaders were leery of the renewal of tension between their two countries. However, Japan's advance in the Pacific and China forced American policy makers to devise a formula that could contain Japan's expansion in China without provoking Japan's hostility. After the Manchuria Incident, many Americans condemned Japan's aggression and violation of the Washington treaties of 1921–1922, but Secretary of State Henry Stimson only issued the Non-Recognition Doctrine that the U.S. government would not recognize the legal existence of Manchuria or any other territories seized by Japan. Washington did not want to assist China or impose economic sanctions against Japan in part because it had to solve its own economic problems created by the Great Depression.

Although the U.S. government saw no reason to do more in the East Asia crisis, American businessmen had great interest in China's aircraft market. Americans' increasing influence in China's military aviation drew Japan's attention in the 1930s. The Japanese government and military had been extremely concerned over American assistance to the Chinese Air Force in the years prior to the reopening of hostilities in 1937. The expansion of Chinese military aviation was likely to contribute directly to the effective unification of China, which was certain to be detrimental to Japanese interests. Also, in the event of Sino-Japanese hostilities, a powerful Chinese Air Force could deal a serious blow to the Japanese and might even attack Japanese cities from the air. More important, in Japanese eyes, American air assistance to China represented American influence, and perhaps control, which was interpreted as a U.S. effort to prevent Japan's expansion and to extend American military influence in East Asia. Therefore, Japan appeared to discourage American air assistance to China.

After Japanese airplanes shot down Robert Short's airplane, the Japanese government issued a strong protest that the United States had breached its neutrality. The arrival of the Jouett Mission in Hangzhou exacerbated this situation. On January 16, 1933, Toshio Shiratori, director of the Bureau of Information and Intelligence of the Japanese Foreign Office, announced that the U.S. Army had loaned China 40 officers on the active list "to take part in hostilities against Japan," apparently referring to the Jouett Mission.[214] The Japanese War Office issued a statement on the same day, which claimed that the United States was supplying automobiles and airplanes through Shanghai merchants. This statement implied that the U.S. government was itself supplying China airplanes under the cover of private firms. The next day, after asking the Japanese War Office to explain "the United States" in the January 16 statement, a United Press correspondent in Tokyo received the reply that "they were referring to the American commercial firms" rather than the U.S. government.[215]

In the following months, some Japanese newspapers continued to spread the rumors that "the United States has arranged to lend the Nanking Government 20 million yuan for warlike preparations against Japan. . . . Japan will warn the United States, Germany and other countries against supplying arms to China under present circumstances, as such action may lead to the outbreak of a world war." Obviously, the Japanese press was pointing to the sale of American military airplanes to China. In response to these rumors, the Department of State orally informed the Japanese press that there was not "any understanding or agreement of this nature between the United States and China."[216]

During a meeting of the military councilors at the Japanese War Office on September 6, Japanese military leaders became anxious about American air assistance. As one officer said, the United States had established "a strong foothold in aviation" in China and extended its political "predominance" in Central and South China. Yahei Ohba, the major general of the Imperial Army, claimed, "If the above-mentioned air force expansion plan of China has been completely backed by the United States, it will prove a great menace to Japan."[217] The Japanese War Office spokesman later even warned that Japan would monitor American assistance to the Chinese Air Force, although at that moment the Chinese Air Force was not powerful.[218]

American air assistance to Canton in South China also irritated Japan. The Japanese press in January 1934 was filled with Admiral Suetsugu's article in the jingoistic magazine *Gendai*, which stated, "They [the Americans] are now bringing airplanes to Canton in large numbers. . . . At the present moment they are bringing more to Canton and then to Amoy and Foochow, constructing air bases along the coast up to Shanghai. What are the Americans doing now? They will surround Japan by air raids in three directions."[219]

Indeed, the story did become more complex. In January, the Foreign Office alleged that Chiang Kai-shek was engaged in the construction of elaborate airdromes on the Fujian coast at Amoy and Fuzhou, which were financed from proceeds of the cotton and wheat loan that China had negotiated with the United States several months previously.[220] Several speakers in the Imperial Diet spoke of American aid to the aviation school and airdromes. Vice Admiral Saburo Yasumi, a reserve naval officer and a member of the Seiyukai Party, addressed the Lower House on January 25: "Since the spring of 1932 the United States has sent many aviators to China as instructors. . . . A loan contract is to be signed between China and the United States with the object of enlarging the Chinese Air Force."[221]

In order to put more pressure on the Nanjing government, Minister Akira Ariyoshi arrived in Nanjing on February 21, 1934, to discuss the issue of the Fujian airdromes with Wang Jingwei, a long-time Nationalist Party official and acting foreign minister. Wang insisted that the airdromes in Amoy and Fuzhou be exclusively for the purpose of the anti-Communist campaigns. He emphatically denied having made a secret treaty on aviation development with the United States.[222] Japan was very concerned about American air influence in China because signatories to the Four Power Treaty signed at Washington in 1921 forced them to maintain the status quo in their respective territories. For Japan, a powerful air force built by Americans in Canton and Fujian was too close to Japan's colony, Taiwan; such operations were contrary to the spirit, if not the letter, of the Washington treaty. In addition, if Americans as well as Europeans continued to provide China with military assistance, the Chinese government would not abandon its anti-Japanese campaign.

In April, the Japanese government complained loudly of American aid to the Chinese military aviation when it learned that the Curtiss-Wright Corporation was building a military aircraft factory in Hangzhou. Irritated by America's air assistance as well as other Western powers' military assistance to China,[223] on April 17, Eiji Amau, chief of the Intelligence Department of the Japanese Foreign Office, perhaps a spokesman for the Japanese Foreign Office, issued a statement to the Japanese press, stating the position of Tokyo with regard to aid from the Western powers to China. This was the so-called Amau Statement, which boldly and openly challenged the Open Door Policy and the Nine Power Treaty System.[224] This statement asserted a Japanese Monroe Doctrine in Asia, which placed China in Japan's sphere of influence, claiming "any joint operations undertaken by foreign powers were bound to acquire political significance." This statement particularly objected to "supplying China with war planes, building aerodromes in China and detailing military instructors or military advisers to China."[225]

Obviously, the Amau Statement opposed any Western military assistance to China in general, and American air assistance in particular. As the respected *Wall Street Journal* remarked, "His 'declaration' or 'warning' or 'manifesto'—or mere indiscretion—stated no ends not long foreseen by every statesman in Europe or America. It presaged—as far as now appears—no new policy or further expansion or tangible action on the part of Japan. It protested a little more strongly than what was courteous American aid to Chinese 'civil' aviation."[226]

To put more pressure on America, the Japanese ambassador to Washington, in the course of an interview with the *Washington Star*, specifically warned the administration, "All new purchases of airplanes are intended by the Chinese to be used eventually against Japan, and we cannot tolerate such things. . . . Any assistance given to the Chinese which might help them to continue their internal wars or to prepare themselves to fight Japan will have to be stopped."[227]

Yosuki Matsuoka, chief of the Japanese delegation to the League of Nations and vice minister for foreign affairs, on April 28 issued a statement, claiming that Japan particularly disliked any activities that thwarted the recovery and development of friendship between China and Japan. He particularly pointed to the actions of "American airplane manufacturers and former American military officers."[228] The Japanese foreign minister Koki Hirota even declared later "that an international embargo should be placed on the exporting of war material to

China" because "much foreign capital, mostly American, has been put into aviation in China."[229]

To clarify the position of the American government with respect to Amau's statement, the State Department delivered an aide-mémoire to Japan on April 29 and made it public the next day. No mention was made of American aid to Chinese military aviation, but the State Department argued loftily that no nation could rightfully endeavor to make conclusive its will in situations where the rights, the obligations, and the legitimate interests of other sovereign states were involved.[230]

Although the State Department justified American air assistance to China, claiming that American pilots and aircraft businessmen in China were acting solely in private capacities, it tried to modify its policy in order not to provoke Japan. The Far Eastern Division of the State Department thereafter recommended no more military assistance to China. Under the pressure of Japan, Nanjing informed Jouett in July 1934 that his contract would not be renewed when it expired in June 1935, in spite of his striking successes. Thereafter Jouett turned to the U.S. government for help. He wrote a letter directly to President Roosevelt, saying that "The re-employment of a large proportion of our group should be the concern of the president of the United States . . . because such a show of interest on your part would be of infinite value."[231] The White House referred Jouett's letter to the State Department. Having warned the U.S. Commerce Department that the existence of the Jouett Mission could cause a serious problem in the U.S.-Japan relations and never greatly appreciating Jouett's activities, the State Department showed no interest in Jouett's request at that time.

The State Department not only declined Jouett's help, but also rejected the Chinese government's request to send another U.S. "civil" air mission to China because it would "give the impression that the American Government was inciting the Chinese Government to military preparedness," and such activities "would have a net effect disadvantageous to both countries."[232] Here, the State Department implied the potential protest of Japan for American air assistance to China.

The Chinese foreign minister issued a statement on April 19, in reply to the Amau doctrine, claiming that China opposed any foreign intervention in its domestic affairs and that the foreign assistance to China was only for the purpose of China's defense.[233] The Chinese government, however, did not renew its contract with John Jouett when it expired in June 1935, partly due to the objection of the Japanese. The Japanese were not satisfied, though, and put more pressure on Nanjing.

In October 1936, Japanese prime minister Hirota summoned the Chinese ambassador to his office and presented him a three-point plan, Hirota's Three Principles. The plan required that (1) China had to abandon its policy of playing one "barbarian tribe" (foreign country) against another; (2) the Nationalist government must enter an alliance or collaborate with Japan for the suppression of Communists in Asia; (3) the Chinese government must cooperate economically with Japan and Manchukuo. By the phrase playing one "barbarian tribe against another," Hirota explained that Japan objected that China continued relying on European and American resources to retard Japan's expansion in China, and opposed the European military assistance to China in general and American air assistance to China in particular.[234] One of the objectives of Hirota's Three Principles was to force China to close its door to Westerners. As American ambassador Johnson commented, these principles would "place China almost completely under Japanese control."[235] Hirota's Three Principles provoked American public criticism of Japan, but the State Department remained silent in its determination not to confront Japan.

No matter how much the State Department denied any connection with American aviation support of China, Japan still was wary of American air assistance to China; the Japanese press continued to spread rumors that the U.S. government was helping China build its air force. In November 1936, when new Ambassador Wang Zhengting arrived in Washington, the Tokyo press claimed that Nanjing had instructed him to conclude a secret military agreement with Washington to augment development of the Chinese Air Force. The correspondent alleged that the agreement would provide that the Chinese government purchase a large number of American planes "at prices below market value for a speedy build-up of the Chinese air force," in return for which China would allow the United States to expand its military "activities in the Far East." The agreement would take on "the nature of an offensive and defensive alliance," an obvious expansion of U.S. aviation interests in the Pacific.[236]

Thus, U.S.-China policy in the 1930s was to promote China's stability and to provide an open market in China for American trade. The Department of Commerce encouraged American manufacturers to increase their exportation to the Chinese market, which was significant for the U.S. economic recovery during the Great Depression. Assisted by the Commerce Department, American aircraft manufacturers increased their aircraft sales to China, which expanded and strengthened the Nationalist Air Force. The expansion of the Nationalist Air Force had great impact on the political and military developments in China. With strength in air, the Nanjing government was more efficient to suppress the revolts of the local warlords and force the Communists to escape to North China. For the first time since 1926 the Nanjing government extended its influence to South China.

The great demand of the Chinese for aircraft contributed to the intense competition among the Western aircraft manufacturers, especially between Americans and the British. Vying for the aircraft market in South China, American and British aircraft dealers tried to sell warplanes to the local warlords who were independent from the Central government. Their competition forced their governments respectively to modify their arms policies and to criticize each other for selling warplanes to local warlords. These arguments between Washington and London over airplane sales to regional warriors weakened Anglo-American diplomatic cooperation in controlling arms sale to the regional warlords.

American air assistance to China also aroused Japan's suspicion of the American intention to increase U.S. military influence in East Asia, becoming one important cause of friction between the Japanese and the U.S. governments. To stop American air assistance and other Western powers' military aid to

China and for other reasons, the Japanese issued the Amau Statement and Hirota's Three Principles, the boldest and most open challenge to the Open Door Policy, clearly violating the Nine Power Treaty of 1922.

However, to avoid increasing the dangers already existing in U.S.-Japan relations after the Manchurian invasion in 1931, American policy makers tried to appease Japan when the Japanese expressed strong objections. Reaction from Washington to Amau and Hirota was weak. The State Department first repeatedly denied any government sponsorship of American military aviation support of China and then discouraged American air assistance to China. For the U.S. policy makers, American economic interests did not justify a bold policy to provoke antagonism from other powers, especially from Japan. Washington's attitude on aircraft sales to China was in line with U.S.–East Asia policy: to avoid friction between the United States and Japan. American appeasement policy on this issue would make the Japanese more aggressive in forcing Americans to withdraw their air support of China in the following years.

Despite the expansion of the Nationalist Air Force assisted by Americans, the Chinese Air Force (CAF) was not an effective fighting force in comparison with other powers' air forces during that time. The CAF was not very efficient because of China's limited financial resources to support an expanded air force, a powerful struggle over the air force, and the use of older model American aircraft.

NOTES

1. See the Commission on Aviation Affairs of the Nationalist Government, "Hongjun yangeshi chugao," SHAC.

2. "Current Aviation Activities," June 21, 1932, Report Number 8322, File Number 2078, RG 165, NA.

3. *Zhongguo jindai junshishi lunwenji* [A collection of papers on modern Chinese military history] (Beijing: Junshi kexue chubanshe, 1987), p. 553.

4. Westervelt to Soong, March 28, 1932, Young Papers.

5. Westervelt to Soong, March 10, 1932, ibid.

6. Westervelt, "Effect on Military Aviation of Commercial Aviation," March 28, 1932, ibid.

7. See the Commission on Aviation Affairs of the Nationalist Government, "Hongjun yangeshi chugao," SHAC.

8. *Kongjun* [Air Force] (The Central Aviation School), 8 (1936):65.

9. See "Hanghongshu gongzhou zongbaogao" [General report of the Aviation Administration] 1930, File Number 16971, Record Group 787, SHAC.

10. Julean Arnold to Leighton Rogers, Chief of the Aeronautic Trade Division, Department of Commerce. Memorandum of Rogers to Colonel Young, March 2, 1931, File Number 560, RG 151, NA.

11. Nelson Johnson to secretary of state, March 14, 1931, File Number 893.20/302, RG 59, NA.

12. Rogers to Mackay, March 28, 1932, File Number 893.20/315, RG 59, NA.

13. William R. Castle, acting secretary of state to Edwin S. Cunningham, consul general in Shanghai, March 29, 1932, Department of State, *FRUS, 1932*, 3: 643.

14. Nelson Johnson to the secretary of state, February 26, 1932, File Number 893.222/2, RG 59, NA.

15. "American Train Mission in China," June 2, 1930, File Number 893.20/348, RG 59, NA.

16. Leighton Rogers, chief of the Aeronautic Trade Division, to Thomas H. McConnell, officer of the United Aircraft Exports, May 24, 1932, File Number 560, RG 151, NA.

17. "Plans for Reorganization—Nanking Air Force," May 3, 1932, Shanghai, Commercial Section, Records of the Foreign Service Post of the Department of State, Record Group 84, Federal Record Center, Suitland, Maryland. Hereafter cited as RG 84, FRC.

18. Acting secretary of state Castle to Ambassador Johnson, April 19, 1932, File Number 893.20/327, RG 59, NA.

19. Soon after the Shanghai Incident, the National government asked Young "to handle negotiations with the American government in matters concerning civil and military aviation," see Arthur N. Young, *China's Nation Building Effort, 1927–1937* (Stanford, Calif.: Hoover Institution Press, 1971), p. 351.

20. The Bureau of Foreign and Domestic Commerce, Department of Commerce, to Howard, May 9, 1932, Shanghai, Commercial Section, RG 84, FRC.

21. The Bureau of Foreign and Domestic Commerce, Department of Commerce, to Howard, May 14, 1932, Shanghai, Commercial Section, RG 84, FRC.

22. John Jouett to the President of the United States, July 20, 1934, File Number 893.20/495, RG 59, NA.

23. John H. Jouett's Memorandum, *Hongjun zhandou shilu* [The records of the combats of the Air Force] (Guangzhou: Lizhun chubanshe, 1938), pp. 12–13.

24. Jouett to the President of the United States, July 20, 1934, File Number 893.20/495, RG 59, NA.

25. American Trade Mission in China, June 2, 1932, File Number 893.20/348, RG 59, NA.

26. Rogers to Howard, June 4, 1932, RG 84, FRC.

27. American Training Mission in China, June 2, 1932, File Number 893.20/348, RG 59, NA.

28. *Hongjun* [Air Force] (the Central Aviation School, China), 8:65 (1932); Zhirou Zhou, "Kongjun zhandou shilu" [The records of the combat history of the Air Force], unpublished manuscript in the Department of Asian Collections, Library of Congress.

29. American instructors on duty at the Central Aviation School:

Jouett, John H.	Director
Clark, G. R.	Assistant Director for Engineering
Rowland, Harry T.	Assistant Director for Administration
Holbrook, R. L.	Observation, photographic and communication officer
Kent, W. C.	Instruction librarian
Knight, M. R.	Meteorological officer
Sansburg, R. L.	Pursuit Instructor
Shannon, E. D.	Instructor
Taylor, T. L.	Procurement supplying and accountability officer
Spencer, W. C.	Chief Mechanic
Ponceti, H.	Mechanic
Mathewson, Christopher	Navigation officer
Pennington, W. H.	Mechanic

Cooper, Harold	Flight Surgeon
Miss Alma Wade	Secretary to the Director

Principal Chinese personnel of the school:

P. T. Mow	General, Vice President
Pei-ken Hsu	Lieutenant-General, officer
Te-hsieh Shen	Director of the Aviation Bureau, officer
C. H. Yen	Lieutenant-Colonel, executive officer
Chia-mei Hu	Major, Aviation Instructor

See W. S. Drysdale, "American Aviation Mission," December 18, 1934, Report Number 8465, File Number 2078, RG 165, NA.

30. Edwin S. Cunningham to the Department of State, June 10, 1935, File Number 893.20/541, RG 59, NA.

31. John Jouett, "War Planes over China," *Asia* 37 (1937): 827–30.

32. "Current Aviation Activities," February 13, 1934, Report Number 8766, File Number 2078, Records of the War Department General and Special Staffs, RG 165, NA; "Hanchow Air Graduate Display Marvelous Skill," *China Weekly Review*, January 19, 1935, p. 268.

33. Jouett to Soong, August 1, 1932, Young Papers; Jouett to Major Ralph Royce, the United States Army Air Corps, January 27, 1933, File Number 893.20/421, RG 59, NA.

34. Soong to Jouett, August 15, 1932, File Number 2078, RG 165, NA.

35. A contract between the representative of the National Government of the Republic of China and the principal adviser on military aviation to the National Government, 1932, Young Papers.

36. See the Commission on Aviation Affairs of the Nationalist Government, "Hongjun yangeshi chugao," SHAC.

37. W. S. Drysdale, "Central Aviation School, Hanchow," Report Number 8786, File Number 2078, RG 165, NA.

38. W. S. Drysdale, "China: Current Aviation Activities," January 19, 1934, Report Number 8833, ibid.

39. Chiang to Director Chu, May 13, 1934, *Zhonghua minguo zhongyao shiliao chupian: Duiri kangzhan shiqi,* 1:311.

40. *Xianzongtong Jianggong shuxin yanlun congji* [The collection of President Chiang's speeches and writings] (Taipei: Zhongyang danshi weiyuanhui, 1963), 12: 729. To strengthen that Commission, Madame Chiang served formally as its secretary general from February 1936 to March 1937.

41. Drysdale to the War Department, May 4, 1934, Report Number 8825, File Number 2078, RG 165, NA.

42. *New York Times*, April 29, 1934, p. 1.

43. "Yinianlai hanghong jiuguo yundong" [The movement of aviation saving China in 1933], *Shenbao Year Book* (Shanghai: Shenbao chubanshe, 1934).

44. "The Nation-Wide Celebration of Chiang's Birthday," *China Weekly Review*, November 7, 1936, p. 344; R. G. Lockhart, "Central Aircraft Factory," November 16, 1936, Report Number 388, Intelligence Division, Naval Attaché reports, 1886–1939, Records of the Office of the Chief of Naval Operations, Record Group 38, National Archives. Hereafter cited as RG 38, NA.

45. For a biography of James H. Doolittle, see Quentin Reynolds, *The Amazing Mr. Doolittle: A Biography of Lieutenant General James H. Doolittle* (New York: Appleton-

Century-Crofts, 1953). For his spectacular flights in China, see James H. "Jimmy" Doolittle, *I Could Never Be So Lucky Again: An Autobiography* (New York: Bantam Books, 1991), pp. 193–96; *New York Times*, April 22, 1934, p. 25.

46. "China Buys New Planes," *New York Times*, January 14, 1934, p. 1.

47. Lincoln C. Reyholds, "Aviation in China."

48. W. S. Drysdale, "Jouett Aviation Mission," August 28, 1934, Report Number 8902, File Number 2078, RG 165, NA.

49. "Weekly Confidential Reports," Box Number 30, Case Files, 1934–1935, United States Senate Special Committee Investigating the Munitions Industry, Records of the United States Senate, Record Group 46, National Archives. Hereafter cited as RG 46, NA.

50. Carl A. Nahmmacher to S. A. McClellan, February 10, 1933, Box Number 35, ibid.

51. *North China Herald*, January 10, 1934, p. 45; "Current Aviation Activities," February 13, 1934, Report Number 8766, File Number 2078, RG 165, NA.

52. "Sale of Aircraft to China Since January 1, 1934," File Number 893.113 Airplanes/901, RG 59, NA.

53. "Chiang Kai-shek Pushes China's Aviation Program," *China Weekly Review*, June 22, 1935, p. 134.

54. Jouett to Young, September 18, 1933, Young Papers.

55. The Memorandum of the Conversation between Jouett and Consul General Peck, November 13, 1933, File Number 893.20/463, RG 59, NA.

56. W. S. Drysdale, "American Aviation Mission," Report Number 8974, File Number 2078, RG 165, NA.

57. The secretary of state to the chargė in China (Gauss), January 26, 1935, Department of State, *FRUS, 1935*, 3: 21.

58. Drysdale, "Jouett Aviation Mission," August 28, 1934, Report Number 8902, File Number 2078, RG 165, NA.

59. Grew, American ambassador in Japan, to the Secretary of State, August 14, 1933, Department of State, *FRUS, 1933*, 3: 388.

60. H. H. Kung, "Building China's Air Force, the Nation's Pride," 1939, Young Papers.

61. Chiang to Kung, December 8, 1934, *Zhonghua minguo zhongyao shiliao chupian: Duiri kangzhan shiqi*, 1:311.

62. Chiang to Soong, March 15, 1933, ibid., I:308.

63. Jouett's memorandum, September 15, 1934, Young Papers.

64. Edwin S. Cunningham to the Department of State, June 10, 1935, File Number 893.20/541, RG 59, NA.

65. "Current Aviation Activities," by John Magruder, Major, General Staff, military attaché, Report Number 7483, File Number 2078, RG 165, NA.

66. For example, on July 28, 1932, Chiang Kai-shek instructed He Yinqin, minister of military, to send Wang Pinghang, director of National Aviation Administration, to France and Germany for the purpose of investigating airplane factories. See Chiang Kai-shek to He Yinqin, July 28, 1932, *Zhonghua minguo zhongyao shiliao chupian: Duiri kangzhan shiqi,* 1:303.

67. Westervelt to Soong, April 14, 1932, Young Papers.

68. Young to Soong, April 25, 1932, ibid.

69. F. B. Rentschler, president of the United Aircraft and Transport Corporation of New York, to Soong, February 22, 1933, ibid.

70. E. P. Howard, the former commercial attaché, resigned in 1933 and served as a

representative of the Douglas Aircraft Corporation.

71. Westervelt to Soong, August 19, 1932, Young Papers.

72. Rentschler to Soong, February 22, 1933, ibid.

73. Love to Kung, April 12, 1933, RG 59, NA.

74. Born in Florence, South Carolina, on September 7, 1896, Pawley began working in Latin America for an important New York firm. Those preliminary ventures led him into the infant commercial aviation business. He organized and later became president of the Companies National Cuban de Aviation Curtiss, Havana, Cuba, in 1928. In 1932, the company merged with Pan American Airways. In 1933, he became vice president of the Intercontinent Corporation in New York and the president of the China National Aviation Corporation, China's first airline. See William D. Pawley's biography, William D. Pawley Papers, George C. Marshall Foundation Library, Lexington, Virginia.

75. Westervelt to Young, January 6, 1933, Young Papers.

76. Pawley, "Russia is Winning," Pawley Papers.

77. Pawley to Kung, April 17, 1933, ibid.

78. "Disputes and Rivalries among American Airplane Manufacturers," File Number 893.796/270, RG 59, NA.

79. "Proposed Establishment of American Airplane Factory at Hangchow, Checking Province," February 27, 1934, Report Number 53, Box Number 18, Intelligence Division, Naval Attaché Reports 1886–1939, RG 38, NA. See also "American Airplane Factory to Be Established at Hangchow, Chekiang, China," January 25, 1934, File Number 893.20/471, RG 59, NA.

80. Americans in the factory were C. W. Hunter (factory manager), W. D. Walish (treasurer), H. A. Symonds, G. B. Arnold, Walter Sobol, A. J. Sargeant Jr., W. O. Stoker, Frank Leckell, and Joseph Killen. Salary Payroll of the Central Aviation Factory, 1936, File Number 10, Records of the Central Airplane Factor, Record Group 645, SHAC.

81. R. G. Lockhart, "Central Aircraft Factory," November 16, 1936, Report Number 409-400, Box Number 18, Intelligence Division, Naval Attaché Reports 1886–1939, RG 38, NA.

82. "A Summary of the Intercontinent Corporation's Activities in China From 1929 to Date," January 24, 1939, File Number 893.248/134, RG 59, NA.

83. Memorandum of a Conversation between Holbrook and S. K. Hornbeck, October 31, 1935, File Number 893.20/561, ibid.

84. "Hanchow Fly School and Related Matter," August 24, 1935, File Number 893.20/553–554, ibid.

85. Memorandum of a Conversation between Holbrook and S. K. Hornbeck, October 31, 1935, File Number 893.20/561, ibid.

86. Claire Chennault, *Way of a Fighter: The Memoirs of Claire Lee Chennault* (New York, G. P. Putnam's Sons, 1949), p. 29.

87. American pilots in this school were:

Smith, Sebie B.	Mechanic Instructor
Watson, Ralfe M.	Armament Instructor and Mechanic
Tatum, Stirling S.	Bombardment Instructor
Mcdonald, William C. Jr.	Pursuit Instructor
Williamson, John H.	Pursuit Instructor
Holland, John M.	Mechanic Instructor

"American Air Mission," August 19, 1936, Report Number 276, Intelligence Division, Naval Attaché Reports 1886–1939, RG 38, NA.

88. C. E. Gauss Memorandum, December 8, 1936, File Number 893.20/590, RG 59, NA.

89. Unfortunately, Colonel Reinburg died suddenly of heart failure on the morning of December 5, 1936. The Department of State did not know when Reinburg was engaged by the Chinese government as aviation adviser and director of the Central Aviation School. When Reinburg died, Holbrook informed Gauss, but Gauss "failed to find him registered as an American citizen at the Consul office in Shanghai," see "Central Aviation School," November 16, 1936, Report Number 32-38, 1936, Intelligence Division, Naval Attaché Reports 1886–1939, RG 38, NA.

90. "American Air Mission," August 19, 1936, Report Number 276, ibid.

91. "China Aircraft Factories," August 7, 1937, Report Number 276; "Training of the Flying Personnel of the Chinese Air Force," August 9, 1937, Report Number 277; "National Aviation Bureau Moved to Nanking," January 30, 1936, Report Number 40, Intelligence Division, Naval Attaché Reports 1886–1939, RG 38, NA.

92. Claire Chennault, *Way of A Fighter*, pp. 37–38.

93. *China Year Book, 1931,* pp. 252–54.

94. For an account of German assistance to the Chinese military aviation in the 1930s, see John P. Fox, "The Development of Germany's Far East Policy, 1933–1936," Ph.D. dissertation, London School of Economics, 1972 and Jerry Bernard Seps, "German Military Advisers and Chiang Kai-shek, 1927–1938," Ph.D. dissertation, University of California, Berkeley, 1972.

95. Douglas Jenkins to Secretary of State, October 23, 1928, File Number 893.113 Airplane/15, RG 59, NA.

96. Major Magruder, Field Artillery, "China Aviation," Report Number 7723, File Number 2078, RG 165, NA.

97. Rogers to James H. Smiley, Director Manager, the Bureau of Foreign and Domestic Commerce, December 21, 1932, Box Number 503, File Number 560, RG 151, NA.

98. In fact, at that time, the Canton Air Force had asked the British Far East Aviation Company in Hong Kong to send to Canton an air mission of one pilot and one aircraft engineer. Their monthly salaries would be ranged from $100 to $120. See C. H. Evans, officer of the Air Ministry, to C. W. Orde, officer of the Foreign Office, December 13, 1932, File Number F8613/16/10, General Correspondence, Record Class FO 371, Foreign Office Archives, PRO.

99. S. A. McClean, assistant to the president of the United Aircraft Exports, to P. G. Johnson, president of the United Aircraft, January 3, 1933, The United States Senate Special Committee, *Hearings before the Special Committee Investigating the Munitions Industry* (73d Cong. and 74th Cong.), 6: 1554.

100. Rogers to Smiley, director manager, the Bureau of Foreign and Domestic Commerce, December 21, 1932, Box Number 503, File Number 560, RG 151, NA.

101. That mission included Captain Edward Deeds, Lieutenant Clarence Terrell, and Lieutenant Stuart Baird. S. A. McClean to F. B. Rentscher, January 26, 1933, The United States Senate Special Committee, *Hearings before the Special Committee Investigating the Munitions Industry* (73d Cong. and 74th Cong.), 6: 1554–5; Frank W. Mondell to Leighton Rogers, April 25, 1933, Box Number 503, File Number 560, RG 151, NA.

102. "Body of American Aviator Deeds, Killed in Crash, July 1, Shipped to U. S.," *New York Times*, July 18, 1933, p. 5.

103. W. S. Drysdale, "Survey of Military Aviation in China," December 4, 1933, Report Number 8700, File Number 2078, RG 165, NA.

104. Wang Hanyen, "Guangdong hanghong jianshi" [A brief history of Guangdong aviation], Guangdong wenshi ziliao [Sources of literature and history of Guangdong], 4:179–88.

105. "Guangdong hongjun feijing ji" [The record of the Canton Air Force flying to Nanjing], *Chuanji wenxue* [Biographical literature] (Taiwan), 30, 1 (1959): 66–68.

106. See the Commission on Aviation Affairs of the Nationalist Government, "Hongjun yangeshi chugao," SHAC.

107. W. D. Pawley, "Proposal for the Management of an Airplane Factory," see Pawley to K. Y. Wong, December 13, 1933, File Number F3675/218/10, General Correspondence, Record Class FO 371, Foreign Office Archives, PRO.

108. Love to the Canton Aviation Bureau, February 27, 1934, Box Number 18, RG 38, NA.

109. Edwin S. Cunningham to Secretary of State, June 21, 1934, File Number 893.20, RG 59, NA.

110. C. J. Spiker, American consul general, to N. T. Johnson, May 31, 1935, File Number 893.113 Aeroplane/109, RG 59, NA.

111. C. L. Zakhartchenko's biography, C. L. Zakhartchenko Papers, Hoover Institution on War, Revolution, and Peace, Stanford University, California.

112. "Half a Million Dollars Worth of American Aircraft Machinery Arrives at Canton," *China Weekly Review*, January 26, 1935, p. 123; see also *North China Herald*, January 23, 1935, p. 126.

113. In December, 1935, design of an all-purpose plane was 60 percent completed and design of the bomber was 13 percent completed. Zakhartchenko to Day, December 5, 1935, Zakhartchenko Papers.

114. They were:

Charles H. Day,	Chief engineer
C. L. Zakhartchenko,	Assistant Chief Engineer, in charge of design
William Zebrouski,	Shop Engineer
William Beardslee,	Crafting Engineer
Harry M. Campbell,	Welder

"Kwangtung Aircraft Factory," American consul in Canton to the American embassy in China, July 27, 1936, File Number 893.20/581, RG 59, NA.

115. Marshal Chen Chi-tang, "Order From General Headquarters," June 2, 1936, Zakhartchenko Papers.

116. Zakhartchenko, "History of Fushing Airplane Development," ibid.

117. P. H. Chow, Director of the Shiuchow Aircraft Factory, to Marshal Chan Chi-tong of the First Group Army Corps, June 13, 1936, Zakhartchenko Papers.

118. In June, 1936, Charles W. Hunter, American manager of the Central Aviation Factory at Hangzhou, arrived in Canton. In July, L. R. Dooley, a representative of the Intercontinent Aviation Corporation in South China, and his assistant, J. Y. Lin, a Chinese-American, established an office in the Victoria Hotel, Xiamen, Canton. They made plans for the new airplanes when the materials arrived. See "Shiuchow Aircraft Factory," July 10, 1936, Report Number 233, Box Number 18, RG 38, NA; C. J. Spiker, American Consul General, 'Kwangtung Aircraft Factory," July 27, 1936, File Number 893.20/581, RG 59, NA.

119. Zakhartchenko to P. H. Chow, director of the Shiuchow Aircraft Factory, June 13, 1936, Zakhartchenko Papers.

120. C. K. Gauss, American consul general, "Chart of Organization of Shiukwan Aircraft Works of the National Aviation Commission," November 6, 1936, File Number 893.20/587, RG 59, NA.

121. The United States, Special Committee, *Hearings before the Special Committee Investigating the Munitions Industry* (73d Cong. and 74th Cong.), 6: 1561.

122. The Aeronautical Chamber of Commerce of America, "Sale of Aeronautical equipment to China," 1935, 893.113 Airplanes/117, RG 59; *North China Herald,* January 20, 1937, p. 121.

123. W. E. F. Jones, head of the Aviation Department, the Messrs. Arnold & Company, to C. F. Garstin, the British consul general in Shanghai, September 4, 1929, File Number 4030, Embassy and Consular Archives, China: Correspondence: Series I 1834 to 1930, Record Class FO 228, PRO.

124. The directors of the company were: Messrs. G. N. Tinson (chairman), A. Ritchie, F. R. Smith, R. Voughan Fowler, and D. S. Scott. See W. E. F. Jones, aeronautical expert of Messrs. Arnold & Company, to Cadogan, August 1, 1935, File Number 599/202/10, ibid.

125. The British instructors were J. R. Brown, chief adviser; C. B. Musson, flying instructor; P. R. May, flying instructor; E. Hutchings, wireless and air photography instructor; D. H. Steven, technical instructor; F. Fidler, technical instructor. See R. P. Willock to M. C. Cadogan, April 12, 1935, File Number F3915/202/10, ibid. See also the Report of American Consulate General at Canton, April, 1935, p. 6, Political Reports, 1934–1938, China, Foreign Service Posts of the Department of State, RG 84, FRC.

126. Ibid.

127. The British instructors in the school were: W. F. Murray, commandant and chief instructor; Lord Douglas Hamilton, assistant flying instructor; W. Waldron, chief ground instructor; H. Willock, assistant ground instructor. See Willock to Cadogan, May 18, 1934, File Number 4423/218/10, G. C. Pelham, trade commissioner and commercial secretary, to the British Legation in Peking, February 22, 1934, File Number 1962/218/10; Willock to Cadogan, May 18, 1934, File Number 4423/218/10, General Correspondence, Record Class FO 371, Foreign Office Archives, PRO; Consul General Douglas Jenkins to Department of State, October 12, 1933, File Number 893.113 Airplanes/88, RG 59, NA.

128. See the Commission on Aviation Affairs of the Nationalist Government, "Hongjun yangeshi chugao," SHAC.

129. See Lampson to A. Henderson, July 18, 1929, File Number F3463/120/10, China Confidential Print, Record Class FO 405, PRO.

130. Ingram to Sir J. Simon, October 24, 1932, Ingram to Sir J. Simon, November 17, 1932, *Documents on British Foreign Policy 1919–1939* (London: Her Majesty's Stationery Office, 1970), 11: 60–1.

131. *Times* (London) September 23, 1933, p. 9 and December 28, 1933, p. 15.

132. Ingram to John Simon, January 29, 1934, File Number 1384/218/10, Willock to Cadogan, May 18, 1934, File Number 4423/218/10, General Correspondence, Record Class FO 371, Foreign Office Archives, PRO; "China: Current Aviation Activities," February 13, 1934, Report Number 8766, File Number 2078, RG 165, NA. See also *North China Herald*, January 24, 1934, p. 130.

133. Phillips to Cadogan, October 16, 1934, File Number 7338/218/10, General Correspondence, Record Class FO 37 1, Foreign Office Archives, PRO.

134. For a study of Germany military assistance to the Nanjing government, see William Kirby, *Germany and Republican China* (Stanford, Calif.: Stanford University Press, 1984).

135. Johnson to the secretary of state, May 9, 1930, Department of State, *FRUS, 1930*, 2: 619.

136. Department of State, *Press Releases*, June 7, 1930, p. 273.

137. Multilateral Convention and Statute on Freedom of Transit, signed on April 20, 1921, *League of Nations Treaty Series, 1921–1922* (New York: United Nations, 1946) 7: 11; C. J. Spiker, "Importation of Airplanes and Airplane Parts into Canton," July 17, 1936, File Number 893.113/1642, RG 59, NA.

138. *Hongjun Yuekan* [Air Force Monthly] (The Air Force command headquarters of Guangdong), 1 (July 13, 1931): 17, preserved in the Archives of Guangdong Province, Canton; The Minister in China (Johnson) to the Acting Secretary of State, July 11, 1931, Department of State, *FRUS, 1931*, 3: 1019–20.

139. The consul general at Nanking (Peck) to the acting secretary of state, ibid.

140. Johnson to the State Department, August 1, 1931, File Number 893.113 Douglas Aircraft Co./32, RG 59, NA.

141. The consul general at Nanking (Peck) to the acting secretary of state; Memorandum by the vice consul at Nanking (Reynolds) of a Conversation with Major Wu Hsieh-shek of the Aviation Bureau of Nanking, July 31, 1931; The Minister in China (Johnson) to the Acting Secretary of State, August 19, 1931, ibid., 3: 1020–25.

142. Stanley K. Hornbeck to Castle, July 17, 1931, File Number 893.113/1307, RG 59, NA.

143. Emphasis added. Memorandum of conversation between Hornbeck and Rogers, June 8, 1931, Department of State, *FRUS, 1931*, 3: 1016–17.

144. Hornbeck Memorandum, July 13, 1931, File Number 893.113 /1307, RG 59, NA.

145. See M. Lampson to the Hong Kong government, July 29, 1931, File Number F4177/52/10, General Correspondence, Record Class FO 371, Foreign Office Archives, PRO.

146. Ibid.

147. William Castle to the American embassy in London, July 27, 1931, File Number 893.113/1312A, RG 59, NA.

148. See Department of State to American embassy in Peking, July 31, 1931, File Number 893.113/1315, ibid

149. Mackillop's Memorandum, September 7, 1931, the Hong Kong government to M. Lampson, September 14, 1931, File Number F4853/52/10, General Correspondence, Record Class FO 371, Foreign Office Archives, PRO.

150. The minister in China (Johnson) to the secretary of state, October 23, 1931, Department of State, *FRUS, 1931*, 3: 1026.

151. Leighton Rogers to Stanley Hornbeck, May 31, 1932, File Number 893.113/1387 and Leighton Rogers, "Policy on Exports of Military and Civil Aircraft to China," April 4, 1933, File Number 893.113/1502, RG 59, NA.

152. The secretary of state to the Ambassador in Great Britain (Dawes), November 13, 1931, Department of State, *FRUS, 1931*, 3: 1026; Minister M. Lampson to John Simon, secretary of state for Foreign Affairs, December 10, 1931, File Number F6911/52/10, General Correspondence, Record Class FO 371, Foreign Office Archives, PRO.

153. *Hongjun yuekan* [Air Force Monthly] (The Air Force command headquarters of Guangdong), 2 (December 12, 1931): 17.

154. The chargė in Great Britain (Atherton) to the secretary of state, January 25, 1932, Department of State, *FRUS, 1932*, 4: 581.

155. Thomas C. Barringer, American trade commissioner in China, to Jenkins, American consul general in Hong Kong, December 15, 1932, File Number 893.113/1428, RG 59, NA.

156. Barringer to Jenkins, December 15, 1932, File Number 893.113/1428, RG 59, NA; Douglas Jenkins to the Department of State, September 8, 1932, Department of State, *FRUS, 1932*, 4: 592.

157. Jenkins to Secretary of State, December 19, 1932, Department of State, *FRUS,*

1932, 4: 592–93; Hinke to Mackay, July 1, 1932, File Number 893.113/1396, RG 59, NA.

158. "Review of General Condition: Aviation," *China Trade Monthly Report*, January 1, 1933, p. 11.

159. Rogers to G. C. Westervelt, North American Aviation, December 8, 1932, Aviation File, (China), RG 151, NA.

160. Memorandum by the under secretary of state (Castle) of a Conversation with the British ambassador (Lindsay), December 29, 1932, Department of State, *FRUS, 1932*, 4: 596.

161. Julian Arnold to Bureau of Foreign and Domestic Commerce, June 6, 1933, Julian Arnold Papers, Hoover Institution on War, Revolution, and Peace, Stanford University. See also Frank M. Tamagna, *Italy's Interests and Politics in the Far East* (New York: International Secretariat, Institute of Pacific Relations, 1941).

162. Stanley Hornbeck, "Export of Arms or Munitions of War to China," March 28, 1933, File Number 893.113/1495, Leighton W. Rogers Memorandum, April 4, 1933, File Number 893.113/1502, RG 59, NA.

163. Ibid., III: 560.

164. William Phillips, acting secretary of state, to Minister Johnson, July 13, 1933, Department of State, *FRUS, 1933*, 3: 565.

165. According to the statistics prepared by the U.S. Commerce Department in 1933, the value of export of arms and ammunition from the United States to China was only $23,953.00 and the value of American airplanes to China was $5,634,000.00, see previous statistics and Department of State, *FRUS, 1934*, 3: 516–18.

167. A gold unit was US$0.40.00.

166. Hornbeck to Phillips, April 18, 1934, Stanley K. Hornbeck Papers, Hoover Institute on War, Revolution, and Peace, Stanford University, California.

168. "Britain Ahead of U.S. in Aircraft Import to China in January," *China Weekly Review*, March 17, 1937, p. 103.

169. John Simon to Alexander Cadogan, May 9, 1934, *Documents on British Foreign Policy 1919–1939*, 20: 182, 242.

170. The minister in China (Johnson) to the secretary of state, April 13, 1934, The Chinese Minister (Sze) to the secretary of state, April 20, 1934, *FRUS, 1934*, 3: 490–1; Joseph Green Memorandum, April 26, 1934, File Number 893.113/1543, RG 59, NA.

171. Memorandum of conversation by Johnson, May 14, 1934, RG 59, NA.

172. *FRUS, 1934*, 3: 492–500.

173. Sir J. Simon to Sir R. Lindsay, August 2, 1934, *Documents on British Foreign Policy 1919–1939*, 20: 275–76; Osborne to Secretary of State, August 4, 1934, Department of State, *FRUS, 1934*, 3: 507.

174. *FRUS, 1935*, 3: 719–20.

175. Secretary of State to Osborne, British chargė in China, August 10, 1934, Department of State, *FRUS, 1934*, 3: 510–11; Lindsay to the Department of State, December 12, 1934, Department of State, ibid., 3: 521–2; Osborne to the Secretary of State, August 21, 1934, File Number F5119/15/10, General Correspondence, Record Class FO 371, Foreign Office Archives, PRO; Lindsay to the Department of State, December 12, 1934, Department of State, *FRUS, 1935*, 3:711–13.

176. British ambassador (Lindsay) to the secretary of state, March 27, 1935, Ibid., 3:718.

177. A. L. Paterson, president of the China Airmotive Company, was the special representative of the Bureau of Air Commerce of the Commerce Department of the United States. He was authorized to issue and to renew aircraft and pilot licenses. See *China Weekly Review*, February 2, 1935, p. 325; Paterson to J. Carroll Cone, March 18, 1935, File Number 893.113 Airplanes/101, Wellword E. Beall, Far Eastern representative

of the Boeing Aircraft Company, to Clarence J. Spike, American consul general in Canton, January 7, 1935, File Number 893.113/1591, Department of State Records, RG 59, NA. See also A. L. Paterson to J. Carroll Cone, July 17, 1935, Central File Number 965, Box Number 453, Records of the Civil Aeronautic Administration, Record Group 237, Federal Record Center, Suitland, Maryland.

178. The acting secretary of state to the British ambassador (Lindsay), August 1, 1935, Department of State, *FRUS, 1935*, 3: 733.

179. Hoare to under secretary of state, Foreign Office, December 17, 1935, File Number F7947/60/10, General Correspondence, Record Class FO 371, Foreign Office Archives, PRO.

180. Lindsay to the secretary of state, January 31, 1936, the secretary of state to Lindsay, February 10, 1936, Department of State, *FRUS, 1936*, 4: 554–7.

181. Max S. Polin to George Westervelt, December 17, 1932, Box Number 30, Case Files, 1934–1935, United States Senate Special Committee Investigating the Munitions Industry, RG 46, NA.

182. Douglas to N.V. Nederlandsche Vliegtuigenfabriek, July 9, 1935, Box Number 101, Current Files, ibid.

183. Memorandum by Green, April 18, 1936 and Hamilton to Green, April 18, 1936, Department of State Records, RG 59, NA.

184. "Canton to Expand Air Forces—50 Planes Bought Abroad to Add to Present Fleet," *North China Herald*, July 4, 1934, p. 11; "Canton Aircraft Strength—200 Aeroplane On Order," ibid., January 17, 1934, p. 96; "Guangdong hongjun" [The aviation of Guangdong], *Zhongguo de hongjun* [China's Air Force] (Chengdu), 48 (Sept. 1940): 403. "Guangxi hongjun," ibid., 49: 25.

185. For the study of Canton's rebellion in 1936, see "Guangdong jindai hanghongshi" [Modern aviation in Guangdong], *Guangdong wenshi ziliao* [Sources of literature and history of Guangdong], 63 (1959): 8–30; Jixu Ding, "Wo he Guangdong hongjun" [I and the Guagndong Air Force], *Wenshi ziliao xuanji (Quanguo)* [The selected sources of literature and history (The nation)], 25: 193–212.

186. "Aviation Notes," November 16, 1936, Report Number 388, ibid.

187. Yingqin He, *Heshangjiang kangzhan qiqian junshi baogao* [General He's military report during the anti-Japanese war] (Taipei: Wenxin chubanshe, 1962), p. 38.

188. Keiji Furuya, *Chiang Kai-shek: His Life and Times* (New York: St. John's University Press, 1981), pp. 425–27.

189. Tuonian Wang, *Fangong kanluan*, 4: 111–12.

190. Approximately 12,000 soldiers of the Nineteenth Route Army took part in the battle. During the combat, about 4,000 had escaped to the south, about 4,000 had surrendered, about 2,000 had been killed or wounded, about 2,000 had escaped in small bands to neighboring hills and forests. See W. S. Drysdale, "Summary of Operations of the Nanking Air Force During Fukien Rebellion Jan. 1934," February 16, 1934, Report Number 8786, File Number 2078, Records of the War Department General and Special Staffs, RG 165, NA.

191. "China: Current Aviation Activities," February 13, 1934, Report Number 8766, File Number 2078, Records of the War Department General and Special Staffs, RG 165, NA.

192. Wang Tuonian, *Fangong kanluan,* 3: 232–34.

193. See the Commission on Aviation Affairs of the Nationalist Government, "Hongjun yangeshi chugao," SHAC.

194. *Jiaofei zhanshi,* 2: 237–239.

195. "The Anti-Red Campaign Air Base-New Aerodrome Building at Nanchang," *North China Herald*, September 13, 1933, p. 407; "To Bomb China Reds," *New York*

Times, June 19, 1934, p. 7; "Canton Air Force Ready," *North China Herald*, November 29, 1933, p. 822.

196. Hollington K. Tong, *Chiang Kai-shek* (Taipei: China Publishing Company, 1953), p. 177.

197. Tuonian Wang, *Fangong kanluan,* 4: 110–114.

198. See the Commission on Aviation Affairs of the Nationalist Government, "Hongjun yangeshi chugao," SHAC.

199. The Historical Document Committee of the Fiftieth Anniversary of the Republic of China, *Wuci weijiao zhanshi* [The combat history of the Fifth Campaign] (Taipei: Peace Press, 1967), 1: 326–28.

200. *Hanghong xuexiao* yuekan [Aviation School Monthly] (Canton), 9 (1933): 7–12.

201. Keiji Furuya, *Chiang Kai-shek*, pp. 425–27.

202. Poyi Zuo, Jiangxi *suweiai zhengquan jianli yiqi benghui* [The establishment of the Jiangxi Soviet and its Collapse] (Taiwan: East Asian Institute of the National Political University, 1969), p. 553.

203. Keiji Furuya, *Chiang Kai-shek*, p. 432.

204. George F. Botjer, *A Short History of Nationalist China,1919–1949* (New York: G. P. Putnam's Sons, 1979), p. 139.

205. *Jiaofei zhanshi*, 3: 431–35.

206. Poyi Zuo, *Jiangxi suweiai zhengquan jianli yiqi benghui,* p. 553.

207. Michel Georges, *1909–1949 nian de zhongguo hanghong* [China's aviation, 1909–1949] (Beijing: Hanghong gongye chubanshe, 1992), translated by Yang Changxiu, pp. 43–44.

208. Jin Yao, *Zhongguo hanghong shi*, pp. 69–70.

209. *Guangdong wenshi ziliao* [Sources of literature and history of Guangdong], 63 (1959): 8–30.

210. "Nine Kwangtung Planes Desert South and Land at Nanking," *China Weekly Review*, July 11, 1936, p. 198.

211. "Wholesale Desertion of the Canton Air Force to Nanking," *China Weekly Review*, July 25, 1936, p. 278.

212. *Guangdong wenshi ziliao,* 4: 179–87.

213. May-ling Soong Chiang, "Unifying China by Air," *Shanghai Evening Post,* March 12, 1937.

214. Grew to the secretary of state, January 16, 1933, *FRUS, 1933*, 3: 94–95.

215. *FRUS, 1933*, 3: 81–2

216. Grew, American ambassador in Japan, to the secretary of state, August 14, 1933, *FRUS, 1933*, 3: 388, 102; Joseph Grew to the secretary of state, January 14, 1933, Department of State, ibid., *1933*, 3: 81; The secretary of state to the ambassador in Japan (Grew), ibid., 3: 81–82.

217. "China Air Force Expansion Sows Seeds of Uneasiness in the Minds of the Japanese," *China Weekly Review,* November 4, 1933, pp. 396–98.

218. "If China Creates Big Air Force," *The North*, September 10, 1933, newspaper clipping in Stanley Hornbeck Papers, Hoover Institution on War, Revolution, and Peace, Stanford University, California.

219. Joseph Grew, *Ten Years in Japan: A Contemporary Record Drawn from the Diaries and Private and Official Papers* (New York: Simon and Schuster, 1944), pp. 111–112, 125.

220. In 1933, T. V. Soong, minister of finance of the National government, arranged with the Reconstruction Finance Corporation in Washington, D.C. for a $50 million

credit that was to be used to purchase American cotton and wheat for China. See Dorothy Borg, *The United States and the Far Eastern Crisis of 1933–1938*, pp. 62–63. See also *China Weekly Review*, March 3, 1934, p. 2.

221. Joseph C. Grew, "Reports of American Air Bases in Fukien," File Number 893.20/427, Department of State Records, RG 59, NA. See also *FRUS, 1934*, 3: 44–46.

222. "Real Reason for Ariyoshi's Trip to Nanking," *China Weekly Review*, March 3, 1934, p. 2.

223. For a scholarly study of the Western military assistance to China in the 1920s and 1930s, see William Kirby, *Germany and Republican China*; Anthony Chan, *Arming the Chinese: The Western Armaments Trade in Warlord China, 1920–1928*; F. F. Liu, *A Military History of Modern China, 1929–1949* (Princeton, N.J.: Princeton University Press, 1956).

224. See Harold S. Quigley, *Far Eastern War, 1937–1941* (Westport, Conn.: Greenwood Press, 1973), p. 56. See also Grew, *Ten Years in Japan.*

225. Arnold J. Toynbee, *Survey of International Affairs, 1934* (London: Oxford University Press, 1938), pp. 650–1. See also "Unofficial Statement by the Japanese Foreign Office," April 17, 1934, Department of States, *Papers Relating to the Foreign Relations of the United States, Japan: 1931–1941*, 1: 224–25. Hereafter cited as *FRUS, Japan, 1931–1941*.

226. "Word Risks in the Orient," *Wall Street Journal*, April 28, 1934, p. 6.

227. Keiji Furuya, *Chiang Kai-shek*, p. 439.

228. "Matsuoka Challenges the Critics of the Japanese Policy on China," *New York Times*, April 29, 1934, p. 1.

229. "Authority in China Development of Aviation: Japan Perturbed," *New York Times*, September 3, 1934, pp. 13–14.

230. The ambassador in Japan (Grew) to the secretary of state, April 29, 1934, *FRUS, Japan, 1931–1941*, 1: 232; The secretary of state to the ambassador in Japan (Grew), April 28, 1934, ibid. , I: 231–32.

231. Jouett to the President of the United States, July 20, 1934, File Number 893.20/495, RG 59, NA.

232. The secretary of state to the Chargė in China (Gauss), January 26, 1935, Department of State, *FRUS, 1935*, 3: 21.

233. Qixue Fu, *Zhongguo waijiaoshi* [A diplomatic history of China], 3d ed. (Taipei, 1966), p. 460.

234. Memorandum of the conversation between Minister Chiang and Hirota, October 27, 1936, *Zhongri waijiao shiliao congbian,* 4: 31–32.

235. *FRUS*, 1935, 3: 404, 429, 460.

236. "Japanese Report of Possible Military Agreement between U.S. and China," *China Weekly Review*, November 14, 1936, p. 380.

Chapter 4

The Nationalist Air Force during the War, 1937–1940

After the Japanese occupation of Manchuria in 1932, Japan continued to expand inside the Great Wall, attacking Shanghai, invading Inner Mongolia, and attempting to make North China independent of the Nanjing government. The Nationalist government made a series of compromises with the Japanese in order to solve the domestic problems. The clashes between the Chinese and Japanese forces on July 7, 1937, at the Marco Polo Bridge near Beiping erupted into a full-scale military conflict. Although Nanjing sued for a local settlement and was prepared to make concessions, Japan took advantage of the occasion as an excuse for sending heavy reinforcements to North China, occupying Beiping in 20 days.

War became inevitable when the Japanese launched another attack on Shanghai on August 13. By the end of 1937, Shanghai and Nanjing fell to the hands of the Japanese. The Japanese troops moved across the North China plain southward and along the Yangzi River westward, occupying most of the major cities and railways lines in Eastern and Central China. Chiang was forced to move the Nationalist government to Wuhan first and, after the fall of the city in October 1938, to the southwestern city of Chongqing, a thousand miles up the Yangzi River from Nanjing. Chongqing, a symbolic center for national resistance to Japan, was a poor place for the GMD government to launch counterattacks against the Japanese military forces.

THE NATIONALIST AIR FORCE IN 1937

Although the Japanese troops were superior in weaponry and in organization, the Chinese troops had been resisting Japanese invasion since the beginning of renewed hostilities. When the war started, the Chinese Air Force boasted nearly 700 airplanes,[1] whereas Japan had about 1,530 army and navy aircraft, of which about 400 were deployed in the China theater.[2] The type and number of the Japanese airplanes in the China theater at the outbreak of the war were:

	Army	Navy
Reconnaissance Planes	82	21
Fighters	100	69
Light bombers	24	54
Heavy bombers	24	50
Torpedo planes		27
Total	230	221[3]

Sixty percent of the aircraft of the Nationalist Air Force were of American origin, as shown in the following list. The total number of military airplanes in China and their origins is as follows:

The United States	440
Great Britain	85
Italy	65
France	40
Germany	50
China	12
Total	692[4]

Most of the squadron leaders and group commanders of the Chinese Air Force received their training in the Central Aviation School.[5]

When fighting began, a number of American pilots were serving in the Chinese Air Force, of whom Claire Chennault was the most visible. During his long service in the U.S. Army Air Corps, Chennault strongly advocated the use of the fighter plane as a bomber escort, in addition to its independent operation. Chennault was originally brought to Madame Chiang Kai-shek's attention by his intimate friend, Roy Holbrook, an adviser to the Central Trust Company of China.[6] On Holbrook's recommendation, she persuaded Chennault for a position as air defense adviser at a salary of $1,000 a month. While contemplating this offer, Chennault suffered heart trouble and the U.S. Army Air Corps asked him to retire from active duty. This premature retirement on April 30, 1937, thus freed him to go to China. Chennault sailed for China on May 1, arriving in Shanghai a month later. Madame Chiang, who served on the board overseeing the Chinese Air Force and Chiang Kai-shek's Australian political adviser, W. H. Donald, greeted him enthusiastically upon his arrival in Nanjing. To avoid a violation of American neutrality statutes, Chennault's employer technically was the Bank of China rather than the Chinese Air Force. When the war broke out between Japan and China in July 1937, Chennault commanded the Chinese Air Corps in resisting Japanese air attacks. He flew the latest Curtiss-Wright fighter P-36 over Japanese-occupied areas along the Yangzi River to provide information on Japanese military activities for the Chinese Army. In an effort to bolster Chinese aviation, he introduced new air tactics. He also taught the Chinese pilots night dive-bombing techniques to bomb the Japanese army at night and recommended that China use its limited funds to strengthen the Chinese Air Force.[7] As the Japanese military officer later admitted, "The Chinese planes took advantage of moonlit nights and frequently carried out air raids on Shanghai from air bases at Nanqing, Zhurung, Guangde and Nanxiangzhen."[8]

In spite of overall numerical inferiority and a sad state of affairs in the CAF, the Chinese pilots trained by Americans flew American airplanes and inflicted heavy losses on the Japanese in the course of the second half of 1937. The day after fighting erupted in Shanghai on August 13, 1937, Chinese planes tried to attack the Japanese cruiser *Idzumo* anchored at the edge of Shanghai, but the Chinese planes bombed inaccurately, causing extensive civilian deaths in Shanghai instead of damaging and sinking the Japanese ship. But in a later air battle that day, the Chinese pilots gave a better account of themselves. When 18 Japanese heavy planes flew above the Jianjiao airfield, Colonel Gao Zhihang led two groups of 27 planes to intercept them and shot down three of Japan's Model-96 bombers. In honor of that first major victory scored by the Chinese Air Force, the Chinese government designated August 14 Air Force Day in the Republic of China. Having no bases in Shanghai at the onset, the Japanese flew from carriers and moved their elite Shikya and Kisarazu squadrons to Taiwan and Jizhou Island. This relocation would allow attacks on major Chinese airfields along the coast and cripple the Chinese Air Force in one stroke. Sixty Japanese bombers attacked airfields in Hangzhou, Qiaxi, and Nanjing in mid-August and were engaged by Chinese aircraft. During the battle, the Ninth Group downed four Model-94 bombers, while the Fourth Group shot down 16 over Hangzhou. The Third and Fifth groups and the temporarily organized group of the aviation school destroyed 14 planes over Nanjing.[9]

The next day, when over 20 Japanese bombers again raided the same airfields, the Chinese pilots blocked them over Hangzhou and Nanjing, and shot down another eight bombers. In three days of intense air fighting, the Chinese Air Corps shot down 42 Japanese planes and destroyed the Shikaya and Kisarazu squadrons. The destruction of those two squadrons, defeating a numerically superior enemy, earned fame for the Chinese Air Force and frightened the Japanese Air Force. On August 17, under the leadership of a deputy group commander, 17 Hawk planes, 12 Northrop planes, and 15 Corsair planes raided the Japanese positions at Hungkou. The attack inflicted heavy losses on the Japanese.[10]

Japan massed its naval forces in Bailong Harbor, and the aircraft carriers moved near Sheshan at the mouth of the Yangzi River. On August 19, the Chinese Air Force sent seven Northrop planes under the cover of seven Hawk fighters to bomb the carriers, and sank one Japanese cruiser. When his plane developed mechanical trouble, the Chinese pilot at full throttle dove into a Japanese ship; both the airplane and the ship were destroyed.[11] The good showing of the Chinese Air Force shocked the Japanese Army. The United Headquarters of the Army and the Navy of Japan pointed out that the most urgent task was to destroy the excellent Chinese Air Force and to occupy the airports around Shanghai immediately.[12]

Masatake Okumiya, a former flying officer of the Japanese Navy during the Pacific War, later proved the achievement of the Chinese Air Force in late 1937:

> Despite the obvious quality of our planes and the caliber of our pilots, the Navy's Air Force suffered heavy losses in the early days of the incident. . . .We learned—almost at once, and with devastating thoroughness—that bombers are no match for enemy fighter planes. We lost many men as this lesson was administered. . . . The planes of the aircraft carrier *Kaga* suffered disastrously. The twelve Type 89 carrier-based attack bombers, led by Group Commander Lieutenant Commander Iwai, left the *Kaga* on August 17 for a raid against Hangchou. Bad weather prevented a rendezvous with an expected fighter escort, and near their target the

bombers were attacked by a group of Chinese fighter planes. Eleven bombers including the commander's, were shot down. Lieutenant (JG) Tanaka managed to bring his bullet-riddled and crippled bomber safely back to the carrier; otherwise, the fate of the attacking group would never have been known, and another bomber formation might have suffered a similar fate. . . . The *Kaga* was ordered to return immediately to Sasebo and to receive a full complement of the new Type 96 carrier-based fighters (Claudes).[13]

To destroy the Chinese Air Force, the Japanese began constructing a new airfield in Zhongming near Shanghai and added new fighters there in early September. Because of difficulties in replenishing its squadrons, the Nationalist Air Corps was unable to ensure air supremacy over Shanghai during the daytime. For this reason, the Chinese began night raids in spite of poor night navigation facilities. Twenty-four Chinese planes launched a large-scale attack on Shanghai on September 18, which inflicted heavy Japanese losses.[14] The next day, more than 30 carrier-based bombers, reconnaissance seaplanes, and fighters attacked Nanjing in retaliation for the Shanghai raid while 21 of the Chinese Hawk and Boeing fighters based in Nanjing and Zhuyong engaged in a fierce fight. The Chinese Air Corps shot down one Japanese plane and damaged four while the Japanese Air Force shot down two Chinese planes and wounded four Chinese pilots in the operation.[15] When 14 Japanese bombers and eight fighters attacked Taiyuan on September 21, four of the Hawks from the Twenty-eighth Squadron and three of the Hawks from the temporarily organized group of the Aviation School took off to meet the Japanese. During the engagement, the Chinese pilots shot down one Japanese plane near Taiyuan, and Major Hiroshi Miroshi, known as a Japanese ace fighter, was captured. Chiang Kai-shek was very pleased with what the Chinese Air Force had done since the beginning of the Sino-Japanese war and instructed General Zhou Chirou on September 22 that the Chinese pilots would be awarded greatly because they were brave and excellent in the air operations.[16]

After nearly three months of fighting, although the Japanese Air Force suffered heavy losses, the Chinese Air Force suffered equally heavy losses. By November, when the Chinese Air Corps had lost most of its airplanes without acquiring new airplanes, Japan acquired air supremacy in the Chinese theater. Despite an increasing disparity in strength, Chinese pilots continued fighting bravely against overwhelming odds. But the Chinese Air Force did not regain strength until a number of Russian bombers and fighters arrived in China in December, and entered a new phase of air operations at the beginning of 1938.[17]

From July through December 1937, the Chinese Air Force flew 137 attack missions over the Japanese Army and engaged in 57 air battles with Japanese airplanes. The Chinese shot down 94 planes and damaged 52 on the ground while the Nationalist Air Corps lost 131 of its planes, nearly 20 percent. In the first phase of air operation, despite Japanese numerical superiority in airplanes, the Chinese Air Force successfully resisted the Japanese air raids due to the superiority of the American airplanes and the skill of the Chinese pilots trained by the American instructors.[18]

American aviation support of China, a centerpiece of Sino-American cooperation in the 1930s, mainly derived from the actions of American individuals and corporations. Since American individuals and corporations had been involved heavily in the Chinese military aviation, the U.S. government began to adopt new policies toward U.S. aircraft sales and American aviators to China in an effort to prevent the United States from becoming involved in the Sino-Japanese war.

LIMITATION ON AMERICAN PILOTS AND PARTIAL ARMS EMBARGO

After the eruption of new Sino-Japanese warfare in July 1937, President Roosevelt reponded at first by refusing to invoke American neutrality, thereby permitting valuable trade to continue with the beleaguered Chinese. It was not enough to save China. On October 5, 1937, Roosevelt addressed a Chicago audience, calling vaguely for a "quarantine" on aggressors to check the "epidemic of world lawlessness." He declared that Americans could not be safe in a world of disorder. American isolationists responded to the speech by warning him against any bold action.

To keep Americans from going to the Asian war, the State Department first tried to prevent American pilots from going to China and from serving in the Chinese Air Force. When the war started, Chennault attempted to persuade some American aviators in China to fly Martin bombers over Tokyo on the condition that China reward them $25,000 for the flight and $200,000 for sinking an aircraft carrier.[19] In the United States, some Americans helped the Chinese government recruit American pilots for the Chinese Air Force by using their influence and business organizations. In July 1937, for example, Russell L. Hearn, a soldier of fortune, recruited 182 American pilots, plus two mechanics for each, in the Los Angeles area. Those pilots and crewmen were able to keep 392 planes available for combat. At the same time, North America Aviation, also of Los Angeles, quietly sought out pilots to fly for China. When a State Department functionary inquired of this recruitment with North America, he was assured that "the reports were quite true and that a number of transport aviators employed by Hearn's factory were expecting to leave for China in the near future."[20]

Predictably, news of this recruitment invoked a prompt and adverse reaction from Japan. Tokyo announced that its consul in Washington had brought the matter to the attention of the State Department. And on the same day, August 5, the Japanese news agency Domei stated that the Japanese government was considering lodging a protest to the American government and that Hearn's plan was seen as contrary to the spirit of the U.S. Neutrality Act and not in conformity with current statements of the American government desiring a peaceful settlement of the China crisis. Several weeks later, when the Japanese consul general informed the American consul general in Shanghai, Clarence E. Gauss, that he had received information that American aviation instructors were taking part in the operations of the Chinese Air Corps, Gauss flatly rejected the charge but could hardly have been entirely persuasive.[21]

The enlistment of hundreds of American pilots for the Chinese Air Force would probably have contributed to U.S. involvement in the Sino-Japanese conflict. When China and Japan were engulfed in a massive undeclared war, most Americans tried to avoid involvement in the conflict due to strong isolationist sentiment all over the country. To clarify its stance, the Department of State issued a memorandum on the very day after Japanese complaints, rather formally and coldly insisting that "the prohibitions imposed on American citizens by Revised

Statutes, Sections 4090 and 4102, are still in force."[22] Additionally, the State Department undertook to prevent American pilots from proceeding to China. Beginning in the latter part of August, it refused to issue passports for travel to China except in unusual circumstances, which would then be endorsed as follows: "This passport is not valid for travel to or in any foreign state in connection with entrance into or service in foreign military or naval forces."[23]

In spite of these measures, Chennault and his business friends continued to help the Chinese government try to recruit American aviators, and enjoyed some success.[24] General L. Cherymisin was one of them. Cherymisin, a second lieutenant, had been in the United States Army Air Force for five years. Later, he resigned and was engaged by Chennault in California for the Chinese Air Force. He would receive a salary of $550 per month in payment for his work as an adviser in China. On August 27, 1937, Cherymisin applied for a passport, stating that he desired to proceed to Hong Kong, thence to China as military aviation representative and adviser to the Chinese government. After advising the War Department, the Department of State refused to issue a passport. Later, Cherymisin received a letter from an attorney in Los Angeles who was in contract with the Chinese vice consul in that city. The letter advised Cherymisin that he did not need to apply for a passport and that he only needed to carry his birth certificate and other documents of identification. On October 16, 1937, Cherymisin boarded the S.S. *President Coolidge* without a passport, departing from San Francisco for Hong Kong. After the ship arrived in Hong Kong, Chinese officials arranged for Cherymisin to pass through customs and go on to China.

To close this loophole which allowed American pilots to go to China, Secretary Hull instructed the American Embassy in China to inform American pilots that if they continued to serve in the Chinese Air Force, they would be subject to fine or imprisonment or even loss of citizenship. Four American instructors therefore were forced to leave China, and Chennault was also threatened by the American consul with arrest and deportation from China if he continued to serve in the Chinese Air Force. The Chinese government was displeased, to say the least, with the State Department's action regarding American volunteers serving in the Chinese Air Force. Madame Chiang Kai-shek, chair of the Chinese Aviation Commission, conversed with Ambassador Johnson and advanced the argument that the American action, which deprived Chinese pilots of being trained by American instructors to operate American planes that made up 90 percent of the Chinese Air Corps, was "unneutral."[25]

Less than a week later, Hull felt it necessary to justify the reasons for the U.S. policy in a communication to his ambassador: "the attitude of this Government . . . represents an attitude taken in response to the strong beliefs of the American people; and such an attitude is not applicable to China alone and is not motivated by any desire to injure China."[26]Johnson, however, in sympathy with China's cause, expressed his disagreement with the Department's interpretation, and in private communication argued that Section 4090 was "permissive rather than mandatory" and that he could not "recall that any general writ or decree has ever been issued by the United States court for China."[27] But Hull was determined to keep Americans away from the Sino-Japanese conflict, instructing Johnson to warn the American aviators employed by the Chinese government that they could not expect "any form of protection from their Government." With regard to Section 4090, "it would appear to be illogical to hold that the power . . . was intended to be merely permissive."[28] The State Department carefully and cautiously acted to maintain both the illusion and reality of American neutrality in the Chinese-Japanese conflict, although refrained from categorically applying and enforcing the Neutrality Act on that conflict.

Beyond this new policy of limitation on American pilots, the Roosevelt administration imposed a partial arms embargo toward both China and Japan. Since Roosevelt refused to apply neutrality laws to the undeclared Sino-Japanese war, American ships continued to transport aircraft to China after the war started. The S.S. *Wichita* was one of them. As a steamship operating in the East Asian trade by the American Pioneer Company but owned by the U.S. government, the S.S. *Wichita*, with 19 airplanes and some other arms, was to sail from Baltimore, Maryland, bound for China on August 28, 1937.[29] If the planes reached China, they might play important roles in the war greatly out of proportion to their numbers. First, the Bellanca, powered by Pratt & Whitney engines of 900 horsepower, had a speed of 280 miles per hour, almost ten miles faster than that of Japan's pride, the Nakajami AN-1. Also, the Bellanca could carry 1,060 pounds of bombs and enough gasoline for more than 1,000 miles of flight. These planes were capable, at least, of flying to Japan with bomb loads of half a ton each.[30]

Fearful of this possibility, the Japanese government decided to block the Chinese coasts to cut off war supplies, especially airplanes; on August 25, the Japanese naval authorities proclaimed a blockade of the Chinese coast. The next day, the Japanese Foreign Office issued a statement to the press on this Japanese blockade, a pertinent portion of which is as follows:

> The Japanese naval authorities found it necessary to close to traffic of Chinese vessels the Chinese sea coast . . . beginning at 6 p. m., August 25, 1937.
>
> The above measure is solely one of self-defense against lawless acts of the Chinese, and applies only to Chinese vessels. It may be added that peaceful commerce carried on by the third powers will be fully respected, the Japanese Navy having no intention of interfering with it.[31]

The Japanese Navy thus would make every effort to stop foreign vessels carrying arms and munitions to China. The Japanese government warned that its warships would intercept the *Wichita* when it arrived in the China Sea because it carried airplanes.[32]

The Department of State was at first outraged by this statement. Secretary of State Cordell Hull announced quickly on August 28 that the U.S. government formally had served notice on the Japanese and Chinese governments that it reserved all rights on its own behalf for damages to American lives or property stemming from their military operations, hoping the Japanese government would be responsible for the safety of the 20 Bellanca airplanes which would shortly be en-route to China and could be expected to pass through the Japanese blockade.[33] The grounds for the proclamation were that under international law only belligerents had the right to impose blockades against third powers or their nationals, and neither China nor Japan had formally declared war. Also, Roosevelt declined to declare a state of war to exist in China, which presumably would have brought the Neutrality Act into play. The shipment, the U.S. policy makers insisted, should go to China without interference.

The shipment of the *Wichita* drew public attention in the United States not only from the Japanese statement, but also from a proclaimed strike of seamen on the *Wichita*. On the morning of August 27, Ralph Emerson, a representative of the National Maritime Union of America, a newly organized sailors' union, called the chief of the Office of Arms and Munitions Control, Department of State, inquiring whether the export of airplanes to China was legal or not.[34] Joseph C. Green replied that there were no restrictions on the export to China of such commodities, to which Emerson retorted that he proposed to call a strike on the *Wichita* that afternoon based on the argument that the proposed export was illegal. He also claimed that the ship should not be permitted to sail until each member of the crew had received an advance bonus of $250. The seamen claimed that the company would make great profits from the shipment to Shanghai, and they intended to receive their fair share. That afternoon, the 28th, the company advised the officers of the union that it would pay a bonus, the amount of which would be determined the next week. A strike was averted, but the crew claimed that if the amount of the bonus was not satisfactory, they would reserve their right to strike at San Pedro, California, where the ship would stop to be refueled.[35]

The circumstances were ripe for pacifists to encourage Roosevelt to invoke the Neutrality Act in the Chinese theater. On August 31, five peace societies formed a joint board of strategy and agreed to ask Washington three questions relating to the *Wichita* and its cargo.[36] To make Americans understand clearly the Roosevelt administration's stand on this issue, on September 2, secretary of commerce Daniel C. Roper at a press conference stated:

> The Wichita is a vessel owned by Federal Government but leased to American citizens. It is therefore not under the direct control of the Federal Government. Under their own statement, we have no reason to believe that the Japanese embargo will be enforced against vessels which are bona fide under the supervision of citizens of this Nation.[37]

In Roper's opinion, the *Wichita* should call at Shanghai; otherwise, future American trade relations with China would be jeopardized by an apparent retreat. Roper's statement inflamed the pacifists. The next day the six peace societies shot off a telegram to the president at Hyde Park.

> We appeal to you to stop the Wichita and any other American vessels carrying such cargoes to the war zone. . . . Further, we feel strongly and respectfully urge upon your attention to the fact that only application of the neutrality law will effectively deny the statement by Secretary Roper.[38]

Meanwhile, the attitude of the Japanese government stiffened. On September 5, the Japanese Navy extended the blockade to the entire Chinese coast "with the exception of the foreign concessions and of Tsingtao."[39] In the days following, Japanese warships halted foreign ships in the China Sea for inspection. To avoid conflicts with Japanese warships, the British Board of Trade suggested that British ships permit Japanese naval officers to board. On September 11, the British government even suggested to the Japanese government that if the Japanese stopped British merchant ships, it should report such to the British naval authorities. London decided not to test the Japanese will.[40]

In light of the British response, Washington tried to find some way to solve the problem. On September 11, Joseph Green of the State Department telephoned Kermit Roosevelt, president of the International Mercantile Marine Company, to discuss the problems of the *Wichita*. Green informed Roosevelt's representative that Japan would not allow the *Wichita* with 19 airplanes to reach China, and the U.S. government would not provide naval escorts for American ships carrying arms to China. Green further stated that if the *Wichita* continued its voyage to China, it would cause trouble and the American government would be in "an embarrassing situation," suggesting that the *Wichita* should unload the airplanes in San Pedro before it continued on to China. Green's suggestions were passed on to Kermit Roosevelt, who rejected Green's suggestion, and rather decided that "all that part of the cargo of the S.S. *Wichita* which was consigned to China would be unloaded at Manila, and that the ship would proceed from Manila to Hong Kong carrying nothing but the cargo consigned to that port." Green then discussed the problem with Joseph P. Kennedy, chairman of the United States Maritime Commission, who agreed to inform the *Wichita* that the ship would not be permitted to leave San Pedro until further notice, and that he would consult President Roosevelt and the secretary of state with regard to all American ships bound for China.[41]

In these circumstances, President Roosevelt lost little time in dealing with the *Wichita* case on his return to the White House from Hyde Park on September 14.[42] After a conference with Secretary Hull, the following statement was issued at a press conference:

> Merchant vessels owned by the Government of the United States will not hereafter, until further notice, be permitted to transport to China or Japan any of the arms, ammunition, or implements of war which were listed in the President's proclamation of May 1, 1937.
>
> Any other merchant vessels, flying the American flag, which attempt to transport any of the listed articles to China or Japan will, until further notice, do so at their own risk.
>
> The question of applying the Neutrality Act remains in status quo, the Government policy remaining on a 24-hour basis.[43]

As a result, on September 15, in order to conform with the policy statement of September 14, the United States Maritime Commission ordered the *Wichita* to unload its airplanes and small arms at San Pedro, and so 19 airplanes and two cases of revolvers consigned to Hong Kong and two cases of cartridges consigned to Saigon were duly removed from the *Wichita*.[44]

Needless to say, the statement of September 14 was a great disappointment to the Chinese government. Stanley Hornbeck, key adviser on political relations in the East, held a conversation authorized by the secretary of state, with Chinese Ambassador Wang Zhengting. Wang admonished the United States for "forgetting its moral obligations."[45] Several days later, when the American ambassador in China visited Dr. Wang Zhonghui, foreign minister, the latter represented the Chinese government as deeply disappointed at the interruption of shipment of military materials.[46]

To understand clearly the president's statement and to advance China's position, Wang Zhengting visited President Roosevelt on September 18, a cordial meeting in which Roosevelt stated that the American government had no way to solve the problem other than the way indicated in the statement of September 14. If the U.S. government undertook to provide naval escorts for American vessels attempting to transport arms to China, the United States would be involved in the war, to which the American people were opposed. Roosevelt went on to argue that if his administration had applied the Neutrality Act to the undeclared war between China and Japan, the Chinese government would have been even more disappointed. In the course of the conversation, Roosevelt hinted that, in spite of the official statement, the Chinese government would still be able to purchase arms, munitions, and war materials in the United States. China might first ship them to Hong Kong or Indochina, and transship them to China. Wang asked immediately what action Washington would take if the Japanese Army interfered with such shipments; Roosevelt promised strongly that the American government would protect that kind of shipment.[47] Clearly, Roosevelt was personally reluctant to impose a partial arms embargo and would continue to support the Chinese resistance against Japan by not discouraging arms exports provided they were indirectly routed.

Roosevelt's statement of September 14 came very close to the Neutrality Act that was applied in 1937 in the case of Spanish Civil War. It was a compromise between Roosevelt and the pacifists,[48] but it was also a retreat of American influence in East Asia under the pressure of Japan. Roosevelt imposed a partial arms embargo on China due to Japan's blockade of China's coast. Japan's blockade was not justified and violated international laws because Japan had never declared war on China since the conflict started. So a third country had the right to transport war materials to both belligerents. The U.S. government, however, retreated in the face of Japan's threat in order to avert the possibility of incidents on the high seas that might drag the United States into hostilities.

The partial arms embargo no doubt hurt China's war effort more than it did Japan's. After the fighting started in July, the United States continued to export arms, ammunition, and war materials to both China and Japan. Military airplanes represented the major item of arms exported to China, while other war materials were the principal exports to Japan. China depended on American aircraft, which were badly needed to replenish the Chinese Air Force. The partial embargo thwarted the delivery of American aircraft to China because the Japanese Navy took a firm grip of China's coast and set up a naval blockade as far as Hong Kong. Because of a partial embargo on China, the Chinese Air Force was seriously reduced in its effectiveness, and Japan's goal to reduce the number of U.S. aircraft to the Chinese was realized. The Chinese Air Force had declined in the absence of new aircraft by the end of 1937. Japan, however, would continue to import American oil and scrap iron because it dominated the sea.

Additionally, the U.S. partial embargo made the Japanese much more aggressive toward Americans because they understood that Americans did not want to fight the Japanese. Several months later, on December 12, 1937, Japanese planes provocatively bombed and sank the U.S. gunboat *Panay*. Before the incident, American ambassador Joseph Grew had called on the Japanese minister of foreign affairs, requesting that all necessary steps be taken to prevent the Japanese Air Force from endangering American lives by dropping bombs in the vicinity of the *Panay*, which settled about 28 miles above Nanjing on the Yangzi River.[49] Ignoring the American's warning, the Japanese aircraft attacked the American ship, killing two Americans and injuring several. Washington quickly agreed to accept Japan's apologies and compensation. The *Panay* incident, as Dorothy Borg commented, indicated that the Japanese attempted to destroy American interests in China, while the president did not want to go to war with Japan, taking no measures that would lead to war.[50]

In sum, the American pilots enlisted in the Chinese Air Force and the American aircraft delivered to China drew strong protest from Tokyo to Washington when the Sino-Japanese conflict started. The Japanese protest was unjustified since Japan had never declared war on China. In spite of that, bowing to the Japanese pressure as well as domestic isolationism,[51] Roosevelt was forced to adopt new policies toward East Asia in 1937, which weakened the Chinese war effort in general and the Chinese Air Force in particular. America's new policies not only appeased but also benefited Japan because the Japanese were able to buy and to ship American war materials, which helped the Japanese military force to carry the war to South China. America's new policies might not have been justified, but under pressure from isolationism, the Roosevelt administration had done what it could do.

SOVIET AIR AID TO CHINA

By the end of 1937, Japan had gained complete superiority in the air, and the Chinese Air Force was in urgent need of assistance to combat the Japanese Air Force. The Chinese government had to turn to the Soviet Union because the U.S. government refused to provide air aid to China. Fearing Japan's onward march in Northeastern Asia, Russia had a common interest with China in checking Japan. The Soviet Union came to help the Chinese Air Force both by providing airplanes and by sending pilots. The Russians remained the predominant group advising and assisting China to combat the Japanese in the air until the end of 1940.

Beginning in November 1937, Soviet aircraft were flown to China, just as the Chinese Air Force reached the depths of its misfortune. The first lots of Russian-built planes were flown from a station on the Turkistan railway across Xinjiang to Lanzhou, capital of the western province of Gansu, where delivery was made to China.[52] The Soviet aircraft sent to China varied in quality, including E-15 biplane pursuits, E-16 monoplane pursuits, SB light bombers, and MB heavy bombers. The E-16 was then one of the most advanced fighter aircraft in the world, and the dual-engine SB-2 bomber, a superior aircraft in 1938, was the Soviet Union's first advanced bomber.[53]

The Soviet Union also sent volunteer pilots to fly many of their aircraft. The volunteers were regular air force units commanded by their own officers, but they retained their air force ranks and received automatic promotions upon return to the Soviet Union. Russians played an important role in the air battles in China between 1938 and 1940. They shot down 986 Japanese planes, limited the Japanese Air Force's freedom of action, effectively bombed bases and lines of communication, and greatly boosted Chinese military and civilian morale. About 2,000 Russian pilots served in China, more than 200 of whom died in the air battles. In addition, the Russians established aviation schools in Xinjiang and Sichuan provinces, built a large

Soviet air base at Lanzhou in Gansu province, and supplied the Chinese Air Force with equipment, such as radios, refueling machinery, aviation gasoline, and air weapons.[54]

With the help of the Russians, the Chinese Air Force, though small in number, played an important part in strafing the Japanese. From the beginning of 1938 the Japanese had been unsuccessful in repulsing Chinese air attacks on the airfields at Nanqing, Wuhu, and Hangzhou because the Russian-made planes used by the Chinese were superior to the Type 95 used by the Japanese. In an encounter over Nanchang in January 1938, new Chinese pursuits downed three planes. In a duel over the city of Wuhan on February 18, 1938, Chinese E-15s and E-16s shot down 12 Japanese pursuits in 12 minutes. The bombing expedition to Forma on February 23, undertaken by three Chinese squadrons, destroyed 12 grounded planes, three hangars, and a number of barracks. On the Japanese emperor's birthday, April 29, 21 Japanese planes were shot down over Wuhan. Japanese airfields, military establishments, shipping, and warships were also repeatedly attacked along the Yangzi, on the South China Sea coast, and in North China. Strafing and bombing were effectively carried out by Chinese planes supporting the ground forces during the famous Battle of Taierzhuang in southern Shantung. Chinese-manned SB-2 bombers raided Bengbu on April 30 while E-15s attacked Langeng to hit supply lines and troop concentrations. In mid-March, the Japanese were reinforced by a few Type 97 fighters which proved helpful in shooting down the Chinese planes.

As combat continued through May, the Chinese launched their first mission over Japan in order to convince Japanese citizens of their vulnerability to possible air raids and inform them of the Chinese viewpoint of the conflict. The Chinese decided to use the Martin B-10 because it was capable of reaching Japan, and the planes would carry only propaganda pamphlets. On May 19, 1938, the Chinese pilots flew two B-10s loaded with leaflets to Japan. Leaflets fell over the lands of Kyushu and Honshu, dropping into five cities. The objective was to awaken the self-consciousness of the Japanese people, although there is no evidence of any such significant effect. Japanese citizens of these cities could not doubt therefore that the Chinese Air Force had the capability of bombing Japan's homeland, if only in token raids. Therefore, all Japanese cities underwent emergency blackout procedures for six months.[55]

Japan's reaction to the Chinese leaflets came in ten days. On May 31, 54 Japanese planes raided Hankou. Fifty Chinese aircraft intercepted them, and 13 invaders crashed to the ground. Other encounters were also fought over the Wuhan area, Canton, Hengyang, Kuide, and other points. In the Battle of Wuhan, Chinese bomber squadrons were effective in delaying the Japanese naval advance along the Yangzi. Five out of nine Japanese bombers were shot down over Guangdong in June.

The CAF carried out raids on Japanese shipping along the Yangzi. Daily, twin-engine bombers were sent to attack Japanese warships and transports on the river, sinking 33 Japanese vessels and damaging 109 others. Raids on other Yangzi cities were occasionally made and a number of grounded Japanese planes destroyed. The Japanese quickly applied new tactics learned in combat and maintained a well-ordered logistical base that the Chinese could never equal. In the summer of 1938, the Japanese Army received a new monoplane fighter, the Nakajima, which was equal to the Claude pursuit. These new fighters helped the Japanese to achieve instant air superiority and to speed the drive toward Hankou.[56]

By 1939 Japan had been spending more and more of its manpower and treasury on a war that they wanted to end quickly. Since the Chinese did not capitulate to the Japanese conquest of their eastern seaboard, the Japanese planned a series of intensive air campaigns to shock them into surrender. In January 1939, Japanese airplanes bombed positions in Guangxi, Guangdong, Sichuan, Henan, and Hunan. Japanese bomber pilots began to train for night flying. The first raids over the Western Chinese cities and flight centers were escorted, but the Japanese soon discovered that the Chinese Air Force was so depleted that protective fighters were not necessary.[57]

Japan's first raid on Chongqing took place on May 3, 1939, and did little damage, but the next day the Nells returned with incendiaries which gutted the city and burned some 10,000 people to death. Although the CAF took part in some intercepting with E-15s and E-16s during the air raids in 1939, it generally tried to avoid combat in order to conserve its experienced pilots as well as instructors. In November 1939, the Japanese captured the last unoccupied seaport, then marched inland to Nanning. Further attempts to move inland were blocked by Chinese troops, but a weakened Chinese Army stretched from Mongolia to Canton and the whole nation was subject to air raids. The CAF could only afford to mount an interception every few weeks over Chongqing, Chengdu, and Lanzhou.[58]

The vast majority of Japanese missions were successful, but occasionally the Japanese sustained defeats. On February 2, 1939, a Japanese heavy bomber, a Fiat BR-20, was brought down by the Chinese in western Hubei. Among the crew of six were Captain Watanabe of the Japanese Naval Air Force and Lieutenant-Commander Fuhida, who had established a world non-stop-flight record. On October 3, 12 Chinese SB-2 bombers staged a surprise raid over Hankou, destroying 100 Japanese planes. More than 200 pilots and mechanics were killed or wounded, and large quantities of gasoline, bombs, and accessories were destroyed. The two Chinese raids practically wiped out the 12th and 13th Air Corps of the Japanese Naval Air Force. Chinese squadrons also made a number of raids on Yongzheng, a Japanese air base in southern Shanxi. During the battle over Chengdu on November 4, 27 Chinese pursuits shot down 18 of the 54 Japanese bombers. On three successive days, December 25–27 of the same year, the Japanese attacked Lanzhou with a total of 301 planes. The Chinese brought down 9 of them. The fight for air supremacy in Guangxi during the month resulted in the destruction of 12 Japanese planes over Liuzhou. Chinese planes strafed and bombed Japanese positions and troop concentrations during the Battle of Kunlunkwan in southern Guangxi in December 1939.

In 1940, the Chinese pursuits were engaged in defending strategic cities. A five-month dogged defense of Chongqing by new Chinese E-15s in the summer of 1940 marked the greatest success of Chinese pursuits. The Japanese made 39 raids, each with an average of 100 aircraft, and dropped more than 2,000 tons of bombs on the Chinese wartime capital. More than 70 of the invaders were brought down by Chinese defenders. Many more were so badly damaged that it was unlikely they could return to their bases.

Another "pamphlet raid" was carried out by a long Chinese bomber in a daylight raid over Beiping on October 3, 1940, dropping 200,000 copies of Generalissimo Chiang Kai-shek's July 7 anniversary message to the Chinese Army and people, his September 18 anniversary message to the Japanese soldiers and people, and another one to the Chinese nation as a whole.

AMERICAN WARPLANES TO CHINA

Although Washington refused to aid the Chinese Air Force between 1938 and 1940, American officials still showed America's support for China by attempting to limit the sale of aircraft to Japan. Japan's repeated bombings of civilian populations in China and the partial arms embargo on China placed U.S. relations with Japan in an entirely new light, which led to the questioning of sales of military airplanes to Japan. The State Department was persuaded to enunciate a policy of disapproving sales of aircraft and aircraft armament to Japan. On July 1, 1938, the State Department addressed a letter to all U.S. aircraft manufacturers and exporters, stating that "the Government of the United States is strongly opposed to the sale of airplanes or aeronautical equipment which would materially aid or encourage that practice in any countries." The letter asserted that "the Department would with great regret issue any licenses authorizing exportation . . . to countries the armed forces of which are making use of airplanes for attack upon civilian populations." Although the statement did not specifically refer to Japan, it obviously implied that country.[59]

American manufacturers complied with the request of the department, and an informal embargo on aircraft exportation to Japan was in effect from the latter part of 1938. The consequence of this embargo, which may be considered a "moral" embargo, amounted to a practical end of aircraft exports from the United States to Japan. In 1938, the State Department issued licenses for $8,799,219 worth of the exports of aircraft to Japan but in 1939 for only $761,684, less than 10 percent of the previous year. During the first ten months of 1940, no licenses at all were issued for such shipments.[60] As Japanese ambassador Kensuke Horinouchi stated in early 1940, "it has virtually become impossible for Japanese firms to import any airplanes and airplane parts of American make." The Japanese government was deeply disturbed by this "moral" embargo on aircraft to Japan, claiming that it was "a discriminatory export embargo" and "in derogation of the provision" of the Treaty of Commerce and Navigation between the two countries.[61]

While applying a virtually complete "moral" embargo on Japan, Washington still discouraged Americans from serving in the Chinese Air Force. The training facilities so carefully built in the past were plainly inadequate to the task of supplying sufficient numbers of capable Chinese pilots and crews. In May 1939, H. H. Kung invited William Pawley to his office, together with Captain B. G. Leighton and Edward P. Pawley, vice presidents of the Intercontinent Aviation Corporation, to observe that one of China's most urgent needs was for a crop of American volunteer airmen who would give the Chinese Army air support. Kung reminded them of the magnificent job done by the Lafayette Escadrille in France during the First World War. William Pawley promised that he "would do everything possible to put China's problem before various men of influence in the United States."[62] But there was no quick action.

Then on December 27, 1939, the first secretary of the Chinese embassy informed Joseph W. Ballantine, assistant chief of the Division of Far Eastern Affairs, that the Chinese government was trying to obtain the services of American aviators as instructors for the Aviation Academy at Kunming. After discussing the request with Hornbeck, M. M. Hamilton, Ballantine's superior, told Cui Zuanling, first secretary of the Chinese Embassy, that if the American government received such a request, "our attitude would have to be unfavorable. . . . [T]his was . . . a private matter between the Chinese Government and American individuals" and "the Chinese Embassy might care to get in touch with Colonel [John] Jouett of the Aeronautic Chamber of Commerce." Hamilton also recommended that Colonel Chennault, who was then in the United States, "might be of some assistance to the Chinese Government."[63] It would seem that these officials were not as inflexible as in 1937.

American officials, however, showed their support for China by encouraging military aircraft sale to the Chinese. Since Washington did not officially recognize that a state of war existed between China and Japan, only limited prohibitions were placed upon China's purchase of military aircraft in the United States. The Department of State in these years positively encouraged the sale of American military aircraft to China in order to inhibit Japan's expansion to the South. The primary area of U.S. government involvement in Chinese military aviation concerned the sale of aircraft during that time. One may observe this in the instance of the Patterson contract.

Allen L. Patterson, president of the Consolidated Trading Company, represented Chance-Vought and other American aircraft companies in China. To arrange credit for the Chinese government to purchase a large number of American military aircraft, Patterson first approached the Chase National Bank and the National City Bank, both of which bluntly refused even to discuss the possibility. After the refusal by these two banks, Patterson approached the Hong Kong and Shanghai Bank, a British institution, and arranged credits applicable to the prospective aircraft contracts.[64]

Subsequently, on February 24, 1939, H. H. Kung concluded a private loan agreement with Patterson's Consolidated Trading Company, providing for credit of up to $12,800,000 on the condition that the Hong Kong and Shanghai Banking Corporation would finance the deal. One month later, Kung signed a contract with Patterson, the largest order that China had placed since 1928. The Chinese government agreed to purchase 199 American planes of the kinds and makes detailed in the following list:

4 Seversky biplane pursuits (STC-4)
50 Seversky single-wing pursuits (EP-1)
50 Ryan two-place primary trainers (STC-4)
20 Ryan single-place pursuit trainers (STC-P4)
50 North American basic trainers (NA16-4)
25 Chance Vought dive bombers (SB-21)[65]

These airplanes would, without doubt, enormously strengthen the Chinese air capabilities.

From the beginning of the negotiations between Kung and Patterson, W. D. Pawley, president of the Intercontinent Aviation Corporation, had attempted to block the contract. Rather, in February 1939, he presented a loan scheme to the Chinese government for the purchase of Curtiss-Wright products, a total of $11,700,000. The difference between the two prospective arrangements was that Pawley's contract would require the Chinese government to use the full amount of the loan whereas

Patterson's did not.[66]

After the Patterson contract was completed, Pawley still endeavored to persuade Kung to abrogate it. Only three hours after the Patterson contract was signed, Pawley obtained accurate information about it from the members of the Commission on Aviation Affairs, information that supposedly was confidential. Pawley did succeed in causing Kung to entertain serious doubts regarding the ability of Patterson's company to carry out the contract, and in persuading him that excessive commissions were charged and that the prices were far in excess of those for similar equipment purchased by the U.S. government. When the Hong Kong and Shanghai Bank learned of these charges, it tried to withdraw from the arrangement for credits that it had entered into in connection with the contract.[67] Furthermore, on May 20, Pawley was able to sign an agreement with Kung that the Chinese government purchase 100 Curtiss-Wright airplanes for $4,500,000 from his company.[68]

When the Department of State learned that Pawley's interests were endeavoring to prevent the execution of the Patterson contract and that the Hong Kong and Shanghai Bank was threatening to withdraw its credit, it urged the Chinese government to honor the Patterson contract. On May 19, Hull instructed Ambassador Johnson that the State Department deemed the contract highly desirable from point of view both of Chinese government's interests and of American commercial interests.[69] Steps should be taken promptly and effectively, the secretary admonished, to bring to an end the efforts of an American competitor, Pawley, to prevent carrying out this contract.

The next day, Stanley K. Hornbeck asked the Chinese ambassador, the highly respected Hu Shi, to present himself at the State Department to be told, "We view it important both as regards American interests and as regards Chinese interests that difficulties and delay be reduced to a minimum."[70] Under the instructions of the State Department, the U.S. Embassy then in Chongqing also tried its best to persuade Kung to carry out the Patterson contract. The consul of the American Embassy in Chongqing called on Kung on May 23, informing him that the contract was important both to American commercial interests and to the Chinese government, and that the American government was anxious to see it executed. Kung replied that he greatly appreciated the State Department's assistance and that the Chinese government was fulfilling its part of the contract and was awaiting fulfillment by the other party. Yet Kung showed some resentment that an American company had to get itself financed by a British bank and that the commercial credit was unavailable from the U.S. government, not only for the purchase of military planes but also for the civilian planes needed by the Chinese National Aviation Corporation.[71]

Kung alleged that although the American Army and Navy had tested and ordered the Seversky planes, they were not up to the specifications. Without doubt, the Pawley group inspired some of these misgivings regarding the Patterson contract, especially Kung's suspicions with regard to airplane specifications.

To allay these suspicions, Hull informed the U.S. Embassy in Chongqing, "The War Department purchased in 1936 a considerable number of Seversky P35-EP1 planes which have given entire satisfaction and are still in service," and the War Department had "just made a new contract with Seversky Company for Seversky pursuit planes" similar to the type in the Patterson contract.[72] On the same day, Hornbeck and Green discussed the matter in regard to these Seversky planes with the Chinese ambassador and urged him to encourage the Chinese government to expedite action on the contract.[73]

Green confirmed and amplified Hull's view in a session in his office with Cui Zuanling and H. D. R. Burgess of the Bank of China: "As the carrying out of that contract would appear to be to the advantage of a Government with which we maintained the most cordial relation and to the advantage of an important American industry, we hoped that nothing would occur to prevent the carrying out of that contract."[74] Burgess, however, informed Green that it was his understanding the Chinese government was prepared to discard the Patterson contract and substitute for it a series of contracts negotiated directly with the American manufacturers. Even in the face of this turnabout in Chongqing, State Department officials spared no efforts to persuade the Chinese government to implement the contract. Two weeks later, after the first visit, Cui called again at Green's office, mentioning the difficulties that had arisen in connection with carrying out the contract, but Green suggested that "if the Chinese Government really wished to obtain the planes provided for by the Contract it almost certainly could find some means of overcoming the comparatively minor difficulties."[75]

These efforts could not save the Patterson contract. In August 1939, the Chinese government sent Patterson a notice that it had decided not to proceed with the Patterson contract but to substitute a new contract with the Pawley interests.[76] And in late September, the foreign minister of the Chongqing government informed Washington of the export of 100 airplanes ordered by the Chongqing government from the Intercontinent Aviation Corporation, Pawley's group. The major cause of the canceling of the Patterson contract was Pawley's intervention, certainly. As Arthur N. Young, financial adviser to the Nationalist government, told Green, "Notwithstanding Mr. Pawley's denials, he [Young] was convinced that Mr. Pawley was entirely responsible for the misguided action of the Chinese Government in canceling this contract."[77] Although the Chinese government terminated the contract, the Patterson case illustrates not only the severe competition among American aircraft businessmen but also the State Department's persistent interest in facilitating aircraft sales to China, in spite of the restrictions upon which neutrality legislation and White House policy laid.

American aircraft manufacturers continued to sell airplanes to China between 1938 and 1940.[78] During that time, the United States exported a total of 279 airplanes to China, valued at $22,000,000. Indeed, in 1937, China was in first place among countries purchasing American planes. In 1938, however, China fell to third place, and in 1939, due to reliance on the Russian airplanes, China fell to 21st place. Yet in July 1940, China rose to third place again.[79]

American aircraft, though few in number, posed a threat to Japan. When Chiang launched the April 1939 offensive, the Japanese government took precautions in western Japan against possible air raids by the Chinese. As the *New York Times* commented, Japan had some apprehensions of the Chinese Air Corps because it was "greatly strengthened recently by the arrival of airplanes from the United States, Russia, and other countries" and was able to "raid Japan troops on the 1,300-mile front and on the island of Hainan."[80]

AMERICAN EFFORTS IN CHINESE MILITARY AVIATION

Americans continued not only selling aircraft to the Chinese but also providing assistance on training and aircraft assembly in China. Although the Soviet Union dominated Chinese military aviation between 1938 and 1940, American advisers and technicians employed on an individual basis were still active in military aviation in South China, while the Russians sent by their government were active in North and Central China.

The Central Aircraft Factory

The Central Aircraft Factory, erected by the Central Aircraft Manufacturing Corporation in 1934, was in its day the largest aircraft factory in China. Under the management of William Pawley, president of the company, the factory was engaged in the manufacture of 30 Vultee V-II attack bombers in July 1937 when hostilities broke out. The first one was test-flown on July 12 with very satisfactory results. The factory staff was greatly impressed with its performance, since it appeared much faster than a Northrop of the same type. It was expected that the factory would eventually be able to turn out two planes a week. No less important, the cost of planes assembled in China was only 20 percent of the price of completely assembled planes imported.[81]

The outbreak of the Sino-Japanese war, however, quickly made the plant inoperable. Within three days, the Japanese Air Force bombed the factory, and as a consequence of the continuous air attacks, Pawley made preparations for the removal of the factory. Less than two months after fighting began, the Nationalist government instructed the Commission on Aviation Affairs to remove the factory to the South Lake Airport in Wuchang City.[82] The factory workers then proceeded to tear down the plant and move $2,000,000 worth of machinery and equipment to Nanjing, where they transferred the machines and equipment to ships sailing up the Yangzi River. Soon after the factory moved to Wuchang, Nanjing instructed the factory to move again, this time to Hankou city two miles away from Wuchang, and occupied premises formerly owned by Standard Vacuum Oil Company which adjoined a Japanese concession.[83]

Early in January 1938, the factory surprisingly began manufacturing again and was able to produce three bombers per month. When the Japanese army appeared to threaten Hankou, the plant was again moved to distant Kunming, capital of Yunnan province in the southwest.[84] Financial minister H. H. Kung asked Bruce G. Leighton, vice president of the Central Aircraft Manufacturing Corporation, to make an on-site visit to Kunming, and after completing his investigation, Leighton left Kunming in mid-February for Hanoi and eventually returned to Hankou by way of Hong Kong.[85]

Pawley himself went to Kunming in June to assess the Kunming location, but abandoned it after several days spent there making inquiries. Several factors influenced his decision not to locate the factory in Kunming itself. The location was vulnerable to Japanese air raids, and the crucial Kunming-Haiphong railway over which supplies for the factory must come could be closed down by such attacks. Furthermore, the cost of this rail transportation was too high, estimated at approximately 25 percent of the value of the goods transported.[86] Pawley consequently chose a site about eight miles east of Kunming on the railway from

Indochina. The new factory site was at a village. Pawley approached the French government in Indochina and customs and railway authorities to arrange for the transportation of material. He believed that within six months he could effectively remove the plant and have the new plant in operation. Pawley returned to Hankou in mid-July 1938 to discuss with officers of the Commission on Aviation Affairs all of these issues in moving the plant to the new site. In August 1938, the Central Aircraft Manufacturing Corporation completed a new agreement with the Chinese government, providing that China would guarantee to the corporation production orders of $1,000,000 per annum in contrast to half that in the original 1934 agreement.[87] The company agreed to invest an additional $350,000 in the new factory, and the Chinese would bear the cost of moving of all equipment and materials.[88]

After the beleaguered Chinese government approved Pawley's plan in November, the company's 30 American technicians and 500 Chinese employees, together with their families, began the trek to Kunming, totaling in all about 1,500 people. By the end of 1938, all had arrived. In addition to maintaining offices at Shanghai, Hong Kong, Kunming, and Chongqing, Wallace Pawley, a brother of William and vice-president of the Intercontinent Aviation Corporation, established an office at Rangoon to arrange the movement of equipment, materials, and supplies through Burma to Kunming, a route of incredibly difficult passage. A great deal of the machinery and equipment was thus moved from Hankou to Hong Kong, thence to Rangoon and north by the route later famed as the Burma Road.[89]

Actual construction of the new plant began in January 1939, and while it was under construction, Pawley reached a new agreement with the Chinese government, calling for an expansion of factory activities in the next two years. The agreement consisted of two contracts signed in 1939, involving a total sum of approximately $13,000,000. The first contract provided for the supply of about $5,500,000 worth of materials to the factory; the second called for the sale of approximately $7,500,000 worth of U.S. aircraft materials and parts.[90]

Construction of the factory was completed in June 1940, less than a year and a half after beginning, and cost $670,000, financed jointly by the Curtiss-Wright Corporation and the Intercontinent Aviation Corporation. Two main factory buildings totaled 100,000 square feet, plus a foundry and drop hammer building of 13,000 square feet. In addition, 84,000 square feet were devoted to offices and quarters.[91] When fully operative, the factory had about 3,000 Chinese employees and 12 American technicians at that time.[92] Although few in number, Americans played a key role in the factory. As British air attaché R. S. Aitken commented: "The Americans are the backbone of the aircraft factories and repair organizations. In general they are most efficient and have rendered very valuable service."[93]

While the factory was still under construction, production began in May 1939. The new factory had a production capacity of between 250 and 300 airplanes annually,[94] but each plane took 32,200 man-hours, or about four times as much as needed for similar production in the United States. The factory was able to produce only 30 Hawk III pursuit planes by September 1940.[95] The low rate of production seems to have resulted from several different factors. First, malaria affected the personnel in the factory; the average absentee rate resulting from that illness was 18 percent at one point. Second, the staff had to supervise construction of the factory and securing of materials, as well as presiding over actual production. Third, the British authorities closed the China-Burma road on July 16, 1940, in response to Japan's protest.[96] As a result, approximately 400 tons of aircraft materials, parts, and engines were stored at Bhamo, the river port in Burma on the upper Irrawaddy. In addition, about 300 tons of aircraft materials were warehoused at Rangoon, together with approximately 100 tons of machinery for the Central Aircraft Factory, awaiting shipment by river to Bhamo and thence, by truck, to Kunming. The total value of the aircraft materials and parts stored at Bhamo and Rangoon was at its peak about $5,250,000. Those aircraft materials were scheduled for use in the manufacture of 163 military planes contracted by the Chinese government with Pawley.[97]

Only three months after ordering it, the British authorities in Burma lifted the prohibition on arms, munitions, and other war materials from Burma to China. The flow to Kunming began again. The factory expected to produce weekly five to six fighters or interceptors, or from three to four Vultee bombers in order to complete the contractual 163 airplanes by the end of 1941. The plant also took on the assembly of 50 Northrop and 50 Ryan training planes.[98]

It is not surprising that the Japanese military took notice of these developments, and the factory became a major target of the Imperial air force.[99] On October 26, 1940, for example, 36 Japanese planes bombed the factory, but the damage done was negligible. Pawley, who witnessed the raid, told an officer of the American Embassy in Chongqing that the damage caused by raiders was not heavy and all Americans escaped without injury, although they remained at or near the factory and had no intention of withdrawing despite the advice of the American Embassy to do so.[100]

All of these handicaps notwithstanding, the Central Aircraft Factory succeeded in manufacturing about 400 airplanes for the Chinese Air Force between 1937 and 1940. It also became a factory base of the Aviation Volunteer Group of Americans and a vital factor in the brilliant achievements of the famed Flying Tigers when the Group expanded in the following years.

The Shaoguan Aircraft Factory

The Shaoguan Aircraft Factory in Guangdong, the second largest aircraft factory in China, employed about 500 employees before the Sino-Japanese war. It produced Fuxing airplanes and assembled Hawk III fighting airplanes under the management of the American engineer, chief engineer Charles H. Day, and assistant chief engineer C. L. Zakhartchenko.[101]

After fighting broke out, however, the conditions of production became extremely difficult in the plant due to continuous Japanese air raids or air raid alarms. During one air raid, two bombs hit the factory, which so damaged several buildings that production ceased. From August 31, 1937, until the removal of the factory, the Japanese Air Force sent more than 100 observation and bombing missions, hoping to destroy the factory completely, but miraculously only three bombs actually hit the factory.[102] To continue manufacturing airplanes, Lin Fuyuan, who was appointed general manager of the Shaoguan Factory in August 1937, asked his workers to remove the factory's equipment to nearby forests.[103] Despite truly primitive working conditions, excellent planes were turned out, if in small number, and Zakhartchenko recommended to his supervisor in the autumn of 1937 that the factory proceed immediately "with the design and construction of an airplane capable of out-maneuvering and out-fighting the enemy aircraft."[104] But the factory, he concluded, had to find "the ways and means for speeding up the work" and placing "orders for indispensable new parts."

Conditions deteriorated still further. On January 14, 1938, Day addressed a letter to Madame Chiang Kai-shek, formally director of the Commission on Aviation Affairs, who was in Hong Kong at the time, suggesting that she arrange for the financing and transshipment at Hong Kong of war materials for the Chinese government. Day wrote, "I have been advised that consideration has been given to the question of moving the activities of the Shaoguan Aircraft Factory to a more remote location. If China is preparing for a long war, I strongly advise such a move and have no doubt but that we can thus produce many more airplanes over a period of but a relatively short time." He recommended that the factory move to Kunming, "because of its railroad connections," and "someone, possibly Colonel Lin Fu-yen and myself, visit the new location . . . to determine the facilities available and the means of transportation to that location."[105]

After receiving orders from the Chinese government, Day went to Kunming in February with Lin Fuyuan to fix upon a site for the Shaoguan Aircraft Factory, and they chose a town about four miles west of Kunming. Day predicted to the American consul in Yunnan that the Shaoguan Aircraft Factory would be removed to Kunming by its staff.[106] The Commission on Aviation Affairs subsequently approved the removal of the Shaoguan Aircraft Factory to Kunming and the facilities of the factory were transported by railroad first to Hong Kong, then to Haiphong and finally to Kunming. Workers and technicians of the factory proceeded to Kunming in small groups either by sea (by way of Hong Kong and Indochina) or by land (via Guangxi and Guizhou provinces).[107]

The new factory was located in a small valley about one mile long, consisting of a dozen sheet-iron buildings with an average floor space of 100 square feet. The buildings, the offices, and the dormitories were scattered over several neighboring villages. The buildings of modern construction contained about 80,000 square feet of floor space, and were divided into metal-working, wood-working, wood-assembly, repair-assembly, production-assembly, and miscellaneous workshops. For all this effort, the factory was really able to produce only wings and fuselages. Engines, landing gear, instruments, and most of the metal fittings had to be imported from the United States.[108]

Between August 1937 and September 1939, the factory manufactured, under very difficult conditions, 42 new airplanes, rebuilt 14 airplanes, and repaired 222.[109] From 1939 on, *Fuxing* training airplanes and Hawk III pursuits were the main products. In September 1939, under instructions from the Commission on Aviation Affairs, the factory started research on the possibility of utilizing local materials for the *Fuxing* airplane. Still, the maximum capacity of the plant, and only if it was provided with ample materials, was only 10 planes per month.[110] In 1940, the factory employed a total of 1,300 workmen, of whom 200 were originally from Canton and the rest hired locally. Most of the Cantonese engineers had studied aviation in the United States and had considerable manufacturing experience. Day and Zakhartchenko were in charge of the day-to-day operations of the plant.[111]

After the Shaoguan Aircraft Factory moved to Kunming, it never enjoyed adequate materials for manufacture. Day was disappointed time and time again by the failure of the Commission on Aviation Affairs to arrange for critical supplies.[112] Furthermore, the Commission on Aviation Affairs established a new organization in the factory that limited the freedom of action of factory management. In the new structure, there was no position of chief engineer. The foreign advisers worked directly under the Commission on Aviation Affairs, which directly managed the factory and made appointments to the factory, and transferred employees from the factory without previously consulting or even giving advance notice of such action. In addition, the commission changed the factory organization plan and regulations without consulting the factory as to its practical needs.[113]

Day, now only an adviser, had no line authority in the factory. Under this circumstance, he complained, "I am employed as chief engineer of the Shiuchow Aircraft Works. There is no such position in the new organization plan. I am entirely unconcerned as to what my position or title may be provided."[114] The factory management ignored Day's complaints, leading him to write directly to the director of the Commission on Aviation Affairs, requesting relief:

> Since in my position and under these conditions I am powerless to in any way correct the situation, I believe my service here is superfluous and therefore an unnecessary expense to the Government and a waste of my time. Accordingly it is requested that the Commission release me from my present employment contract.[115]

The Commission on Aviation Affairs sided with the factory management, and predictably, Day, who had been in charge of the factory for over five years, resigned and left for the United States in November 1940.[116]

Zakhartchenko , who also had worked in the factory for five years, decided to come back to the United States at the end of 1939 to validate his citizenship status in accordance with U.S. emigration regulations. On April 24, 1939, the Commission on Aviation Affairs approved his application for leave. Taking the opportunity, the commission suggested that Zakhartchenko visit aircraft factories while he was in the United States to catch up with the latest airplane developments including propeller manufacture, plywood fabrication, and so on. The U.S. War Department interposed no objection to Zakhartchenko's plan to visit the Consolidated Aircraft Corporation, the Douglas Aircraft Corporation, the North American Aviation, and others in February 1940.[117] After Zakhartchenko returned to China, he also expressed his desire to resign when his contract expired. In a letter to the commission, he wrote:

> I beg to decline with thanks the offer of renewal of my contract for a period of one year. However, in order to complete the work on hand, I will be quite willing to stay for a few months under extension of the present contract provided one month of unused leave will be included in it.[118]

The commission renewed his contract, and he did not return to the United States until 1943.

From July 1937 to December 1940, with the assistance of American engineers, the factory successfully manufactured about 500 *Fuxing* training airplanes and pursuit airplanes of the Hawk III design. *Fuxing* airplanes were turned over to the Aviation School for training the students while the Hawk III pursuit airplanes went to the Chinese Air Forces at the front. The factory contributed greatly to Chinese military aviation in the subsequent years.[119]

Aviation Schools

The production of training and fighting aircraft, difficult as it was in the rough and deprived conditions of beleaguered China, was only a part of the tale. Training of adequate pilots was not a bit less critical to effective resistance in the skies over China. In 1937, the Nationalist government had three major aviation schools for training Chinese pilots, the Luoyang Aviation School, the Central Aviation School in Hangzhou, and Nanchang Aviation School. At the Luoyang and the Central Aviation Schools, primary training was conducted in Fleet and Douglas trainers and by American instructors under the command of Claire Chennault. Instruction was given in advanced pursuit and light bombardment. After the outbreak of the Sino-Japanese war, the Chinese government reorganized and relocated all the aviation schools. It moved the Central Aviation School from Hangzhou to Kunming in October 1937, dividing it into a basic school and an advanced school. A few months later, the Guangdong and Luoyang Aviation Schools relocated in Liuzhou in Guangxi province, and were combined with the Guangxi Aviation School. The new school conducted primary training.[120]

In early 1938, Chiang Kai-shek asked Chennault to implement a two-year program of training for Chinese pilots. Authorized to take charge of all training, Chennault went to Yunnan province to assume his duties by the summer.[121] The aviation schools, under Chennault's guidance, selected students from Chinese middle schools between the ages of 18 and 22. Before undertaking air training, these young men were first required to do nine months at a military school, then six months at the Liuzhou Primary School, and finally six months in each of the basic and advanced schools in Kunming, for a total training period of 27 months, a very long time in the light of China's always pressing need for flying crews. By 1938, about 400 students were enrolled in the primary school of Liuzhou and approximately 400 students in the Kunming Schools. On February 20, 1938, the advanced flying school graduated 183 pilots. Of those, 50 specialized in bombardment, 43 in observation, and 85 in pursuit. Pursuit graduates were to remain for an additional six weeks of training.[122]

The courses at all three schools duplicated the instruction at the U.S. Army aviation training schools. Much as in other countries at the time, the primary schools taught take-offs, landings, climbs, climbing turns, gliding turns, steep banks, and spins. Native Chinese instructors provided all instruction at that stage, although American pilots closely observed and checked each cadet. The basic schools, the next level of training, taught cross-country, night, and formation flying, including accuracy spins. Chinese instructors, with Americans again acting as check pilots, did the training. Unlike the primary course, the basic schools used only Douglas training planes instead of Fleet trainers. The Chinese students received instruction in all the maneuvers taught in American primary military aviation schools, but when they graduated, they were considered to be only about 60 percent as skilled as the average American graduating from a primary flying school, at least according to one report. After the primary and basic stages, the successful cadet went on to the advanced flying school, where he was assigned a specialty: bombardment, pursuit, or observation. Chinese instructors offered the ground instruction and some in bombardment and observation, while Americans gave all the instruction in pursuit and air tactics, which required 70 hours of flying with instruction in aerial combat, aerial gunnery, air bombing, ground gunnery, and different tactics used in combat formations. The advanced courses in bombardment and pursuit work used the Curtiss-Hawk pursuit planes, types II and III, for training, in which the cadets logged approximately 200 flying hours. Upon graduation from the advanced school, some students were sent to Xi'an and Lanzhou in the north and west for a short training course in Soviet-built planes. Others were sent to various fields to specialize in the type of foreign planes they were assigned to fly.[123]

Colonel S. D. Jiang was nominally in command of the schools at Kunming,[124] but actually Billy McDonald, formerly with the U.S. Army Air Corps and a member of their acrobatic group, was in charge of the instructor personnel and training at Kunming. According to all reports, American instructors at the schools were extremely capable with years of flying experience.[125] Four ex-American Army Air Corps officers were in charge of the Primary School at Liuzhou, most of whom had been known to Chennault prior to their retirements from the U.S. Army.[126] These instructors were all graduates of Kelly Field and remained reserve officers in the U.S. Army. Forty-two Chinese instructors also served at the school. When the school needed additional instructors, the American instructors recommended to Chennault other American pilots known to them, who had served in U.S. forces. Chennault then recruited them in the United States.[127]

In mid-1939, Chennault submitted to Chiang a long and detailed report concerning the training program, in which he suggested that American instructors should enjoy greater authority in the aviation schools.[128] As chief adviser, Chennault was successful in keeping the American training program going in the schools in the succeeding years.[129] From July 1937 to December 1940, the American-supervised aviation schools trained about 900 excellent Chinese cadets, who thereafter constituted the bulk of the flying personnel of the Chinese Air Force.

From the outbreak of the Sino-Japanese conflict in July 1937 to the end of 1940, China fought alone. The Chinese received sympathy, moral support, and some loans from the Western countries. The Soviet Union was the only state that gave China material aid, especially air aid. During this period, the U.S. government did not provide air aid, but it promoted aircraft sales to China, and Americans helped to assemble and manufacture aircraft and to train the Chinese pilots. By the end of 1940 one of the most serious problems Chongqing had was its almost total lack of air power. Due to the events in Europe, Russia slowed down its air aid to the Nationalists and began to withdraw its pilots. Beginning in May 1939 the Japanese Air Force undertook destruction of China's cities, especially Chongqing, by air bombardment. As a result, Chiang turned to the Americans for air aid again.

NOTES

1. Estimates of the number of Chinese military airplanes at the beginning of the Sino-Japanese war range from 300 to 700, see Yingqin He, *Heshangjiang kangzhan qiqian junshi baogao,* p. 38, and Long-hsuen Hsu, *History of the Sino-Japanese War, 1937–1945* (Taipei: Chung Wu, 1985), p. 266.

2. "The First Year of the Chinese Air Force in the War," *Zhongguo de hongjun* [Chinese

Air Force], 13 (1938): 1.

3. *Air Operations in the China Theater, 1937–1945* (Japanese Monograph No. 76, prepared by Headquarters, USAFFE and Eighth U.S. Army, 1956), p. 17.

4. A. L. Paterson's Report, File Number 893.796/216, RG 59, NA.

5. John H. Jouett, "War Planes Over China."

6. Beginning in February, 1936, Madame Chiang was the director of the Commission on Aviation Affairs and was reorganizing the Chinese Air Force for the coming war.

7. Claire Lee Chennault, *Way of a Fighter*, pp. 29–54; Paul Frillman and Graham Peck, *China: The Remembered Life* (Boston: Houghton Mifflin, 1968), p. 79.

8. *Air Operations in the China Theater, 1937–1945,* p. 28.

9. Yingqin He, *Rijun jinhua banian kangzhanshi* [The eight-year history of Chinese resistance to Japan's invasion] (Taipei: Guofangbu shizhenju chuban, 1983), pp. 299–304.

10. Weiguo Jiang, *Kangri yuwu* [Resisting Japanese invasion] (Taipei: Liming wenhua chubangongsi, 1979), 2: 111.

11. Yingqin He, Heshangjiang kangzhan qiqian junshi baogao, p. 38.

12. *Zhongguo shibian: lujunzouzhanshi* [China Incident: The combat history of the Japanese Army] (Beijing: Zhonghua shuju, 1981), originally issued by the Department of History, Defense Institute of the Defense Ministry of Japan, translated by Ji Fulin, p. 19.

13. Masatake Okumiya and Jiro Horikoshi, *Zero!* (New York: E. P. Dutton, 1956), pp. 24–25.

14. Weiguo Jiang, *Kangri yuwu*, 4: 95–101.

15. Long-hsuen Hsu, *History of the Sino-Japanese War, 1937–1945*, pp. 266–69.

16. Chiang to Zhou, September 22, 1937, *Zhonghua minguo zhongyao shiliao chupian: Duiri kangzhan shiqi,* 2: 76.

17. Jiang Weiguo, *Kangri yuwu*, 4: 95–101.

18. Long-hsuen Hsu, *History of the Sino-Japanese War, 1937–1945*, pp. 266–69; Air Major-General C. J. Chow, "Two Years of Undeclared War in the Air," *China Weekly Review*, January 6, 1940, pp. 204–7.

19. Clarence E. Gauss, American consul general at Shanghai, to the Department of State, September 29, 1937, File Number 893.20/627, RG 59, NA.

20. "182 U.S. Aviators Are Eager to Join China Air Service," *Washington Post*, August 2, 1937, p. 9; "War Recruiting Here Banned," *Los Angles Times*, August 3, 1937, p. 2; Yost Memorandum, August 9, 1937, File Number 893.20/604, RG 59, NA.

21. "Japan Disturbed by Report 182 Americans Have Enlisted to Fly Warplanes for China," *New York Times*, August 6, 1937, p. 3; "Considers Protest," *Washington Post*, August 6, 1937, p. 9; "7,000 Americans Volunteer for Brigade to Fight for China," *New York Herald Tribune*, August 6, 1937, p. 5.

22. "American Citizens Engaging in Military Activities in China: Prohibition of," by Division of Far Eastern Affairs, August 6, 1937, File Number 893.20/616, RG 59, NA.

23. The secretary of state to the Consul at Hong Kong (Donovan), August 21, 1937, *FRUS, 1937*, 4: 522.

24. Addison E. Southard, American consul general in Hong Kong, to State Department, November 16, 1937, File Number 893.20/636, Department of State Records, RG 59, NA; the State Department to the American consular officer in chargé, Hong Kong, November 17, 1937, ibid.; Addison E. Southard, "American Aviators for Chinese Government," November 17, 1937, ibid.

25. The ambassador in China (Johnson) to the secretary of state, September 1, 1937, *FRUS, 1937*, 4: 871–72,523–4. See also Claire Lee Chennault, *Way of a Fighter*, p. 51.

26. The secretary of state to the ambassador in China (Johnson), September 7, 1937, ibid., 4: 524–5.

27. The ambassador in China (Johnson) to the secretary of state, September 10, 1937, ibid., 4: 526–7.

28. The secretary of state to the ambassador in China (Johnson), September 15, 1937, ibid., 4: 528–30.

29. These airplanes, manufactured by the Bellanca Aircraft Corporation of New Castle, Delaware, originally were designed for carrying mail in the United States. The U.S. government prohibited their domestic use, however, because the airplanes were judged unsafe for long-distance flight. In November 1936, the Hanover Sales Corporation of New York bought these 19 planes for Air France, a commercial operating company, but when Washington revised its neutrality law to include countries engaged in civil war, the Department of State refused to issue an export license for their export to France because the Spanish loyalists might ultimately use the planes. Then, the Chinese government purchased the airplanes, and on August 9, 1937, the State Department issued a license to the Huahang Trading Company, an agent of the Nanjing government, authorizing the company to export 20 Bellanca airplanes as commercial airplanes to China. With the permit from the Department of State, the Bellanca factory loaded 19 on the S.S. *Wichita*; the remaining plane was under construction and would be shipped to China later. See C. W. Yost Memorandum, August 9, 1937, File Number 711.0011 Lic Wah Chang Trading Corp./52, RG 59, NA.

30. Newspaper clipping, File Number 711.0011 Lic Wah Chang Trading Corp./170, RG 59, NA.

31. The ambassador in Japan (Grew) to the secretary of state, August 28, 1937, *FRUS, 1937*, 4: 436–7.

32. Later, Japan informed the *Wichita* that the Japanese Navy would examine it when it arrived in China sea. See C. T. Wang to the minister for foreign affairs, September 18, 1937. See *Zhonghua minguo zhongyao shiliao chupian: Duiri kangzhan shiqi*, 3: 409.

33. "To Reserve Rights: State Secretary Calls Nations' Attention to Damages Invaded," *New York Times*, August 29, 1937, pp.1, 31.

34. Joseph C. Green's Memorandum, August 27, 1937, 711.00111 Lic.Wah Chang Trading Corp./43, RG 59, NA.

35. "19 Bombers off for Shanghai: As Strike of Crew Is Averted," *Baltimore Sun*, August 28, 1937, p. 18.

36. They were the World Peace Ways, the National Council for the Prevention of War, the Emergence Peace Campaign, the Women's International League of Peace and Freedom, and the Fellowship of Reconciliation. See "Five Groups Open Neutrality Drive," *New York Times*, September 1, 1937, p. 3.

37. "Roper and Hull Differ," ibid., September 2, 1937, p. 1.

38. See "U.S. Policy Backs Joint China Moves," ibid., September 3, 1937, p. 1.

39. The ambassador in Japan (Grew) to the secretary of state, September 5, 1937, *FRUS, 1937*, 4: 449.

40. Arnold J. Toynbee, *Survey of International Affairs, 1937* (London: Oxford University Press, 1938), I: 226–7.

41. Memorandum of conversations, by the chief of the Office of Arms and Munitions Control (Green), September 13, 1937, Department of State, *FRUS, 1937*, 4: 527–8.

42. "U.S.-Owned Ships Under Arms Ban," *New York Times*, September 15, 1937, p. 1.

43. Department of State, *Press Release*, September 18, 1937, vol. 17, no. 416, p. 227.

44. Edgar P. Allen Memorandum, September 15, 1937, File Number 711.00111 Lic Wam Chang Trading Corp./43, Department of State Records, RG 59, NA. On October 28, 1937, the *Wichita* arrived in Hong Kong without having been challenged by Japanese warships. It unloaded a considerable amount of barbed wire and other materials that could not be used for military purposes. See *New York Times*, October 28, 1937, p. 4; Memorandum by the adviser on political relations (Hornbeck) of a conversation with the

Chinese ambassador (C. T. Wang), September 15, 1937, Department of State, *FRUS, 1937*, 4: 530–1; The ambassador in China (Johnson) to the secretary of state, September 18, 1937, ibid., 4: 535–6; C. T. Wang to the minister for foreign affairs, September 18, 1937, *Zhonghua minguo zhongyao shiliao congbian,* 3: 409.

45. Memorandum by the adviser on political relations (Hornbeck) of a conversation with the Chinese ambassador (C. T. Wang), September 15, 1937, Department of State, *FRUS, 1937*, 4: 530–1.

46. The ambassador in China (Johnson) to the secretary of state, September 18, 1937, ibid., 4: 535–6.

47. C. T. Wang to the minister for foreign affairs, September 18, 1937, *Zhonghua minguo zhongyao shiliao chupian: Duiri kangzhan shiqi,* 3: 409.

48. Roosevelt's statement failed to satisfy the pacifist organizations. On September 16, the six organizations issued a statement requesting the explicit application of the Neutrality Act to China and Japan. See *New York Times,* September 16, 1937, p. 1.

49. *FRUS, Japan, 1931–1941*, 1: 520.

50. Dorothy Borg, *The United States and the Far Eastern Crisis of 1933–1938*, p. 518.

51. An account of American isolationism is found in Manfred Jonas, *Isolationism in America* (Ithaca, N.Y.: Cornell University Press, 1966), which analyzes the various shades of isolationist ideology. For the best study of congressional isolationists, see Wayne Cole, *Roosevelt and the Isolationists, 1932–1945* (Lincoln: University of Nebraska Press, 1983) and Robert A. Divine, *The Illusion of Neutrality* (Chicago: University of Chicago Press, 1962). For a study of pacifist opinion, see Lawrence S. Wittner, *Rebels Against War* (New York: Columbia University Press, 1969).

52. Chennault to the adjutant general, U.S. Army, June 1, 1938, Pawley Papers.

53. Enzo Augelucci, *The Rand McNally Encyclopedia of Military Aircraft, 1941–1980* (New York: Military Press, 1983), pp. 217, 223, 286.

54. Chi-jou Chow, "The Chinese Air Force," *China Year Book, 1944–1945*, p. 315; *Zhonghua minguo zhongyao shiliao chupian: Duiri kangzhan shiqi,* 2: 509–512, 526; Aleksandr Ya. Kalyagin, *Along Alien Roads* (New York: Columbia University East Asian Institute, 1983), pp. 43, 46; Arthur N. Young, *China and the Helping Hand, 1937–1945* (Cambridge, Mass.: Harvard University Press, 1963), pp. 10, 54–5.

55. Long-hsuen Hsu, *History of the Sino-Japanese War, 1937–1945*, p. 270.

56. See *China Handbook, 1937–1943,* pp. 33–34.

57. *Air Operations in the China Theater July 1937–August 1945*, pp. 71–72, 96.

58. Lawrence K. Rossinger, *Deadlock in China* (San Francisco: Institute of Pacific Relations, 1940), p. 27.

59. "The Chief of the Office of Arms and Munitions Control, Department of State (Green), to 148 Persons and Companies Manufacturing Airplane Parts," July 1, 1938, Department of State, *FRUS, Japan: 1931–1941*, 2: 201–2.

60. Elton Atwater, *American Regulation of Arms Exports*, p. 233.

61. The Japanese ambassador (Horinouchi) to the secretary of state, January 6, 1940, Department of State, *FRUS, Japan: 1931–1941*, 2: 205–7.

62. W. D. Pawley, "A Brief History of the Flying Tigers," Pawley Papers.

63. Memorandum by the Chief of the Division of Far Eastern Affairs (Hamilton) of a Conversation with the First Secretary of the Chinese Embassy (Tsui), January 3, 1940, Department of State, *FRUS, 1940*, 4: 251.

64. H. D. R. Burgess, Joint Manager of the Bank of China, New York, to Tsuyee Pei, Assistant General Manager, Head Office of Foreign Department, Hong Kong, April 3, 1936, Young Papers.

65. Major M. W. Pettigrew, "Purchase of Airplanes by China," April 16, 1939, File

Number 2078-152, RG 165, NA.

66. Ibid.

67. Peck to the secretary of state, May 24, 1939, File Number 893.248/120, RG 59, NA.

68. Green's Memorandum, September 13, 1939, File Number 893.248/168, ibid.

69. Hull to the American embassy in Chungking, May 19, 1939, File Number 893.248/117, ibid.

70. Memorandum of conversation among Hu Shih, Hornbeck, and Green, May 20, 1939, Hornbeck Papers.

71. The Chinese government had been endeavoring to use the American commercial credit for the purchase of four airplanes that were badly needed by the China National Aviation Corporation, a joint America–Chinese enterprise. But the American government stated that the credit could not be used for the purpose because airplanes were regarded as war materials. See Peck to Secretary of State, March 18, 1939, File Number 893.796/262, RG 59, NA.

72. Hull to American embassy in Chungking, May 29, 1939, File Number 893.248/119, ibid.

73. Memorandum by Green, May 29, 1939, File Number 893.248/136, ibid.

74. Memorandum by Green, June 13, 1939, File Number 893.248/140, ibid.

75. Memorandum by Green, June 28, 1939, File Number 893.248/149, ibid.

76. The Foreign Ministry to the Chinese Ambassador in the United States, September 25, 1939, *Zhonghua minguoshi ziliao conggao* [The historical records of the Republic of China] (Beijing: Zhonghua shuju, 1983), 3: 25.

77. Memorandum by Green, July 27, 1940, File Number 893.248/179, RG 59, NA.

78. H. D. R. Burgess, Joint Manager of the Bank of China, New York, to Tsuyee Pei, Assistant General Manager, Head Office of Foreign Department, Hong Kong, April 3, 1936, Young Papers.

79. The secretary of state to the ambassador in China (Johnson), October 24, 1940, Department of State, *FRUS, 1940*, 4: 679–82.

80. "Japan Reveals Her Fear of Air Raids by Chinese," *New York Times*, April 13, 1939, p. 18.

81. "Chinese Aircraft Factories," Report Number 276, August 7, 1937, Box Number 18, Intelligence Division, Naval Attaché Reports, 1886–1939, Records of the Office of the Chief of Naval Operations, Record Group 38, NA.

82. See the Commission on Aviation Affairs of the Nationalist Government, "Hongjun yangeshi chugao," SHAC.

83. The Standard Oil Company sold the premises to the Central Aircraft Manufacturing Company on November 20, 1937. See P. R. Josselyn, American consul at Hankow, to Secretary of State, January 14, 1938, and Secretary Hull to Josselyn, January 19, 1938, File Number 393.115/132, RG 59, NA.

84. "Moved Chinese Plane Factory Four Times," *New York World-Telegram*, September 22, 1939.

85. Memorandum of Convention between Bruce G. Leighton and Paul W. Meyer, February 15, 1938, File Number 893.796/243,RG 59, NA.

86. "Removal of Central Aircraft Manufacturing Company to Yunnanfu," by Paul W. Meyer, American consul at Yunnanfu, July 11, 1938, File Number 893.796/249, RG 59, NA.

87. Memorandum of Conversation on purchase of military aircraft and equipment by the Chinese government, June 23, 1939, File Number 893.248/154, RG 59, NA.

88. Nelson T. Johnson, "Removal of Central Aircraft Manufacturing Company from Hankow to Yunnanfu," June 30, 1938, File Number 893.796/248, ibid.

89. Paul W. Meyer, "Removal of Central Aircraft Manufacturing Company, Federal Inc.,

U. S. A. to Yunnan," December 5, 1938, File Number 893.796/259, ibid.

90. The terms of these contracts were to be fulfilled within a period not longer than two years. The material to be furnished was to be used chiefly for the manufacture of pursuit and interceptor planes. The figure included the value of materials and parts, their transportation charges from the United States to Yunnan via Rangoon, and fabrication and assembly costs. Supplies were to be obtained by the Intercontinent Corporation from Curtiss-Wright interests that would furnish the factory some of their own materials and undertake to secure additional parts and equipment from other American manufacturers. See Austin Brady, "Construction of and Plans for the Operation of A New Aircraft Manufacturing Factory at Loiwing, China," November 18, 1939, File Number 893.796/276, RG 59, NA. See also Austin C. Brady, "Reported Terms of Contracts for Aircraft Materials to be Furnished by Intercontinent Corporation to Chinese Government and Imported through Rangoon for Assembly at Loiwing, China," February 14, 1940, Regional File 1933–1944, Military Intelligence Division, Records of the War Department General and Special Staffs, Record Group 165, The Federal Record Center, Suitland, Maryland. Hereafter cited as RG 165, FRC.

91. "China's Aircraft Factory Plant: Modern Industrial Center Established after Many Removals to Escape Japanese Bombers," *North China Herald*, October 18, 1939, p. 128.

92. George B. Arnold was the factory manager and C. W. Hunter was the general manager. The American technicians in the factory were: A. L. Anderson, Frank Leckell, W. D. Stoker, R. M. Bell, C. K. Miller, M. D. Walsh, H. E. Hoteko, M. K. Shipherd, and C. F. Killelea. See "Central Aircraft Factory at Loiwing, Yunnan," September 19, 1940, Report Number 9505/751, File Number 2078-147/3, RG 165. See also W. D. Pawley, "A Brief History of the Flying Tigers," Pawley Papers.

93. R. S. Aitken, "Report on the Chinese Air Force," March 27, 1939, File Number 421, Embassy and Consular Archives, China, Correspondence: Series II, 1875 to 1956, Record Class FO 676, PRO.

94. "Thrice-Moved Airplane Plant Now Turns Out One Plane Daily," *China Weekly Review*, January 6, 1940, p. 199.

95. "Central Aircraft Factory at Loiwing, Yunnan," September 19, 1940, Report Number 9505/751, File Number 2078-147/3, RG 165, NA.

96. The ambassador in China (Johnson) to the secretary of state, July 18, 1940, Department of State, *FRUS, 1940*, 4: 53.

97. Austin Brady to the secretary of state, October 22, 1940, RG 59, NA.

98. "Central Aircraft Factory at Loiwing, Yunnan," September 19, 1940, Report Number 9505/751, File Number 2078-147/3, RG 165, NA.

99. Austin C. Brady, "Possibility of Japanese Air Raid on the Aircraft Factory at Loiwing, China Considered, and Some Protective Measures Taken; Thirty-nine Americans Now at Loiwing, and Number Will Be Increased," March 19, 1940, Regional File 1933–1944, Military Intelligence Division, Records of the War Department General and Special Staffs, RG 165, FRC.

100. Brady to Secretary of State, November 16, 1940, ibid.; *New York Times*, October 28, 1940, p. 4.

101. Day to James Moy, Director of the Shiuchow Aircraft Works, May 25, 1937, Zakhartchenko Papers.

102. "Removal of Central Aircraft Manufacturing Company to Yunnanfu," July 11, 1938, File Number 893.796/249, RG 59, NA.

103. *Zhongguo de hongjun* [Chinese Air Force], 100 (1947):13.

104. Zakhartchenko to Day, November 14, 1937, Zakhartchenko Papers.

105. Day to Madame Chiang Kai-shek, January 14, 1938, File Number 893.248/97, RG 59,

NA.

106. Memorandum of conversation between Day and Paul W. Meyer, February 16, 1938, File Number 893.796/243, RG 59, NA.

107. "Removal of Central Aircraft Manufacturing Company to Yunnanfu,"July 11, 1938, File Number 893.796/249, RG 59, NA.

108. After the factory moved to Kunming, the Commission on Aviation Affairs of the Nationalist government renamed it the First Aircraft Factory in January 1939. The budget for the factory was $140,000 Chinese yuan. See "Kuomin zhengfu hanghong weiyuanhui hongjun diyi feiji zhizaochang shimo" [The history of the first aircraft factory of the Commission on Aviation Affairs of the Nationalist government in Yunnan], 1986, unpublished manuscript preserved in Guangdong Aviation Association, Canton, China.

109. "First Aircraft Factory (Shiuchow Aircraft Works) Output of Airplanes from January 1936 to September 1939," Zakhartchenko Papers.

110. Day to Lin Fuk-yuen, director of the First Aircraft Factory, August 4, 1939, ibid.

111. Zakhartchenko, "History of Fushin Airplane Development," April 3, 1941, Zakhartchenko Papers; "Chinese Military Aircraft Factory at Kunming," by William Mayer, Field Artillery Military Attaché, September 23, 1940, Report Number 9931, File Number 2078-147/2, RG 165, NA.

112. Day to General C. J. Chow of the Commission on Aviation Affairs, October 3, 1939, Zakhartchenko Papers.

113. Day to Lin Fu-yen, August 18, 1939, ibid.

114. Ibid.

115. Day to General C. J. Chow, October 3, 1939, ibid.

116. "Notes on China's First Aircraft Factory," November 5, 1940, Report Number 9505, Regional File 1933–1944, Military Intelligence Division, RG 165, FRC.

117. Colonel J. A. Crane, foreign liaison officer of the U.S. War Department, to Colonel Teh-chuan Kuo, military attaché of the Chinese Embassy in the United States, January 25, 1940, Zakhartchenko Papers.

118. Zakhartchenko to General C. J. Chow, Director of the Commission on Aviation Affairs, December 19, 1940, ibid.

119. See "Kuomin zhengfu hanghong weiyuanhui hongjun diyi feiji zhizaochang shimo."

120. "Notes on Aviation in China," April 1, 1940, File Number 2078-161, RG 165, NA.

121. Chiang to Chou, January 21, 1938, *Zhonghua minguo zhongyao shiliao chupian: Duiri kangzhan shiqi*, 2: 69.

122. "Aviation School at Kunming (Yunnanfu)," June 8, 1938, File Number 1002-400, Intelligence Division, Naval Attaché Reports, 1886–1939, RG 38, NA.

123. "Chinese Air Corps," May 12, 1939, File Number 2078-155, RG 165, NA.

124. S. D. Jiang was a distant relative of Generalissimo Chiang Kai-shek. He was a graduate of the Whampoa Military Academy and received his flying training in the Hangzhou Air School, in which he became chief instructor later. In the summer of 1936, he helped fly the Cantonese Air Force away from Canton, which resulted in the collapse of the "Canton Rebellion."

125. American instructors at the schools were Johnny Preston, Frank Higgs, C. B. Adair, Jim Bledsoe, L. G. Hoston, Emil Scott, Harold Johnson, Billy Cherymisin, Harold Mull, and Boadtner Carney. See Claire Lee Chennault, *Way of a Fighter*, p. 74.

126. "Chinese Aviation School at Yunnan," June 6, 1938, Report Number 50, Intelligence Division, Naval Attaché Reports, 1886–1939, RG 38, NA.

127. American instructors at the school were H. R. Mull, E. B. Adair, L. G. Heston, and J.

L. Preston. See "Aviation School at Kunming (Yunnanfu)," by Captain R. A. Boone, the U.S. Marine Corps, May 12, 1938, Regional File 1933–1944, Military Intelligence Division, RG 165, FRC.

128. Chiang to Zhou, January 21, 1938, *Zhonghua minguo zhongyao shiliao chupian: Duiri kangzhan shiqi*, 2: 69; Chennault to Chiang, June 7, 1939, ibid., 2: 79.

129. Claire Lee Chennault, *Way of a Fighter*, pp. 74–79.

Chapter 5

U.S. Government Commitment to the Chinese Air Force, 1940–1941

After the failure of the major counterattacks by the Nationalists in late 1939, a stalemate persisted between the Chinese troops and the invading Japanese armies, with a vertical line extending from Beiping to Canton as the division between Nationalist China and the Japanese-occupied areas. The Japanese were caught in a "China quagmire." In this situation, they had several choices: to further their conquest of China, which would exceed their limited resources upon Japan; to retreat from China, which would bring humiliation upon Japan; or to negotiate peace with Chinese leaders, which would end the war. Wang Jingwei, Chiang's deputy in the Nanjing government, left Chongqing in December 1938 and then went to Shanghai to negotiate a separate peace with Japan. In March 1940, he established a new "Nationalist government" in Japanese-occupied Nanjing. Since most Chinese regarded Wang as a traitor and Wang's government failed to end the war, the Japanese had to abandon their plan to ignore the Chongqing government.

At that time the GMD government maintained a difficult existence in Chongqing. Having been cut off from prewar political and financial bases in the rich coastal provinces, Chiang had to cooperate with the warlords and landlords in Southwest China and survive on limited resources. The Nationalists could only "trade space for time," waiting and hoping for Western aid.

AMERICAN WARPLANES TO CHINA

To force the Chongqing government to surrender, Japan increased its air attacks on Nationalist China. In early 1940, Japanese attacks increased to 100 plane raids over the Chongqing area daily. In February 1940, Japanese airplanes carried out reconnaissance missions over advanced Chinese airfields throughout Hunan in order to attack a suspected buildup of the CAF. The Chinese could not offer massed resistance, but staged surprise air guerrilla raids over Shanxi and Hunan to attack the supply centers. Through the spring and summer of 1940, the air war over China continued in the same manner, as the CAF tried to generally regroup, stage surprise raids when possible, and rebuild. The Russian E-16s and Curtiss Hawk IIIs that were flown by skilled pilots could occasionally match the best Japanese fighters until August 1940.[1]

Armed with machine guns and cannon, the Mitsubishi A6M2 Zero fighter appeared over China on August 19, 1940. The Zero fighter, 50–100 miles per hour faster than the best fighter the CAF possessed, rapidly destroyed any slight advantage that the CAF enjoyed earlier when one of its skilled pilots might destroy a Nate or Claude pursuit.[2] After Japanese pilots had gone through transition training, the Zero fighters began missions from Hankou. On September 12, they attacked parked Chinese aircraft but discovered that they were dummy aircraft, specifically constructed to draw enemy attacks. On September 13, the Japanese replied with a ruse of their own. Nell bombers escorted by Zeros were launched toward Chongqing. As they headed toward the city, the Nells turned around and returned to Hankou. Reconnaissance planes then reported that Chinese planes had appeared over Chongqing to launch their own patrols. The Zeros proved their capacity to execute long-range missions by flying on to Chongqing, catching the E-15s and E-16s over the city, and shooting down 27 planes during a 30-minute fight. The Japanese lost no planes.[3] The Zero gave the Japanese the ability to use all aircraft that had been endangered by Chinese pursuits, and its arrival coincided with an even more intense air offensive against Chongqing. From May through September 1940, the Japanese flew 168 day and 14 night missions, with an aggregate total of 3,717 bombers over the city. The CAF was able to mount interceptions against unescorted bombers and against the Nate fighters, but its initial strength was drastically depleted by the Zeros. By the beginning of October, 25 fighters fell before Zero-escorted raids over Chongqing and Chengdu, and 10 more were destroyed on the ground.[4] The CAF could not stand these losses, and its rebuilding program was too new to offer any replacements of men or machines. Chiang Kai-shek feared that the air force and the nation could not take much more of this pressure. The Chinese were in urgent need of fighter planes to defend the temporary capital when the Japanese were undertaking destruction of China's cities by air bombardment. Chiang hoped that China might receive American aid because the war in China had now taken on expanded international meaning.

In the European theater, Germany defeated France on June 25, 1940, and reached the English Channel. As a result, France and England began losing control of their Asian spheres of influence, and Japan now entered the power vacuum. It forced the French to allow them to establish bases in southern Indo-China to bomb Southwestern China. With France's surrender, the Soviets gradually began pulling their aid out to deal with a war against Finland and intense defense preparations for their own country. One consequence of these preparations was that the Chinese were unable to buy more advanced Russian fighters and the Soviet Union could not continue to supply China with large quantities of airplanes.

The number of planes sent to China fell greatly after July. By then China had only 37 fighter planes and 31 old Russian bombers while Japan had 968 warplanes in China. From January to November 1940 the Russians had sent China only 80 E-15s and E-16s while the United States sent China the following:

1 Hawk 75
3 Hawk 21
30 NA16/4
11 Ryan 5TC4
1 Curtiss/22 Falcon PT
3 Vultee VIA
1 Stinson 105[5]

The shock of rapid German and Japanese advances created a more aggressive, fearful mood in the United States, and now Chiang, hoping to receive U.S. help, resolved to place more orders for American aircraft. He asked Chennault to return to the United States to buy the latest American planes and to recruit trained American pilots to fly them. T. V. Soong was already in the United States in July 1940 to ask for arms. Chiang Kai-shek instructed him to purchase 150 fighting planes as well as other war materials. Several days later, Chiang further informed Soong that if the United States was able to provide 300 fighting airplanes of the latest type and 50 to 100 long-distance heavier bombers, victory over Japan would be in sight, certainly an exaggeration but an understandable hope. Soong informed Roosevelt that the Chinese Air Force was so weak that the Chinese Army was not able to win victories in every battle.[6] But nothing immediate came of this attempt.

The next attempt was not apparently until October 18, when Chiang met with Ambassador Johnson, stressing that airplanes and economic assistance were urgently needed for maintaining the Chinese spirit and optimism. In view of the importance of airpower, Chiang pointed out that American planes, with Chinese pilots and designations of course, might utilize splendid air bases located in Zhejiang province to attack Japanese naval bases in Japan and Formosa. China, he insisted, could impact negatively on the Japanese naval strength and reduce the possible Japanese pressure on the Philippines if it acquired only one-twentieth or one-tenth of the American planes Great Britain had ordered.[7]

To meet China's requirements for airplanes, at Hornbeck's suggestion, Joseph Green met with the Chinese ambassador at the end of October 1940 to discuss how to obtain American military airplanes for China. Green suggested that "small orders of twenty-five or fifty of each of these types could be added to the large orders of planes of similar types already placed by the British, Canadians, and American Governments."[8]

A week later, Navy Secretary Frank Knox told Soong that in the first cabinet meeting after the presidential election a plan submitted by him for assisting China was approved.[9] The plan included a loan of $100 million for the purchase of 500 airplanes, the time of delivery of those airplanes not set. To urge the swift accomplishment of the cabinet's plan, Chiang Kai-shek addressed a letter on November 28 to Roosevelt, urging the cabinet's plan. The letter emphasized the importance of American planes to China, arguing that the planes would certainly contain an enemy air force of four times their number and constitute a threat to Japan proper.[10] All these heady plans and hopes to the contrary, the State

Department promised at the beginning of December to make available to China only 50 modern airplanes in the near future.[11]

Hope springing eternal, Chiang sent an air mission in November to the United States, under Major Mao Bangchu, director of the Operations Division of the Chinese Air Force, accompanied by Claire Chennault. This mission was authorized to purchase the latest American fighters and to hire American pilots. Chennault was initially pessimistic since the infant American aircraft industry was hard-pressed to meet its own rapidly expanding needs and was shipping any surplus aircraft to England, now alone against the European Axis. He also knew that no American fighter could offer much resistance against the Zero, but he consented to undertake the assignment.

After arriving in Washington, D.C., Major Mao presented the request for 500 combat planes to the President's Liaison Committee, the civilian agency coordinating foreign arms purchases in the United States.[12] Mao found that Chennault's pessimism was well founded. The primary American concern was England's survival. Newspaper columnist Joseph Alsop told Chennault that the chances of receiving first line aircraft were just about impossible for China. Alsop's opinion was confirmed within the War Department when Chennault visited contacts he knew from previous military service. In the face of German and Japanese victories, American military leaders did not want any of their few modern aircraft and trained pilots to go to China. They were desperately needed for an expanding American military. The Chennault-Mao mission, however, discovered that it had powerful allies among a small influential group in the White House. Led by President Franklin Roosevelt, this group included Secretary of the Treasury Henry Morgenthau, Secretary of the Navy Frank Knox, and Secretary of War Henry Stimson. These men all believed in the immediate importance of buttressing the enemies of the Axis powers with the available means at hand. America's war industry was just beginning to produce weapons, and it was hard pressed to meet American demands, much less those of other nations, but the Roosevelt administration directed that all available material be provided to anti-Axis nations.

For the moment, this meant only England, since the United States wanted to refrain from offending the Japanese or goading them into war, but Morgenthau and others lobbied for the Chinese. Chiang sent Morgenthau a more forceful memorandum which was based on strategic plans that Chennault had worked out while in Washington. It stated that if China were given enough planes, it could retake Canton and Hankou and force the Japanese to concentrate on China and away from the Pacific. The CAF could attack Hainan, Formosa, and the Japanese home island.[13]

Although Washington still tried to practice neutrality, the Axis victories had a profound impact on it. Morgenthau was deeply impressed with the first bombing of the Japanese home islands. Chiang's claim that Japan and Germany had synchronized plans to capture Gibraltar and Singapore took on ominous tones as the Japanese were 600 miles from the latter base.[14]

On November 30, Soong submitted to Secretary of Treasury Henry Morgenthau a plan developed by Chennault. This proposal suggested that if China were given a force of 500 planes flown by American pilots, China could totally wipe out the Japanese forces in China and neutralize Japan's naval striking ability. An air attack on Japan also would undermine Japanese civilian morale. To provide American aircraft for the Chinese to contain Japanese expansion in Southeast, Morgenthau discussed this issue on December 3 with British ambassador Lord Lothian because the relocation of any aircraft to the Chinese resulted in reducing aircraft to Great Britain. The secretary told Lothian that given these bombers and trained pilots, China would use them to attack Tokyo and other industrial centers in Japan. Lothian, however, did not agree with this proposal and encouraged Morgenthau to discuss the matter with Soong again. After talking with Roosevelt during lunch on December 8, Morgenthau suggested to Soong that at the time 500 airplanes were simply not available for China because of the limited capacity of airplane production. If China were willing to bomb Tokyo and other cities in Japan, however, Morgenthau would ask President Roosevelt to divert a number of the newest Flying Fortress bombers to China from the allocation of planes committed to Great Britain. Morganthau repeated that if China were to use those airplanes to bomb Japan, Roosevelt probably would approve the reallocation of airplanes to China and only the president had authority for allocation of airplanes. On December 9, Soong lost no time in passing the important offer to Chiang.[15]

Inspired by Morgenthau's suggestion of using American airplanes to bomb Tokyo and other big cities, Chiang telegraphed Roosevelt on December 13, once again pointing out the crucial importance of airplanes for China in the current war, and promising that if China had 500 airplanes of the latest type and a number of Flying Fortress bombers in the near future, China would have used those airplanes to bomb Japanese cities to discourage the spreading of the war to Southeast Asia. Chiang's telegram incited Roosevelt to approve this proposal. On December 19, Morgenthau discussed the proposal with Roosevelt and his cabinet. Convinced that this program would have great impact on the Sino-Japanese war, Roosevelt finally approved additional military aid for China and ordered the State, War, Navy, and Treasury Departments to develop a program to that end. American involvement had come a very long way from the *Wichita* affair of 1937.[16]

On behalf of Washington, Morgenthau, on Christmas Day, 1940, held a conference with Soong, Major Mao, and Claire Chennault. Roosevelt, Morgenthau said, had agreed to deliver 12 Flying Fortress bombers from Manila with two aviators for each airplane with the understanding that the Chinese Air Force would use these planes to bomb Tokyo in February or March of next year.[17] During the meeting, Chennault provided some specific ideas for the use of Flying Fortresses while Soong wished to receive all kinds of American aircraft as soon as possible. Morgenthau, however, found himself in opposition to other high-ranking officials on his proposal. Some doubted whether or not the Chinese pilots had the ability to fly advanced American bombers to attack Japan. Henry Stimson believed that Morgenthau, Cordell Hull, and Roosevelt were misled by Soong and Chennault, and this operation would incite Japan to attack America. George Marshall, army chief of staff, also strongly opposed this plan and suggested that the War Department would transfer only 100 P-40 fighter planes, which had not been delivered from Britain to China. When Morgenthau, Hull, Frank Knox, Stimson, Marshall, and Admiral Harold Stark attended a meeting on December 23, Morgenthau was asked to inform Soong and Chennault that the proposal had been modified.[18]

On January 1, 1941, Morgenthau informed Soong that the Roosevelt administration had allocated only 100 P-40 fighters from the British order to China.[19] The Chinese government was disappointed with this allocation because it needed the Flying Fortresses while Chennault had all but given hope for the creation of a new fighter group for China. Granted permission to buy 100 aircraft, Chennault began a long search to find a source of supply. Most factories were unprepared to manufacture additional aircraft, but Chennault was able to purchase planes with the help of Burdette Wright, vice president of Curttiss-Wright Corporation in Buffalo, New York. Wright's company had had a long and profitable relationship with China, and Wright arranged to have 100 P-40s that were allocated for the British shipped to China. Operated as an official Chinese purchasing agent, the Universal Trading Company (UTC) on behalf of the Chinese government reached an agreement on February 10, 1941, with the Curtiss-Wright Corporation with regard to the purchase of those 100 fighters.[20]

At the beginning of 1941, President Roosevelt strove hard to mold public opinion and encourage Congress to initiate legislation aiding Britain in its fight against Germany. In his State of the Union address in early January 1941, Roosevelt demanded congressional authority and funds to supply war materials to those nations fighting against aggressors.[21] On March 11, 1941, the Lend-Lease Act, designed primarily to ensure Britain's survival, was passed in Congress, giving the president power to use $7 billion to aid "any country whose defense the President deems vital to . . . the United States."[22] This act signaled a radical departure in U.S. foreign policy and ended American isolationism and neutralism. With this authorization, Roosevelt would approve aid to China under Lend-Lease. On May 6, Roosevelt made an important step in the commitment to provide massive government-to-government aid, declaring that China was eligible for Lend-Lease aid. Roosevelt assured Chiang Kai-shek that aid to China constituted an integral part of the Lend-Lease bill.[23] The Chinese request for aircraft for China's defending and for bombing Japan's cities persuaded Roosevelt to approve additional military aid for China which became one of the Lend-Lease Act's recipients.

To receive expanded U.S. aid, Soong created a new corporation in April 1941. China Defense Supplies (CDS), replacing the UTC, which became China's designated Lend-Lease agent. At that time, the CDS had 37 persons, including Chennault and William Pawley who devoted themselves to purchasing aircraft supply. Although by the end of May about $150 million worth of military equipment had been allocated to China, only a small quantity of U.S. equipment arrived in China in the following months due to transportation difficulties and the lack of transportation means.[24] Since Washington was pursuing a Europe-First strategy, more airplanes were shipped to Britain. By June, only 100 P-40 pursuit planes produced by the Curtiss-Wright Corporation had been delivered to the China Defense Supplies.

In the second half of 1941, few American warplanes were delivered to China. By August, 70 Ryan PT-22 aircraft from the U.S. stock had been transferred to the Chinese, thus partially meeting the demand of the China Defense Supplies.[25] These American airplanes helped to strengthen the Chinese Air Force, which began to challenge the Japanese fleet again in the skies over China.[26] From

August to November, however, only a small number of aircraft were delivered to China. Soong complained in Washington about the lack of air support for China, claiming that the Washington government had promised the Chinese much but had given them little in comparison with the U.S. aid to Russia and Great Britain. Soong appealed to American high officials to speed up delivery of aircraft to China.[27]

AMERICAN VOLUNTEER GROUP

The Chinese need for modern fighting aircraft was thus partially satisfied, but pilots were no less in short supply. When Russia stopped sending aviators to China in July 1940, Chongqing was in urgent need of pilots, although once again no action quickly resulted. Only on October 18 did Chiang Kai-shek in a conversation with N. T. Johnson express the hope of the Chinese government that American pilots would soon come to China. Johnson sadly replied that the U.S. government was unable to send American aviators to Great Britain, not to mention to China. Although some wished individually to venture to China, they could do so on their own. Chiang understood that Washington itself could not dispatch volunteers to assist the Chinese Air Force, but he was not confident that the Chinese government could recruit. Johnson ended by promising he would try his best to pass the message to his supervisors.[28]

In Washington, officials differed on the matter of American pilots proceeding to China. Frank Knox and Stanley Hornbeck were the most active in the matter. On October 19, the secretary of the navy handed a letter to the secretary of state, informing him that "a considerable number of American aviators" were trying to go to China to fight Japan, and asked, "Is it at all possible that we can handle this matter of American flyers going to China as we have handled the same situation with respect to young men volunteering for service in Great Britain in the present War there?"[29]

Within days, Hull had a memorandum dispatched to the Department of the Navy, reminding Knox that "the present policy of the Department is to refuse passports to persons who state that they desire to go abroad for the purpose of enlisting in the armed forces of a foreign state."[30] With regard to American volunteers serving in British forces, the statement explained that American pilots engaged in Great Britain did not violate the laws of the United States because they "proceeded to Canada without passports and have enlisted there."[31] It is not wholly certain whether Hull was simply reiterating earlier policy or intimating that Americans wishing to serve in China might enlist in Canada or elsewhere.

Hornbeck had quite a different opinion from that of Hull. In a letter to Hull of October 28, 1940, he forthrightly stated, "In my opinion, it is highly desirable that we endeavor to make available to the Chinese such facilities, including personnel as we make available to the British."[32] Knox took the first palpable step to permit U.S. reserve pilots to serve in the Chinese Air Force. On November 5, after a conversation with Soong, he agreed that 10 dive bombers then in Manila would be turned over to China and the Chinese government would engage American reserve pilot as instructors to fly those planes as participants in the war.[33] When Soong learned of Knox's decision, he immediately telegraphed Chiang Kai-shek, predicting that if the plan were to be carried out, American pilots would be available to China in the near future.[34] During the conference between Chiang Kai-shek and Major James Marshall McHugh of the U.S. Marine Corps, assistant naval attaché, McHugh confirmed to Chiang that Secretary Knox favored American pilots in China, but added that the Chinese government itself would have to recruit them. As for the passports, these pilots should claim that they were businessmen bound for China for their business only.[35] Finally, Secretary Hull went along with these developments, at least in part. At the beginning of December 1940, under Hull's instruction, Hamilton and Hornbeck informed Ambassador Hu Shi and Soong that "the Department of State would probably issue passports to American citizens who desire to proceed to China for the purpose of serving as aviation instructors," although presumably not as combat pilots.[36]

Hornbeck then encouraged Hamilton to "do something toward helping China in the matter of aviation instructors" because Washington had sent "officers of the Army and of the Navy as instructors to certain Latin American countries and recently sent eleven officers to the Netherlands East Indies. . . . [A]n investment of the services of a few of our officers by way of helping the Chinese prepare for an ultimate offensive would . . . be a sound and a profitable investment."[37] Hamilton suggested, at the end of December, that "Colonel Chennault would undertake the employment of American personnel" and that the War and Navy Departments would assist largely in providing lists of qualified personnel and the State Department would finally issue passports for them to go to China.[38]

With the acquiescence of some policy makers, Chennault, who was working in Washington, D.C., for CDS, immediately proposed a plan on January 2, 1941, for organizing a group of American airmen in China, calling for 100 pilots and 150 technicians. William Pawley, after conferring with Knox on a volunteer scheme, began implementing a plan to recruit American military personnel who would resign their commissions and sign contracts with a private corporation, the Central Aircraft Manufacturing Corporation (CAMCO), which operated with funds from CDS and originated as U.S. Lend-Lease aid. As president of CAMCO, Pawley recruited pilots and crews at military bases all over the United States as well as the Philippine Islands, Hawaii, and Puerto Rico, where there were many pilots with P-40 experience.

In April, President Roosevelt orally authorized reserve officers and enlisted men on active duty to change to inactive status, to go to China, and to serve in the Chinese Air Force as fighter pilots or mechanics.[39] The American Volunteer Group, later to achieve exceptional fame as the Flying Tigers, came into being in April 1941. With President Roosevelt's approval, the first group left for China via Rangoon on June 9, boarding a Dutch freighter escorted by American naval ships. The escort was under the orders of Admiral Harold Stark, who declared that American pilots' safe arrival was "essential to the United States' support of China."[40] Legal arbitration had delayed plane shipments, and the Chinese training centers for the Americans became swamped by summer monsoons. The British offered facilities at Toungoo, Burma, north of Rangoon, and the AVG established its headquarters there. Chennault began an intensive transition program for the pilots. After the first AVG contingent arrived in Burma in June 1941, McHugh worked closely with Chennault to solve many problems. At

Chennault's request, he wrote Secretary Knox in August asking him to seek presidential help in getting .50-caliber machine gun ammunition from the reluctant Army. McHugh also sought Knox's assistance in encouraging MacArthur to release urgently needed spare parts for Chennault's P-40s. McHugh played a significant role in helping Chennault establish the AVG.[41]

Due to Washington's permission and active assistance, the American Volunteer Group, under the command of Chennault, was formally constituted as a unit of China's armed forces by an order issued by Generalissimo Chiang Kai-shek on August 1, 1941. By early September, 101 American pilots had signed contracts and arrived in China, the greater majority of whom had resigned from active duty in the Navy. Three-fourths of the 157 enlisted mechanics and administrative personnel had come from the Army Air Corps.[42]

The successful establishment of the AVG encouraged the AVG sponsors to create a second volunteer force. In October, President Roosevelt authorized the establishment of a second contingent of volunteers by asking the War Department to promote the employment of 106 additional pilots and 181 ground crew personnel.[43] Unlike the first one composed of fighter planes, the second group consisted of a heavy bomber force which was to bomb Japanese cities. Secretary of War Robert Lovett authorized Richard Aldworth, former Air Force captain, to recruit volunteers and instructed the commanding officers at each airfield to assist him with recruitment. After the recruitment was finished, volunteers for the second contingent left for China in November.[44]

Since most of the pilots came from varied branches of military aviation, Chennault requested that they have to enter a three-month training program to learn fighter tactics, understand Japanese combat skills, and form a cohesive unit. Because Burma, under English control, remained neutral, the AVG was obligated to perform training flights only. Chennault did not want them to participate in combat until they were thoroughly trained.

The influx of American pilots to China, promoted by the U.S. government, indicated a dramatic change in U.S. policy toward the undeclared Sino-Japanese war. Apart from delivering aircraft to China under the Lend-Lease program, the Roosevelt administration had modified its policy by allowing Americans to fight alongside the Chinese against the Japanese. The AVG arrived just in time to restore some hope to the Chinese, as their only full-pledged ally, the Soviet Union, had pulled out of their nonaggression pact by signing one with Japan on April 13, 1941.[45]

American civilian and military leaders not only helped to create the AVG, but also established programs to train the Chinese pilots in the United States. During Lauchlin Currie's visit in China in January 1941, American military officials considered that conditions of CAF training facilities were very poor: inadequate equipment and insufficient flying hours. At the request of Chiang Kai-shek, Washington sent to China an air mission, led by Brigadier General Henry B. Clagett, commander of the Philippine Air Force. In March, an air mission composed of an Air Corps officer and a naval flying officer arrived in Chongqing to survey civilian and military airfields as well as sites for new airfields. During their visit, they paid attention to CAF training activities and facilities in and around Chongqing, Chengdu, and Kunming, where the aviation training took place. Believing the aviation schools had inadequate equipment for air and ground training, Clagett recommended that if American instructors and advisers were able to train the Chinese pilots, the Chinese Air Force's strength would be enhanced. In the following conferences with Madame Chiang and Brigadier General Zhou Chirou, acting head of the CAF, Clagett's suggestions won their agreement.[46]

As a result, Chinese military attaché Chu Shiming sent a letter to the U.S. War Department, requesting to send Chinese students to the United States for aviation study. The Chinese also requested American advisers to create a modern Chinese Air Force, to maintain an efficient line of communication in China, and to arm 30 Chinese divisions.[47] With Roosevelt's approval, the Lend-Lease Administration appropriated $1 million for training 150 Chinese pilots.[48] In October, the first group of 50 Chinese students began to arrive in the United States to take the standard U.S. Air Force training course for pilots at Thunderbird Field in Arizona. Asked by reporters, Secretary Stimson had to admit that the Chinese pilots were in the United States for aviation training. In the following weeks, the U.S. government attempted to hide their presence so as not to provoke Japan.[49]

By 1941 the Chinese convinced Roosevelt that China needed American air aid because it had a plan to attack the Japanese homeland. This enticement plus a desire on Washington's part to stanch Japanese expansion in Southeast Asia had persuaded some Washington officials to provide a large number of the latest airplanes, to allow American pilots to serve in the Chinese Air Force, and to train the Chinese pilots. The Roosevelt administration had committed itself to the revitalization and enlargement of the CAF.

ABORTION OF THE BOMBING OPERATION

When Hitler launched his attack on the Soviet Union in June 1941, the Russians totally stopped sending aviators and aircraft to China. A sense of urgency gripped the Chongqing military leaders because at that time the CAF was still outmoded and inefficient. Its strength as of June 1, 1941, was only 89 Russian E-15s and E-16s, 70 Russian two-engine light bombers, and 89 assorted but obsolete American aircraft. At that time, 654 Japanese aircraft in the central and southern regions of China were undertaking more intensive bombing of Chongqing to coerce the Chinese government to surrender.[50] Japan became more and more aggressive in the second half of 1941. Japanese air raids posed a constant and serious threat to that city, as evidenced in the Chongqing tunnel tragedy of June 1941 resulting in the death of thousands of Chinese.

To receive more American airplanes, Chiang attempted to revive the bombing proposal that had been dismissed by Washington in December 1940. When Lauchlin Currie, a coordinator of Chinese aid and a specialist for President Roosevelt, was sent by Roosevelt to China in spring 1941 to investigate the economic conditions and to make suggestions about U.S. aid to the Chongqing government, Chiang reminded Currie of the plan to bomb Japan and Washington's promise to deliver 12 Flying Fortresses. Currie obviously did not know the details of the discussion of the bombing scheme but agreed to push for this program again when he returned home, advising the Chinese to request additional fighter planes and long-range bombers from the United States. Chiang was pleased, believing that Currie, an advocate of U.S. air support for China, would "be of the greatest assistance" to the Chinese Air Force.[51]

After coming home, Currie submitted a new proposal in May to the Joint Aircraft Committee, and the committee sent it to the Joint Board. The proposal stated that a powerful Chinese Air Force of 500 planes could defend Singapore, the Burma Road, and the Philippines and stop Japanese advances in Southeast Asia. It could pose a threat not only to the Japanese military forces in China but also to the industrial centers and cities in Japan.[52] The Chinese Air Force, to which 100 P-40 fighters had been allocated, urgently needed 250 fighters and 150 bombers to launch an air attack on Japanese factories in November 1941. This bombing would weaken Japan's production of munitions and its ability to continue expansion. This proposal was similar to that developed by Chennault in December 1940. While working with Chennault to secure the Joint Board approval of their Chinese air aid program, Currie presented their revised proposal on June 11 to the Joint Planning Committee representing the secretaries of war and navy and the Joint Planning Board. The proposal, "Short Term Aircraft Program for China," suggested that the U.S. government provide China with 269 fighters, 66 bombers, and 35 transports and start to train Chinese aviators.[53] While waiting for approval of his program, Currie tried to convince American military and civilian leaders that U.S. air support for China was significant because the Chinese Air Force would attack the Japanese cities from the airfields in China.

Currie's proposal to bomb Japan was partially influenced by Lieutenant Commander E. O. McConnell, an American navy officer, who was on Clagett's air mission to China in May. After returning home, McConnell wrote a letter to Currie, stating that American bomber aircraft allocated to China were not sufficient and the U.S. government should allocate to China the longer-ranged B-25 which could attack Yokohama and Tokyo from East China. He also suggested that the U.S. government provide Chennault with more staff officers so as to increase the strength of the AVG.[54]

When the Joint Planning Committee was considering Currie's proposal, it was hard for the committee to meet the Chinese needs because America's aircraft production had to meet the British demand first, and then satisfy MacArthur's military forces in Philippines. While the program was waiting, Germany invaded Russia in June and the Japanese started military campaigns in Southeast Asia, which posed a threat to Singapore and the Burma Road on which America depended for transporting war materials to China. Since Japan refused to stop marching on French Indochina and Thailand, American policy makers decided to increase U.S. aid, especially air aid, to China in order to prevent the Japanese from extending their influence to the south. The Joint Planning Committee agreed to the Currie-Chennault proposal and sent it to the Joint Board for approval on July 9, suggesting that American planes and advisers be sent to China as early as possible. On July 12, the Joint Board approved this program to promote the "bombing of Japan" with American aviators and airplanes.

Secretary Knox and acting secretary of war Paterson approved this plan on July 18 and President Roosevelt secretly ratified it on July 23. Roosevelt was determined to assist China with air reinforcements to deter the southern expansion of the Japanese military.[55] Three days later, Roosevelt publicly announced a freeze on Japanese assets and imposed an oil embargo on Japan. The embargo left Japan with only an 18-month stockpile of oil which was very important for the Japanese military machine. Washington believed these two policies—an oil embargo and a secret air attack—might stop the Japanese advance in the southeast and force Japan to withdraw its military forces from China and Indochina.

To examine the supply requirements of the Chinese Army as well as the plan for air attacks on Japan, Roosevelt sent a military mission to China in July. The American Mission to China, headed by Brigadier General John Magruder, a former military attaché in China, arrived in Chongqing in September. During the conferences with Chiang in the following months, Magruder offered to provide air assistance to China. Chiang hoped that Washington would send a high-ranking aviation officer with a group of staff officers to recognize and to enlarge the Chinese Air Force. Magruder strongly endorsed this suggestion, proposing to strengthen the AVG with active-duty Army Air Corps units.[56]

Chennault returned to Burma before the committee approved the proposal, helping to assemble the recently arrived P-40 fighters and to train the AVG pilots. In July Chennault received Lauchlin Currie's letter that 24 bombers would be delivered immediately and 66 would be available to China in December. Chennault later informed Chiang Kai-shek that he was recruiting and training a second volunteer group to be prepared for flying 66 bombers which the War Department would allocate to the Chinese.[57]

In Washington, Currie and Soong were helping Chennault to recruit American pilots to fly Lockheed and Douglas bombers. The second volunteer force arrived in China in November, but the bombers were not available by that time. For this reason, Currie wrote to Chennault on November 22 that 24 American bombers would be delivered to China in December. Understanding the urgent need of the British government and General MacArthur in the Philippines, Chennault hoped that the bombing operation supported by some high-ranking officials in Washington would be carried out before the end of the year.[58] Their plan, however, was not realized before December 7, 1941.[59]

By December 1941, key U.S. government personnel had instituted a series of programs designed to help China resist Japan's air attacks with the use of American bombers and American volunteers, and even to carry the war to the Japanese homeland. The Lend-Lease program offered considerable promise to the Chinese in the way of providing pursuit and bombardment aircraft. The U.S. government aided the dispatch of planes and volunteers and trained the Chinese pilots in order to revitalize the Chinese Air Force. Adopted by the Roosevelt administration, a new policy of furnishing China with modern combat aircraft and surreptitiously allowing Americans to join the Chinese Air Force had an impact on the U.S.-Japan crisis in 1941.

JAPANESE ATTITUDE

The years 1940 and 1941 saw still closer cooperation between the air and ground forces of the Chinese military machine. Chinese pursuit and bomber squadrons played a part in attacking Japanese positions and troop concentrations. On February 5, 1941, a Japanese transport plane was shot down in southern Guangdong. Among the nine passengers and crew killed was Admiral Ozumi, former Japanese minister of the navy and at the time of his death a member of the Supreme War Council. He was on an inspection trip to Hainan Island to arrange details for Japanese activities in the Southwestern Pacific. In the Second Battle of Changsha in September–October 1941, Chinese bombers twice bombed Japanese concentrations when the Japanese were approaching the city of Hunan. In the Chinese counterattack on Yichang in October 1941, a Chinese bomber squadron made a moonlight raid on a Japanese airfield and supplies at the Yangze city.[60] With U.S. government air aid, the Chinese Air Force became more efficient.

Japan contended that the purpose of U.S. government's aid, mainly air aid, to China was to boost the Chinese morale to fight the Japanese and to mire millions of the Japanese Army in China. Although it would take years before the Chinese Air Force could challenge the Japanese fleet in the sky over China, Japan was angered by U.S. government's air support. Saito Yoshie, adviser to the Japanese Foreign Office, claimed in July 1941 that the United States had "continued to aid Chiang until the present time; and they are planning to obtain an advantageous position in China in the future."[61] During a meeting with U.S. Rear Admiral R. K. Turner on July 20, 1941, Ambassador Nomura complained that the United States was constantly providing greater support to China. It was supplying "airplanes and pilots to be sent to Chongqing . . . there are over a hundred American pilots now en route, who have been supplied from the armed forces of the United States."[62]

In August 1941, after exchanging their views at the Liaison Conference, three secretaries, Tomita Kenji, chief secretary of the cabinet; Muto Akira, chief of the Military Affairs Bureau of the War Ministry; and Oka Takasumi, chief of the Naval Affairs Bureau of the Navy Ministry, prepared reference materials for answering questions at the Imperial Conference on September 6 regarding "The Essentials for Carrying Out the Empire's Policies." In answering the question, What about the military alliance between the United States, Great Britain, the Soviet Union, China, and Netherlands? the reference material said:

> It is also reported that the air force specialists of the United States, Great Britain, and China met in Chungking at the end of July to discuss the use of existing air bases in the hinterland of China.
>
> It is widely known that the United States has been supplying China with military goods. In addition, it is worth noting that the movement of United States Army personnel, who are going to China a few at a time and bringing weapons with them, particularly airplanes, shows that a military coalition actually exists between the United States and China.
>
> However, partial agreements are being reached on specific matters concerning military alliance between the United States, Great Britain, and China. They are as follows: (1)

between the United States and China, assistance to the Chinese air force; (2) between Great Britain and China, formation of mobile guerrilla units; (3) against Japan's advance southward, use of Chinese bases by American and British air units.[63]

The above reference material suggested that, to the Japanese leaders, the American government's air support of China meant the Americans were fighting with the Chinese Air Force against the Japanese, and the United States had involved itself in the war against Japan. The Japanese militarists might justify their attack on Americans because the U.S. government refused to stop military aid to China. Japanese Ambassador Nomura, in a final memorandum to the secretary of state several hours before the Pearl Harbor attack, claimed that the U.S. government refused to "cease from aiding the Chungking regime, which indicated clearly the intention of the American Government to obstruct the restoration of normal relations between Japanese and China and the return of peace to East Asia."[64] American aid to China included both military and economic aid, but American air assistance constituted the major U.S. support of China in 1941. As military historians Charles Romanus and Riley Sunderland wrote, in an effort to aid China, by December 1941, the U.S. government had been helping the Chinese "to (a) create a modern Chinese Air Force, (b) institute and maintain an efficient line of communications into China, and (c) arm thirty divisions."[65]

On November 26, Secretary Hull informed the Japanese representatives that the final terms for peace required Japan's withdrawal from China and Indochina first, and the U.S. then would lift oil embargo and restore trade. With its oil running out, Japan decided to attack the American Navy at Pearl Harbor.

U.S. Asia policy was to resist further Japanese expansion while sustaining and strengthening Chiang through military and economic aid in order to prevent the collapse of the Nationalist government. The first military assistance to China sponsored by the U.S. government came as part of U.S. air aid for China. The Roosevelt administration began to give military aid to China in 1941. Little of the equipment intended for China's ground forces under this program reached its destination. The United States was more successful in furnishing China with assistance in the air, taking steps to put into effect a program for building a strong and well equipped Chinese Air Force in 1941. This new stage of American military aid to China started with aircraft.[66] From the Japanese point of view, U.S. government air support of China was one of the most important causes of friction between the United States and Japan in 1941. Many factors, certainly, contributed to the 1941 attack on Pearl Harbor. U.S. air support of China probably influenced the Japanese decision to initiate hostilities. The Japanese militarists could justify their attack on Pearl Harbor as a part of a war already underway.[67] Japan's attack on Pearl Harbor signaled not only full-scale American involvement in the war with Japan, but large-scale U.S. aid, especially air aid, to China.

Table 5.1
Value of U.S. Arms, Aircraft, and Petroleum Exports to China and Japan, 1937–1941 (in $ million)

	To China			To Japan		
Years	Firearms	Aircraft	Petroleum	Firearms	Aircraft	Petroleum
1937	6.6	4.0	10.5	0.4	2.5	64.0
1938	2.6	6.4	1.5	0.1	11.1	51.2
1939	0.7	1.1	2.7		3.3	45.3
1940	2.0	11.1	4.2		0.9	55.1
1941	5.0	9.5	5.1		0.04	28.2

Source: U.S. Commerce Department Bureau of the Census, *Foreign Commerce and Navigation of the United States.*

NOTES

1. Long-hsuen Hsu, *History of the Sino-Japanese War, 1937–1945*, p. 510.
2. *China after Five Years of War* (New York: China News Service, 1942), p. 68.
3. Jiro and Masatake, *Zero!* p. 21.
4. Ibid., pp. 18–24.
5. Allan Nevins, *The New Deal and World Affairs* (New York: United States Publishing Association, 1950), p. 137; John Morton Blum, *From the Morgenthan Diaries: Years of Crisis 1928–1938* (Boston: Houghton-Mifflin, 1959), p. 509; *FRUS Far East 1940,* 8: 685.
6. Chiang to Soong, July 6, 1940; Chiang to Soong, July 10, 1940, *Zhonghua minguo zhongyao shiliao congbian,* 3: 410.
7. Memorandum of conversation between Chiang and Johnson, October 18, 1940, ibid., 3: 100–2; The Ambassador in China (Johnson) to the secretary of state, October 20, 1940, Department of State, *FRUS, 1940*, 4: 672–4.
8. Department of State, *FRUS, 1940*, 4: 683.
9. The cabinet meeting was on November 8, 1940. Soong to Chiang, November 9, 1940, *Zhonghua minguo zhongyao shiliao congbian,* 3: 123.
10. Memorandum of Conversation, by the Secretary of State, November 28, 1940, Department of State, *FRUS, 1940*, 4: 698–700.
11. The Department of State to the Chinese Embassy, ibid., 4:705–8
12. Claire Lee Chennault, *Way of a Fighter*, pp. 90–95.
13. *FRUS 1940 Far East*, 4: 673–4; John Morton Blum, *From the Morgenthau Diaries: Years of Urgency 1939-1941* (Boston: Houghton-Mifflin, 1965), p. 365.
14. *FRUS 1940 Far East,* 5: 712; Blum, *From the Morgenthau Diaries: Years of Urgency 1939–1941,* p. 361.
15. Soong to Chiang, December 9, 1940, *Zhonghua minguo zhongyao shiliao chupian: Duiri kangzhan shiqi*, 3: 423. Chiang was "Most grateful for" Morgenthau's message and asked him to present "these views to Present Roosevelt so that a decision will be made in time for us to block the Japanese move." See The President of the Chinese Executive Yuan (Chiang) to the Secretary of the Treasury (Morgenthau), December 16, 1940, Department of State, *FRUS, 1940*, 4: 712.
16. Chiang to President Roosevelt, December 13, 1940, *Zhonghua minguo zhongyao shiliao chupian: Duiri kangzhan shiqi*, 4: 427–9.
17. Mao to Chiang, December 25, 1940, ibid., 3: 432.

18. Transcript of Treasury, Department Group Meeting, December 23, 1940, Morgenthau Papers, Franklin D. Roosevelt Library and Museum, Hyde Park, New York.

19. *Morgenthau Diary* (Washington: U.S. Government Printing Office, 1965), 1: 334–36.

20. Arthur Young, *China and the Helping Hand,1937–1945* (Cambridge, Mass.: Harvard University Press, 1963), p. 256.

21. *U.S. Congressional Record*, 77th Cong., 1st Sess., LXXXVII, Part I, pp. 44–7.

22. U.S., *Statutes at Large*, LX, Part 1, pp. 31–33.

23. Franklin D. Roosevelt to Henry L. Stimson, May 6, 1941, AG 400.3295, Sec 1, Records of the Office of United States Army, National Archives, Washington, D.C. Hereafter cited as RG 407, NA.

24. Transcript of Treasury Department Groups Meeting, May 12, 1941, Morgenthau Papers.

25. Letters initiated by Chief of Air Staff, August 19, 1941, China Defense Aid Program, 145.91-584, July– August, 1941, USAFA.

26. General George C. Marshall to General Moore, June 3, 1941, Adjutant General File, AG 400.3295, Sec. 1, RG 407, NA.

27. *FRUS, 1941, The Far East,* 5: 707–709.

28. Memorandum of a Convention, October 18, 1940, *Zhonghua minguo zhongyao shiliao chupian: Duiri kangzhan shiqi*, 3: 100–2.

29. The Secretary of the Navy (Knox) to the secretary of state, October 19, 1940, Department of State, *FRUS, 1940*, 4: 671.

30. Hamilton Memorandum, October 23, 1940, File Number FW 740.00 111A RECRUITING/183, RG 59, NA.

31. The Department of State to the Department of the Navy, October, 1940, Department of State, *FRUS, 1940*, 4: 677-8.

32. Hornbeck Memorandum, October 28, 1940, File Number FW 740.00 111A RECRUITING/183, RG 59, NA. Actually the War Department began helping China acquire the services of American citizens for the Chinese Air Force at that time. As a State Department officer said, "a number of months ago officers of the War Department had informally made available to the Chinese Embassy here a list of American qualified in aviation matters." See "The Chief of the Division of Far Eastern Affairs (Hamilton) to the Adviser on Political Relations (Hornbeck)," December 23, 1940, Department of State, *FRUS, 1940*, 4: 712–3.

33. The State Department previously approved those ten North American dive bombers for export to Thailand. After the pro-Japanese militarists took over the Thailand government, the United States decided to cancel the grant. On October 9, Hamilton informed the Thai minister in Washington that due to the needs of the American national defense program, the U.S. government was obliged to take steps toward requisitioning the airplanes. See Department of State, *FRUS, 1940*, 4: 176–8.

34. Soong to Chiang, November 5, 1940, *Zhonghua minguo zhongyao shiliao chupian: Duiri kangzhan shiqi*, 3: 145.

35. Memorandum of a Conversation between Chiang and McHugh, December 11, 1940, ibid., 3: 424–7.

36. The Department of State to the Chinese Embassy, Department of State, *FRUS, 1940*, 4: 705–8.

37. Hornbeck to FE, December 12, 1940 File Number 893.208/222, Department of State Records, RG 59, NA.

38. Hamilton to Hornbeck, December 28, 1940, File number 893.248/222, ibid.

39. W. D. Pawley, "A Brief History of the Flying Tigers," Pawley Papers; Martha Byrd, *Chennault: Giving Wings to the Tiger* (Tuscaloosa: University of Alabama Press, 1987), pp. 116–117; Daniel Ford, *Flying Tigers* (Washington, D.C.: Smithsonian

Institution Press, 1991), p. 54.

40. Chennault, *Way of a Fighter,* pp.101–104; Charles F. Romanus, *Stilwell's Mission to China* (Washington, D.C.: Office of the Chief of Military History, 1953), p. 18.

41. For a detail of McHugh's key role as a naval intelligence officer in Sino-U.S. relations between 1937 and 1943, see William Leary, "Portrait of an Intelligence Officer: James McHugh in China, 1937–1942," in *Naval History* (Wilmington, Del.: Scholarly Resources Inc., 1988), edited by William B. Gogar.

42. Memorandom of Major H. L. Boatner, September 2, 1941, Decimal File, 1941–42, Army Service Forces, International Division, Mission Branch 336.2, Chinese Air Force, Record of Headquarters Army Service Forces, Record Group 160, National Archives, Washington, D.C. Hereafter cited as RG 160, NA.

43. Robert A. Lovett to Lauchlin Currie, October 10, 1941, Army Air Force Central Files, 1939–1942, Chinese Training, 350.2, Records of the United States Army Air Force, National Archives, Washington, D.C. Hereafter cited as RG 18, NA.

44. Lovett to Lauchlin Currie, October 10, 1941, Army Air Force Central Files, 1939–1942, Chinese Training, 350.2, RG 18, NA; Lovett to Currie, October 17, 1941, Lovett to Captain Richard Aldworth, October 17, 1941, Decimal File, 1940–1946, Assistant Secretary of War for Air, Records of the Office of the Secretary of War, National Archives, Washington, D.C. Hereafter cited as RG 107, NA. See also Donald F. Drummond, *The Passing of American Neutrality, 1937–1941* (Ann Arbor, Mich.: University of Michigan Press, 1955), pp. 320–23.

45. Fong-cheng Tien, *A History of Sino-Soviet Relations* (Washington, D.C.: Public Affairs Press, 1951), pp. 215–219.

46. Brig. Gen. H. B. Clagett to Chief of Staff, June 12, 1941, "Report of Air Mission to China," Adjutant General's Files, 380.3, 4-4-41, RG 407, NA.

47. Maj. Gen. Shih-ming Chu to Chief, Foreign Liaison Section, War Department, July 22, 1941, China Defense Aid Program, 145.91-584, July–August, 1941, Air Force Historical Research Agency, Maxwell Air Force Base, Alabama. Hereafter cited as AFHRA.

48. Adjutant General to Maj. Gen. Arnold, October 15, 1941, Chinese Training, 350.2, RG 18, NA.

49. *The China White Paper* (Stanford, Calif.: Stanford University Press, 1967), pp. 28–9.

50. Brig. Gen. H. B. Clagett to the Chief of Staff, June 12, 1941, "Report of Air Mission to China," Files of Adjutant General, 380.3, RG 407, NA.

51. *Zhonghua minguo zhongyao shiliao chupian: Duiri kangzhan shiqi*, 1: 593. See also Owen Lattimore, *China Memoirs: Chiang Kai-shek and the War against Japan* (Tokyo: University of Tokyo Press, 1990), pp. 87–88; Chiang to Soong, February 27, 1941, Laughlin Currie Papers, Hoover Institution on War, Revolution, and Peace, Stanford University.

52. General H. H. Arnold to Stimson, May 10, 1941, Joint Board Paper #355 File, Serial 691, Records of Joint Army and Navy Boards and Committees, Record Group 225, National Archives, Washington, D.C. Hereafter cited as RG 225, NA.

53. H. H. Arnold to Lovett, June 11, 1941, ibid.

54. Lieutenant Commander E. O. McConnell to Lauchlin Currie, June 24, 1941, Adjutant General Files, 380.3, 4-4-41, RG 407, NA.

55. Joint Planning Committee to Joint Board, July 9, 1941; Notice of Joint Board approval, July 12, 1941, Patterson and Knox to Roosevelt, July 18, 1941; Roosevelt approval of July 23, 1941, Joint Board Paper #355 File, Serial 691, RG 225, NA. See also Report of Joint Planning Committee, July 14, 1941, China Defense Supplies, Inc., 336.9, RG 160, NA.

56. Magruder to Adjutant General, October 28, 1941; Magruder to Marshall, November 8, 1941, Magruder to Adjutant General, November 1, 1941, Adjutant General Files, AG 400.3295, RG 407, NA.

57. Currie to Chennault, July 22, 1941; Chennault to Chiang, August 8, 1941, Chennault Papers.

58. Chennault to Soong, September 4, 1941; Currie to Chennault, November 22, 1941; Chennault, "Notes on CAMCO," November 23, 1941, Chennault Papers.

59. For more details of American air assistance to China in 1941, see Michael Schaller, "American Air Strategy in China, 1939–1941: The Origins of Clandestine Air Warfare," *American Quarterly*, 28 (1976): 2–19.

60. *China Handbook, 1937–1943*, pp. 333–34.

61. See *Japan's Decision for War: Records of the 1941 Policy Conferences* (Stanford, Calif.: Stanford University Press, 1967), translated and edited by Nobutaka Ike, pp. 94–95.

62. *FRUS, Japan, 1931–1941*, p. 518.

63. *Japan's Decision for War: Records of the 1941 Policy Conferences*, pp. 153–4.

64. *FRUS, Japan, 1931–1941*, pp. 791–2.

65. Romanus, *Stilwell's Mission to China,* pp. 48–9.

66. For a scholarly study of the beginning of the military aid to China of the United States government, see Charles F. Romanus, *Stilwell's Mission to China*, pp. 1–10. See also Chiang to President Roosevelt, December 13, 1940, *Zhonghua minguo zhongyao shiliao chupian: Duiri kangzhan shiqi*, 4: 427–9.

67. As a document prepared for the Imperial Conference of Japan on September 6, 1941, stated, a "military alliance" actually existed between China and the United States because of "assistance to the Chinese air force." See *Japan's Decision for War: Records of the 1941 Policy Conferences*, pp. 153–4.

Chapter 6

U.S. Air Support for China and Sino-U.S. Disagreements, 1942–1945

The Pearl Harbor attack in December, changing the military situation in Asia, forced the United States and its allies to face the Axis powers along two fronts. The U.S. declaration of war against Japan and China's declaration of war on the Axis powers turned the war in Asia into part of a worldwide war against fascism. The Japanese advances throughout the Pacific encouraged Americans to establish a military alliance with the Chinese, who had been fighting the Japanese alone for more than four years. China's situation after Pearl Harbor was entirely different. China was now incorporated as part of a regional war zone, covering China, Burma, and India, and was expected to cooperate with its Allies in forwarding the war against Japan. The Chinese, in fact, privately greeted the news of the Japanese attack because Americans were finally involved in war with Japan. China's strategic position became more important, and Americans would support China.

When the war started in the Pacific, American strategic planning was clear: Europe first and Asia next. American policy makers regarded Germany as the most dangerous enemy, and the European theater enjoyed a top priority in American military aid and intervention. China should be kept in the war in order to hold hundreds of thousands of Japanese troops which might be dispatched to Southeast Asian countries or islands in the Pacific. China should be kept in the war because the United States needed to establish bases in China for large numbers of American troops for future military campaigns against Japan. Therefore, Washington assigned to China the lowest priority of any other theaters and provided limited amounts of American aid to Chongqing during the war between 1942 and 1945.

American military policy toward China consisted of two major objectives, the first of which was to supply the Chinese Nationalist Army against Japan. The principal means of doing this was to send to China Lend-Lease war materials, especially aircraft, which carried the highest priority in all Chinese requests. American military aid became an increasingly important factor in the war. The second major objective was to coordinate operations with the Chinese Army, especially with the Chinese Air Force. Therefore, the United States played a more important role in aviation matters in China during the wartime. The U.S. War Department established an American air force in China, provided assistance to the Chinese Air Force, and transported Lend-Lease materials by air from India to the Chinese government.

U.S. AIR SUPPORT FOR CHINA

Under the supervision of Claire Lee Chennault, the American Volunteer Group (AVG) was the first group of Americans involved in support of the Chinese in the war with Japan. Formed into three squadrons, they had experienced barely three months of training as fighting units before the Japanese air attack in December 1941. On December 20, when 10 Japanese planes took off from Hanoi and attempted to raid Kunming, 27 P-40 planes of the AVG went up and engaged the Japanese in intensive air combat. They destroyed four Japanese bombers over the city of Kunming. This was the first air battle fought by the AVG against the Japanese Air Force. Accompanied by pursuit planes, 54 Japanese bombers tried to bomb Mingaladon in Burma on December 23. Because there were no air raid warning facilities available, 14 P-40s of the 3rd Pursuit Squadron and 16 Brewsters of Britain's Royal Air Force (RAF) were ordered immediately to intercept the Japanese aircraft and shot down six of them. When a large formation of Japanese airplanes, 18 Zero pursuit planes and 60 bombers, crossed the Burma border for Rangoon on December 25, 12 P-40s of the 3rd Pursuit Squadron and several RAF Brewsters went up to intercept the Japanese over Thongwa. After chasing them for 15 miles over the Gulf of Martaban, the AVG shot down 13 fighters and four bombers. In three air battles, Chennault's fighters, with the assistance of the British airplanes, destroyed 21 Japanese aircraft by the end of 1942.[1] The AVG not only survived the Japanese assault but repulsed them with heavy losses, exciting the Allied side of the war, which had been repeatedly defeated by the Axis powers.

The success of the AVG aroused the attention of American military leaders who tried to absorb it into the U.S. Army Air Force (USAAF). At the beginning of 1942, General George Marshall, Army chief of staff, instructed General John A. Magruder to create a U.S. Army Air Force unit in China—a pursuit group—by incorporating the AVG pilots and technicians into it. But Chiang Kai-shek was not interested in this proposal and did not reach an agreement with the Americans during the negotiation of January 1942. Chennault also objected to this plan to induct the AVG into the American Army on the grounds that this would reduce the combat effectiveness of the AGV, and most AVG pilots also opposed the merger.[2]

The AVG, however, was facing some problems in the spring of 1942. With Chiang's approval, Chennault divided the AVG between Burma and China in an effort to resist the Japanese Air Force attacks in both areas. This division, without doubt, affected the effectiveness of the AVG. To make things worse, when Japan invaded Burma, the AVG had to evacuate a number of air bases in Burma, abandoning a lot of needed supplies. The poor living conditions and hot temperature of Burma also impacted the AVG's morale, which was sinking lower. The decline of combat effectiveness, supplies, and morale pressed both Chiang and Chennault to change their minds about transfering the AVG to the U.S. Army.[3]

Since the presence of AVG pilots in Chongqing was sufficient to heighten the morale of the Chinese and to force the Japanese Air Force to reduce its raids on Chongqing, the U.S. War Department decided to induct it into a U.S. air unit. The U.S. Army Air Force returned Chennault to active duty with the rank of colonel and immediately nominated him for promotion to brigadier general to command the unit that would succeed the AVG. After this agreement, General Marshall praised the AVG's combat record and stressed that President Roosevelt had agreed to increase supplies to the AVG.[4]

The AVG had become an object of admiration and acclaim during its seven months of fighting in Burma, China, Thailand, and French Indochina. Outnumbered as much as eight to one and fighting under primitive conditions with shortages of both food and supplies, the AVG held its airplanes together by the determination and resourcefulness of ground crews and halted the Japanese invasion of China. From December 1941 to July 1942, the AVG created a most impressive record in aviation history while firmly establishing Chennault's reputation as a brilliant tactician and leader. It provided an effective air defense for Southwest China and rendered invaluable assistance to the hard-pressed Chinese and other forces in Burma.[5] Shooting down 297 Japanese aircraft, this group of 87 pilots with a fleet of only 100 airplanes destroyed another known 240 planes. In seven months of combat, the group lost four pilots in combat, 11 more in strafing or bombing actions, 45 airplanes by combat attrition, and 45 more by accidents, bombing, or capture by Japanese ground forces.[6] Chiang praised the AVG: "The American Volunteer Group of the Chinese Air Force has acquired a world-wide reputation for greatest courage."[7]

In July 1942, the AVG was disbanded. The China Air Task Force (CATF), a segment of the U.S. Tenth Air Force in the China-Burma-India theater, took over its 54 P-40s, which were about worn out. Chennault was named head of the China Air Task Force and his personnel were incorporated into the CATF. Only four AVG pilots and 16 communication, armament, and airplane mechanics wanted to serve in the CATF. Most of the AVG refused to join the CATF because they claimed that they had been treated badly by the U.S. Army Air Force.[8] Chennault thereafter enlarged the CATF to 1,000 American officers and men, and brought its aircraft strength to 105 fighters, 12 medium bombers, four reconnaissance planes, and several transports. The Chinese Air Force furnished the CATF with about 1,000 semiskilled mechanics, engine repairmen, radio mechanics, and other less skilled airmen.[9] The primary mission of the CATF was to defend the eastern segment of the route between India and China, especially the bases, and to destroy the lucrative Japanese targets in South China.

After acquiring eight B-25 medium bombers, in addition to some P-40s, the CATF started destroying Japanese airfields and coastal and river shipping in East China. Using Guilin Air Base and other auxiliary fields which were about 500 miles east of Kunming, the bombers, supported by CATF fighters, did considerable damage in proportion to their number. Within only four months, they conducted 35 raids to keep the Japanese Air Force on the defensive and executed a number of air strikes which damaged Japanese military installations in Hankou, Canton, and Hong Kong. By the end of 1942, the CATF comprised four P-40 squadrons of the 23rd fighter Group plus the B-25s of the 11th Bomber Squadron.[10] By February 1943, this air force grew to 114 fighters and 12 medium bombers, together with 817 officers and enlisted men.[11]

In recognition of its increasingly important role, the U.S. Army Air Force formally activated the CATF as the Fourteenth Air Force in March 1943. On October 1, 1943, as one of the wings of the Fourteenth Air Force, a Chinese-American Composite Wing (CACW) of the Chinese Air Force was established. Ably commanded by Chennault, this Composite Wing, which was composed of Chinese and American airmen and ground units, was equipped with P-40 fighters, B-25 light bombers, and C-47 transports. Engaging in joint operations with other American units, the CACW supported Chinese ground forces and struck at Japan's shipping through advanced bases in East China while protecting the northern end of the Hump route.[12] It destroyed the Japanese aircraft during the air encounters and airfields in bombing operations. The six leading American fighter pilots of the CACW amassed 51 aircraft while the top six Chinese fighter pilots destroyed 24. The Chinese in eight fighter squadrons destroyed about 200 enemy aircraft from November 1943 to November 1944.[13]

In late 1944, the Chinese succeeded in destroying and damaging a considerable amount of Japanese military equipment. Periods of intense activity in offensive strikes characterized the Chinese operations. In 1945, when the Japanese were on the defensive and did not offer lucrative targets, the Chinese bombers started destroying Japanese shipping in the inland and coastal waters of China. The Chinese B-25 pilots sank over 6,000 tons and probably and damaged another 2,000 tons.[14]

The destruction of Japanese ground troops along with their supplies and equipment perhaps entailed the most important mission for the units of the CACW. During Japanese ground offensives in 1944 and 1945, the CACW units flew intensively against targets close to their air bases, destroying rail lines, hundreds of locomotives, and boxcars. Low-flying fighters and bombers attacked heavily the Japanese trucks and trains bringing supplies and reinforcements along the roads and railroads to the front.[15] As historian Joe G. Taylor wrote:

> A small air force of limited logistic and tactical means proved able to carry out an effective interdiction campaign in China. Once air superiority had been won, the Fourteenth Air Force was able to interdict effectively, if not completely, the rivers and railroads of the theater. There were errors of omission and commission, but by and large a remarkably small number of sorties inflicted a remarkably great amount of interdiction upon the enemy communications. This campaign not only reduced supplies to the battlefield to a trickle, but also reduced significantly the flow of strategic raw materials to Japanese factories. Mainly by virtue of its interdiction campaigns, the Fourteenth Air Force had made the most significant contribution to victory in China. With its anti-shipping and railroad interdiction operations, it had made a less significant but nonetheless very real contribution to the final defeat of Japan.[16]

Since the CACW mixed its air crew personnel at a ratio of approximately two Chinese for every American, the Chinese in the CACW not only conducted highly effective sorties against Japan, but also received ample opportunities to observe and to learn techniques the Americans used. The Chinese airmen acquired experience and skills necessary in aerial warfare. As the Chinese more competently conducted their operations, Chennault withdrew more and more American personnel. In May 1945, Stimson and Marshall were recommending that the CACW revert to the control of the CAF, and that the United States withdraw its remaining personnel.[17] Not until mid-October 1945 did the War Department order the deactivation of the CACW and the transfer of its equipment and Chinese airmen to the Chinese Air Force. The establishment of CACW had provided a means to train the Chinese in combat against the Japanese and trained the cadre of combat-tested, American-oriented Chinese aviators for the postwar era.

One month before V-J Day, the CACW's strength reached 107 P-40s and P-51s and 57 B-25s.[18] From 1943 to 1945, Chennault did not even receive the minimum of supplies for which he asked, but he succeeded in realizing the better part of his plan, with the exception of the bombing of Japan. This force, becoming the nucleus for a strong Chinese Air Force, kept control of the air over unoccupied China, engaged in expanding operations against the Japanese, and performed the vital mission of protecting the terminal bases of the air transport route into China.

The achievements of the Chinese-American Composite Wing were without precedent. After three years of fighting, it lost 500 aircraft but destroyed 2,600 Japanese aircraft and sank or damaged 2,500,000 tons of Japanese shipping, plus 44 naval vessels and 13,000 river boats under 100 tons.[19] In fact, by V-J Day, the CACW had played a more important role than the Chinese Air Force in fighting the Japanese. In 1944, the CACW made twice as many sorties against Japan as the Chinese Air Force made; in 1945, the CACW flew seven times more sorties than the Chinese Air Force flew.[20] U.S. air assistance made the United States–Chinese combined air operation on the Asian continent both possible and effective in the later stage of the Sino-Japanese war.

In addition to the establishment of an American air force in China, the Roosevelt administration also provided assistance to the Chinese Air Force. After 1941, the U.S. government increasingly furnished China with aircraft. Almost half of the total number of American aircraft delivered to China during the war reached the hands of the Chinese only after June of 1944; the strength of the Chinese Air Force was greatly increased in the last stage of the war. By the end of the war, this force had a total of about 500 aircraft, including 200 fighters, 60 medium bombers, 30 heavy bombers, 15 reconnaissance planes, and 120 transport planes.[21]

Although the U.S. Army Air Force did not provide China with a large number of aircraft under the Lend-Lease, it did sponsor training to a large number of Chinese pilots and aviation technicians. During the war, the United States took steps to train the Chinese pilots so as to build a strong and well-equipped Chinese Air Force. The U.S. Army started to train Chinese aviation personnel in India in 1941, and by the end of 1942, 79 pilots had completed their training satisfactorily and returned to China proudly.[22] The general pattern of training Chinese students, which had evolved in late 1942, continued until V-J Day. In mid-June 1943, the U.S. Army agreed to give pilot training to a second group of 500 Chinese.[23] By the beginning of 1945, approximately 1,300 students had completed flying training and 320 had attended technical training. The program reached its peak in mid-March 1945 when 2,120 Chinese were attending some phase of instruction in the United States. This number represented almost one half of the total number of foreign nationals training at that time. Fulfilling its commitments, the U.S. Army Air Force offered a variety of courses to train the Chinese Air Force from 1943 to V-J Day.[24] The program for training Chinese aviation personnel had an important impact on operations against Japan.

The U.S. Army Air Force not only helped to strengthen the Chinese Air Force, but also helped to transport war materials by air to the Chinese. During the war, one of the major problems both the Chongqing and Washington governments faced was how to transport war materials to China. When Japan penetrated into Burma in early 1942, China's military situation was becoming increasingly difficult because the Japanese attempted to close the Burma Road to truck traffic, upon which the Chinese primarily depended for receiving U.S. war materials. It was necessary for the U.S. government to open a new route to keep China in the war and to contact China's capital. The difficult mountainous terrain and the heavy jungle foliage made it very difficult to open a highway route immediately. Another route open to American supplies was to pass through Soviet territory, but the Russians refused to allow American goods bound for China to transit their territory because they did not want to provoke Japan. Both Chinese and American leaders began to worry about the complete isolation of China.

As a result, in early February 1942, General H. H. Arnold, chief of the U.S. Army Air Force, suggested to Roosevelt that the development of a new air transport route from India into China should be made possible and immediate in view of the imminent loss of the chief port of Rangoon in Burma. Arnold estimated that if American commercial airlines provided the airplanes, the service could be started in a few days because the China National Aviation Corporation (CNAC), a joint Chinese-American commercial airline, had taken exploration flights two weeks before Pearl Harbor was bombed. Set up by Pan American World Airways to fly commercially in China, CNAC had been in operation since 1929. When the Sino-Japanese war erupted, CNAC began flying supplies and personnel to areas that Japan isolated from land connections. Later, it flew into Burma and China from India, becoming the main source for supplies for the AVG between 1941 and 1942. Sponsored by both the Chinese government and Americans, CNAC was a small company. After receiving C-46 aircraft as Lend-Lease, CNAC soon began flights from Chongqing and Kunming to Calcutta via Lashio into northwest Burma, becoming China's major contractor with the outside world.

The president supported Arnold's suggestion on the grounds that without receiving American assistance China would falter in its resistance, which would lead to serious consequences for the Allied forces in Asia. He thereafter lost no time informing Chiang Kai-shek on February 9 that a new airline into China would be developed if the land route from India to China via Burma was cut off.[25]

With the fall of Rangoon in March and northern Burma in May, the Japanese closed the Burma Road to truck traffic, cutting off China's last link with the Allies. China-Burma-India (CBI) theater air units and the CNAC's freight service over the Hump of the Himalaya Mountains began in March 1942 and regular service started about June 1. From July 1 on, well-organized and efficient CNAC flew one flight every day, pouring an ever-increasing amount of important materials into starving China. For several months, those Hump pilots were flying day and night on the world's roughest and highest terrain, in all kinds of weather, providing almost the entire airlift for all American forces in China.

Because of CNAC's pioneering successful flights, Washington decided to build up an efficient airlift to provide war materials to China. Established in March, the American Ferry Command (AFC) began the mass airlift over the Hump and its haulage started to surpass that of CNAC. Problems erupted in every aspect of the operation and were compounded by a lack of a single controlling force to coordinate the effort. Washington soon realized the importance of keeping China in the war and ordered the ATC to take over. In December, the Air Transport Command (ATC) took over the AFC. The new command was headed by Major General Harold George with Brigadier C. R. Smith as deputy. The ATC was to grow to 85,000 officers and men with countless major air routes, branches, and feeder systems. Perhaps its finest hours were in guiding the aerial resupply of China over the Hump of the Himalayas.[26]

The new air route started from the ports in India. Native laborers transferred equipment to barges and river boats and onto a railroad for hauling to Chabua, the terminal of the Bengal-Assam Railroad. Airplanes transported these materials from Chabua to Kunming, flying about 520 miles. The geographic terrain and weather conditions between Chabua and Kunming were highly variable. Airplanes had to fly over towering mountains rising to 21,000 feet, where violent winds and icing conditions occurred at all levels. Cargo transport flying above the ranges was dangerous. Buffeted by the world's worst weather en route, the pilots often found destination bases socked in upon arrival with little in the way of ground aids to assist them to safe landings. In addition, the presence of Japanese fighters and the lack of navigation assistance contributed to extremely hazardous flying conditions. In the second half of 1943, 135 aircraft and 168 pilots were lost because of mechanical problems, terrible weather, and pilots errors. By mid-1944, Japan began to collapse in Burma, which opened lower and more direct routes to China and made possible the use of the C-54.

When Brigadier General Earl Hoag assumed command in October 1943, the airlift to China was ripe for a period of expansion. New Hump tonnage records were set almost daily and 12,590 tons were airlifted in December 1943. The monthly airlift rose to 20,000 tons in the following years. Until the summer of 1945, China remained primarily dependent on air transport for the sustaining of its military operations. In spite of high casualty rates, a total of 650,000 tons of gasoline, munitions, other commodities, and men traveled the air route over the Hump into besieged China. At the time the war ended, the ATC had 330 C-46s, 167 C-47s, and 132 C-54s flying the Hump. This airlift became the greatest effort of its kind in the war.[27]

These materials, without doubt, greatly strengthened the power of the Nationalist Army and Air Force fighting the Japanese. American air aid and airlift to China played a significant role helping the Chinese to keep more than a million Japanese occupied until the end of the war. However, disputes between the American and Chinese leaders on air aid and air supply rose during the war.

SINO-U.S. DISAGREEMENTS OVER AIR SUPPORT

The Sino-U.S. controversy on air support for China developed over two issues. The first issue was the nature and size of an American air force in China and the quantity and delivery schedule of aircraft to the Chinese Air Force. The second issue was the volume of American air supply to China. The U.S. War Department promised to increase air transport volume of war materials and equipment from India to China, but it never fully kept its promise. The Chongqing government was unhappy with the amount of U.S. air supply over the Hump to China during the war. The relationship between Stilwell and his subordinate Chennault who was supported by Chiang became embittered allegedly over Stilwell's opposition to strengthening the American and Chinese air forces. Both sides repeatedly brought complaints against one another to Roosevelt. In Washington, civilian and military leaders were divided into factions which disagreed over the allotment of the Lend-Lease supplies to the air forces or the Chinese Army. Those in favor of building up American and Chinese air power soon came to blame Stilwell and other American leaders for the neglect of the air forces.[28] The Sino-U.S. argument on air support for China continued until the end of 1944.

After China became one of the Lend-Lease Act recipients in May 1941, the Chongqing government urged the U.S. government to send 1,000 more bomber and pursuit planes to China in an effort to maintain a 500-plane air force for the Chinese Air Force. Washington promised to deliver more aircraft to China, but Chongqing was disappointed with the Lend-Lease allocations of American aircraft, far short of the number required to meet the needs of the operation of the Chinese Air Force.[29]

To solve the problem of allocation of the Lend-Lease materials to China and the problem of command of all U.S. forces in China, at the request of Chiang Kai-shek, the U.S. government decided to send to China a high-ranking U.S. officer of general rank, together with a dozen staff officers. On January 29, 1942, Secretary Henry Stimson informed the Chinese government that an American Army officer was to go to China.[30] To retain the services of Chennault, Madame Chiang Kai-shek encouraged the U.S. War Department to appoint Chennault as chief air adviser because he was familiar with Chinese military aviation and would contribute to the success of the Chinese Air Force.[31] The War Department therefore appointed Chennault as chief air adviser and selected Major General Joseph W. Stilwell as chief American military officer.

Having spent many years in China as an American officer and as an army military attaché between 1937 and 1938, Stilwell, the only high-ranking American officer who could speak the Chinese language fluently, was an indomitable fighter and a capable field operations commander.

Stilwell arrived in China in early March 1942. His mission, according to the instructions of the U.S. War Department, was to lead and to coordinate all phases of U.S. military activities in China, India, and Burma, to increase the effectiveness of the U.S. assistance to the Chinese government for the prosecution of the war, and to assist in improving the combat effectiveness of the Chinese Army. Stilwell was not the ideal man Chiang wanted; Chiang wanted Washington to send a U.S. air officer of high rank who would help to build up China's air force.[32] In addition, Stilwell's natural tendency to give primary attention to military, instead of political, considerations did not make him the ideal theater commander in the most difficult theater of war. As a result, the conflict between Stilwell and Chiang began with Stilwell's appointment, and contention between the two governments intensified in the following years.

In aviation matters, conflict over the Chennault plan between American and Chinese leaders persisted during the war. In July 1942, with the commencement of CATF combat operations, a campaign was initiated by those who wanted a large American air force in China. Chennault, the leading proponent of the plan to bomb Japan from bases in China, was anxious to bring up his air strength to a total of 105 fighters and 42 bombers. Chennault boasted that with this force he would defeat the Japanese Air Force, paralyze all Japanese rail and river communications in China, and bomb the Japanese islands.

Chennault admitted that Americans had gained air supremacy over the Japanese by mid-1942 but could not maintain this position without securing the most up-to-date fighters. Believing that in no other theater could a "comparatively insignificant number" of planes do so much damage and disrupt the enemy air force so much, he called Arnold to supplement his small force with 100 P-47 fighters and 30 B-25s. In return for this allocation, he grandly promised Arnold that he would then assume responsibility for destroying the morale of the Japanese Air Force and their aircraft production facilities. He convinced Stilwell that his air force could destroy much of the Japanese shipping, inspire the Chinese Army to fight, and neutralize the Japanese Air Force in Southeast Asia and China. He persuaded President Roosevelt that if given complete authority and a force of approximately 100 fighters and 40 bombers, he could "bring about the downfall of Japan," secure the lasting friendship of the Chinese, and create a "friendly trade market for generations."[33]

American ambassador to China Clarence Gauss supported Chennault. He informed the State Department in August that "an American air force has a distinct offensive mission in China" and "Chennault's operations during past months" had demonstrated this. He concluded,

> Thus the entry of the American Air Force into the China theater produces three results, all of which are important. First, Japan's earlier tactics of menacing China with an unchallenged small air force is no longer feasible. Second, the Chinese Army will be enabled to achieve better results in immobilizing an increasing number of Japanese

troops. . . . Third, the American Air Force will perform its own work of immobilizing constantly an important portion of the Japanese Air Force in China.[34]

Believing that an efficient American air force in China not only would deal a great blow to the Japanese but also would promote Chinese morale in fighting Japan, Chiang strongly supported Chennault and encouraged him to discuss his plan directly with Roosevelt. To Chiang, building Chennault's air force would probably prevent Stilwell from accumulating stockpiles for Chinese Army reform and the Burma campaign. Impressed by Chennault's plan, Roosevelt advised General Stilwell to give his support to the American air force in China, proposing to accelerate the delivery of transport aircraft for the India-China route in order to increase the flow of supplies to Chennault and to the Chinese Air Force. Roosevelt justified his support of Chennault on the basis that the Chinese had been complaining about the trifling number of aircraft allocated to China.[35]

Stilwell, Marshall, and Arnold opposed the Chennault plan. General Arnold stated, "There was no use sending planes to the Far East without their ground service crews."[36] American military leaders feared that American use of air bases in Southeast China would provoke retaliating Japanese offensives against these air bases, in spite of Chennault's claim that the Chinese Army aided by air support could protect such bases adequately. In addition, an increase of supplies to the Chennault air program would certainly reduce supplies to Stilwell's reform program to reorganize the Chinese Army. They argued that an efficient and strengthened Chinese Army was important to resist the Japanese attacks and to establish a logistical land route which would increase the flow of supplies to China through northern Burma. Their strong opposition to the Chennault plan mainly came from military and strategic considerations.

Therefore, the China theater did not gain enough aircraft from the United States in the second half of 1942. The AVG received only 20 newly improved versions of the P-40 to supplement its strength which consisted of 36 combat-ready aircraft. Other plans to strengthen the AVG with a substantial number of aircraft were never realized in 1942. The Chinese dissatisfaction with Stilwell was rising because he was responsible for the allocation of aircraft and war supplies to China.

The U.S. government not only gave fewer aircraft to China but diverted the aircraft from the China theater to other theaters. Roosevelt issued an order to transfer the U.S. Tenth Air Force from India to Egypt and to stop the shipment of a force of B-24 heavy bombers that was on its way to China. On June 23, the U.S. War Department informed Stilwell that General Lewis Brereton, commander of the 10th Air Force, was to transfer the 10th Air Force's heavy bombers and several transport airplanes of the Assam-Burma-China airline, together with pilots and technicians. At that time the Egyptian situation was critical as Italian and German military forces drove British forces from Cyrenaica into Egypt. The Suez Canal, the Middle East, and the lines of communication to China and India via the Cape of Good Hope were in danger of falling to the hands of the Axis forces. To relieve the pressure in the Middle East, the Anglo-American forces tried to launch an attack on North Africa. For this reason, the U.S. Army Air Force diverted these aircraft to Africa and the Middle East for the British use. When Stilwell informed Chiang Kai-shek on June 26 of this transfer and diversion, Chiang was angered because the removal of the U.S. 10th Air Force did not have his approval. At that time, Chiang was in charge of the U.S. 10th Air Force, which had been assigned by Stilwell for use in the China theater and Burma. In addition, the situation of Zhejiang and Jiangxi was not better than that of the Middle East when the Japanese force launched attacks on Central China while Chiang had selected significant strategic targets for those heavy bombers.[37] Madame Chiang complained, "Every time when the British suffered a defeat, they took away our war equipment or that which had been promised to us. Such being the case there is no need for China to continue the war."[38]

Air transport was another source of conflict between China and American military leaders. The allocation of Lend-Lease supply to China had been a problem since Roosevelt announced on May 6, 1941, that China would receive U.S. Lend-Lease aid. With regard to air aid, President Roosevelt sanctioned on July 23 equipping, manning, and maintaining of a 500-plane Chinese Air Force, but China had not received a total of 500 airplanes by the arrival of Stilwell. Since one of his duties was "to supervise and control U.S. defense air affairs for China," Stilwell became the sole person to control Lend-Lease supplies to China.[39] After the defeat of Burma, the Lend-Lease materials depended entirely on the air route. Transport aircraft thus played a significant role in delivering war supplies to the Chinese. Since the China theater was listed as the third and last priority by the Munitions Assignment Committee, China received only 80 tons of supplies in May, 106 in June, and 73 in July.[40]

Stilwell's reluctance to require Washington to send more transport planes to China also dissatisfied the Chinese. In a conversation with Stilwell, Mrs. Chiang asked him, "Another telegram from Dr. Soong states that replying to the War Department's inquiry whether you need more transport planes, you said that you do not need any more." Stilwell replied, "In this regard, I remembered having wired that for the present we can make use of 100 transport although as a matter of fact the fields here in China can accommodate only 50 planes."[41] Stilwell's reply suggested that he did not deny that he had sent no request to the War Department for more transport aircraft to be shipped to China. By June 1942, China had only six transport airplanes left.

In the meantime, the War Department refused in July to airlift supplies to China, further disappointing the Chinese. By June 1, more than 45,000 tons of Lend-Lease for the Chinese had piled up in India, but the air ferry line was scheduled to take only several hundred tons each month in the next few months, a tenth of what the Chinese urgently needed. For his part, Soong sent a message to Stilwell on June 17, pointing out, "I have put up to the president most effectively [*sic*] immediate aids to China are: 1. American air force unit operation in China. 2. Large scale air transport from India to China." On June 20, Soong sent a note to Roosevelt and Churchill during a meeting in Washington. Soong claimed that if Washington would not deliver 3,500 tons of supplies to China, the Allies' military campaigns in the China theater would cease. However, the Munitions Assignment Committee agreed to release only

400 tons for the Chinese in July while Henry Aurand suggested that no supplies be delivered in July. Angered by this decision, Soong informed Chiang that all Chinese missions in Washington would be stopped very soon.[42]

In this situation, tension increased between the Chinese and American leaders. Chiang pressed Stilwell to ask Roosevelt whether America and Britain still considered China one of the Allied theaters. Understanding that the situation was very serious, Roosevelt cabled Chiang on June 27: "The urgency of the situation demanded that any and all means immediately available be dispatched to pressure our line of communications to the China Theater." He promised that the U.S. government still regarded China as an important ally in the joint operation against Japan and would return to China those heavy bombers immediately after the conclusion of U.S. air aid to Egypt.[43]

Chiang did not allow the matter to drop there. Immediately after this incident, Chiang gave Stilwell a note that defined three requirements for maintenance of the China theater: (1) U.S. dispatch of three army divisions to India in August to help the Chinese restore the supply to China via Burma; (2) maintenance of a force of 500 combat planes in the China theater; and (3) delivery of 5,000 tons of supplies monthly via India. Stilwell, as U.S. military adviser, naturally was unwilling to support any ultimatums to his own government.[44] To seek a solution to the Three Demands crisis, Roosevelt sent his representative Lauchlin Currie to China in July 1942 and promised Chiang a combat strength of 260 planes for the Tenth Air Force, instead of the requested 500 combat planes.[45] To solve the crisis, Currie recommended during his visit to China that the 3,500-ton emergency air transport program be restored, and the War Department later agreed to Currie's proposal.[46] The crisis seemed to be over after Currie's visit to China, but the problem of aircraft to China still existed.

Stilwell still repeatedly denied the Chinese requests for more aircraft in the second half of 1942. In July 1942, the CAF commanders asked for 200 fighter planes to replace their accident and combat losses. These requests were rejected bluntly by Stilwell, who claimed, "The Chinese Air Force has been petted and favored in pay, uniform, prestige, quarters, etc., from the start. It was Madame's plaything."[47] Denouncing the CAF in his diary, Stilwell further suggested to the U.S. War Department that no aircraft should be delivered to the Chinese Air Force until they proved they could make effective use of what they had.[48]

When Chiang asked Stilwell to forward to Washington China's aircraft requests of 180 fighter aircraft and 60 B-25 bombers before the beginning of 1943, Stilwell suggested that the War Department provide aircraft to China "on paper," stopping actual delivery of all aircraft to the CAF. Stilwell justified his suggestion on the grounds that several recent aircraft accidents in China indicated the Chinese inability to employ modern pursuit aircraft in combat. Stilwell, as supervisor and controller of Lend-Lease supplies to China, had the support of American high-ranking officials such as Stimson and Marshall, who had given him authority to recommend diversions of Lend-Lease aircraft for China and had followed his suggestions in most cases.[49] When Chiang's request of air support from the United States was rejected by Stilwell, Chiang thus lost faith in Stilwell. The contradiction between Stillwell and Chiang was clear and reached its first peak in 1942.

To solve the problem of air aid to China, Roosevelt sent his air chief Arnold in early February 1943 to Chongqing to discuss air matters with the Chinese government. Chiang claimed that China would not successfully prosecute the war without these three essential concessions, for which he expected Arnold's firm commitment: (1) An independent air force controlled by Chennault; (2) 500 planes to be operated in China by Americans or Chinese no later than November 1943; and (3) 10,000 tons of cargo a month delivered over the Hump.[50]

In response to Chiang's demands, Roosevelt quickly made a three-point decision. First, he ordered the establishment of the U.S. Fourteenth Air Force with Chennault as commander, which was independent of Brigadier General Clayton Bissell, the commanding general of the Tenth Air Force in India. Second, Roosevelt promised to build up Chennault's air force as rapidly as possible to a strength of at least 500 planes. Roosevelt assured Chiang that the United States would equip China's air force with one fighter squadron and later a light bomber squadron. Third, the air transport volume from India to China would be increased to 10,000 tons per month.[51]

Roosevelt believed that if the Chinese had a powerful air force, they would bomb Japanese cities, which would have a tremendous morale effect on the Chinese people. The bombing would help Americans advance quickly instead of moving forward inch by inch, island by island, and would save the lives of hundreds of thousands of American soldiers and sailors in the final defeat of Japan. To keep his promise, Roosevelt applied more pressure on the War Department to give much more attention to the Chennault plan. In a letter to Marshall, Roosevelt questioned whether the War Department was giving enough weight to bombing Japan from China and hinted that Stilwell was acting unwisely in his dealing with Chiang Kai-shek.[52]

At this time, the Washington Conference was coming up and the Chinese situation was to be examined. Instructed by Chiang, Chennault accompanied Stilwell to Washington in April 1943 for further discussions of the plan. During the conference, Chennault presented General Marshall with a plan for stepping up the strategic bombing program, but Stilwell insisted that Chennault's plan was only a dream because the Chinese would never defend their air bases. Stilwell went on to say:

> My mission is to increase the combat effectiveness of the Chinese Army, and it can be done if we have a little time, if we have a minimum of supply, and if the Japanese are not stung into a major reaction. Morale will rise with the creation of this ground force in a far more lasting and effective way, than it would by an increased tempo in the air, unaccompanied by foundation work on the ground.[53]

In contrast, believing that Stilwell was indifferent to the possibilities of China-based air power, Chennault promised that his program would help to raise the morale of the Chinese attacking Japanese cities.[54] Stilwell's argument was not accepted by the Washington Conference. As he recalled, "So everything was thrown to the air offensive."[55] The conference decided to support an air war in China at the expense of a buildup of the combat effectiveness of the Chinese Army, which meant that Stilwell's program had to be carried on with an inadequate share of the Hump tonnage.

After the conference, Chiang continued to encourage the U.S. government to keep its promise, repeating the long-standing demand for a 500-plane air force and for 10,000 tons of cargo a month delivered to China over the Hump.[56] Roosevelt ordered a drastic change in the Hump priorities. Starting on July 1, 1943, the first 4,700 tons of supplies per month flown into China were to be allotted to Chennault's air force.

With regard to a strength level of 500 combat-ready planes, after considering that the chance for aid might be enhanced if the U.S. Air Force dealt with Chennault as an equal, Chiang obtained Roosevelt's approval to appoint Chennault as the chief of staff of the CAF in July. Chennault then submitted his plan in August for the modernization of the Chinese Air Force.[57] During the Cairo Conference of November 1943, Chennault as well as Chiang constructed a plan for modernizing the CAF, to include 12 squadrons of fighters, four squadrons of heavy and light bombardment planes, one reconnaissance squadron, and one transport squadron. Chennault sought to gain an American commitment to build an air force with 198 aircraft of all types.[58]

Chennault urgently and immediately needed bombers and fighters, but 250 fighters and 20 medium bombers in India were waiting for delivery. Chiang urged Roosevelt to deliver those airplanes to China as early as possible to enable Chennault to carry out the 1943 plan approved by the president. As a result, Roosevelt angrily informed his army chief of staff, "I am still pretty thoroughly disgusted with the India-China matters. Everything seems to go wrong. But the worst thing is that we are falling down on our promises every single time. We have not fulfilled one of them yet." Marshall explained that cargo delivery over the Hump had not reached sufficient proportions to provide more aircraft to Chennault. He "hoped that the Generalissimo [would] understand that China is better off in the long run."[59]

In the meantime, when Chennault launched air attacks in late 1943, the Japanese counterattacked and dealt great blows to the Chinese armies, and the Chinese could not defend Chennault's air bases in South and Central China. After overrunning the air bases, the Japanese quickly attacked previously unoccupied West China and posed a threat to Chongqing in the spring of 1944. This military disaster proved Stilwell's predictions correct. But Chennault and Chiang attributed China's defeat to a lack of weapons, arms, and aircraft. As a result, Chiang asked Roosevelt for more economic and military aid at the beginning of 1944.

In the aviation matter, by the end of 1943, the U.S. War Department fell far short of the promised aircraft and 10,000 tons of supplies a month over the Hump. The Chinese leaders complained that the War Department was not delivering promised equipment. In a letter to Roosevelt on December 23, 1943, Chiang wrote,

> While sharing your view that very long-range air operations against Japan proper from China should do much to heighten the morale of both our nations, I am strongly of the opinion that in order to assure the success of these operations, the present American air force stationed in China should be immediately increased to such an extent as to be able to protect our air bases from attack or destruction by the enemy.[60]

The Chinese accused Stilwell of intending to block Chennault's plan to strengthen the CAF in favor of a concerted offensive in Burma, failing to support the president's decision, and seriously damaging Sino-American relations with his thinly disguised contempt for Chiang. In this way, the conflict between Stilwell and Chiang became very serious. As a result, Chiang requested the recall of Stilwell. Stilwell, however, expressed his willingness to cooperate with Chiang; the second Stilwell-Chiang crisis was over.

At the beginning of 1944, Chennault continued to complain about the lack of aircraft and supplies. In a letter of April 19 to Roosevelt, he wrote, "little had been done to strengthen the Chinese Armies in the interior, and for the same reason the 14th and Chinese Air Force are still operating on a shoe string."[61] Chennault maintained that Washington had agreed at the Cairo Conference to modernize the CAF and that the Chinese had expected the United States to provide 579 fighters and 134 medium bombers in 1944. The U.S. Army Air Force staff, however, denied Chennault's allegation that the United States had failed to carry out its commitments, arguing that the Chinese had been told at Cairo not to expect deliveries earlier than July 1944, and Chennault's correspondence in March was premature.[62]

The fact was that the U.S. Army Air Force planned to reduce allotments to the CAF accordingly. Stratemeyer and other air officers in India sought to prevent Chongqing from receiving the fighters and heavy bombers, claiming that the CAF lacked sufficient gasoline, ammunition, and bombs to use these planes in combat, although the Chinese had hoarded a sufficient quantity of such supplies at most stations. General Arnold, accepting the view of Stilwell and Stratemeyer, decided to seek a drastic revision of the allocations and a delay of the CAF modernization program. In explaining the Air Force's position, Arnold repeated Stratemeyer's statement that the Chinese could not handle American aircraft efficiently.[63] Arnold defended his decision on the basis that the CACW engaging in combat lost many aircraft and thus needed replacement of aircraft while the CAF was not losing as many airplanes. General Marshall backed this decision because support of the Chinese Air Force would use a lot of American resources. Convinced by Arnold's judgment, Roosevelt decided to suspend the reinforcement and modernization of the CAF by limiting the number of aircraft delivered to China. In spite of China's protest, American air staff prepared a reduced schedule of aircraft deliveries for the Munitions Allocations Board's approval. When 102 fighters from the CAF's previous allocation arrived in India and were turned over to the CACW, the Chinese strongly objected.[64]

This revised policy for the delivery of aircraft had worked to the disadvantage of the Chinese, seriously obstructing China's plan to modernize the CAF. The Chinese regarded this Air Force decision as a deliberate effort by the War Department to scale down its commitment to the CAF. Chinese frustration was rising in September when the U.S. Army Air Force rejected a request for 40 aircraft for which the CAF had tried very hard to obtain General Stilwell's approval.[65]

By that time, the issue of air support for China as well as other issues, such as power of command, reorganization of the Nationalist Army, dispatch of the Chinese troops to Burma, and assistance to the Chinese Communists, had affected seriously the relationship between the U.S. and Chongqing governments. Unhappy with Chiang's delayed action in the military campaign, Roosevelt requested Chiang in a letter of September 18 to reinforce his troops in North Burma immediately. Otherwise, Washington would completely stop all aid to China. Angered by this letter, Chiang decided to request Stilwell's recall as a way to resist the U.S. policy, and cabled H. H. Kung, his brother-in-law, with instructions not to seek additional American aid and to return to China as early as possible. The collision between Washington and Chongqing reached a new high. Under this circumstance, Roosevelt had only two choices: to sustain Stilwell and break with Chiang, or to fire Stilwell and support Chiang. If he broke with Chiang, he would be blamed for China's collapse. Roosevelt quickly backed down. He recalled Stilwell and divided the China-Burma-India theater into two parts. Major General Albert Wedemeyer would be responsible for China as chief of staff and General Daniel Sultan would be responsible for India and Burma.[66] The recall ended the period of overt friction between Nationalist China and the United States. The appointment of Wedemeyer, who advocated an increase in air aid and air supply to China in the last stage of the war, ended the Sino-U.S. dispute on air support for China.

THE CAUSES OF SINO-U.S. DISAGREEMENTS

At the end of 1944, Sino-U.S. relations were in crisis when Chiang recalled Kung from the United States and stopped the requests for U.S. aid to China. Sino-U.S. military cooperation began to collapse when Chiang refused to send additional Chinese troops to Burma while the Japanese Army was stepping up its campaigns against the Allies. Arguments between the Chinese and American leaders on air support for China were one of the major factors affecting Sino-U.S. relations and military cooperation during the war. Why were the Chinese and American governments bogged down in these arguments? What were the major differences on the issue of air support for China? Was China's demand for more air aid and air supply reasonable? The following paragraphs will answer these questions.

The Issue of the China Theater

After the outbreak of the Pacific War, the U.S. basic war strategy which emphasized "Europe first and Asia next" was to protect defenses in the Pacific zone. U.S. military goals in China were to encourage the Chinese troops to launch offensives and to cause enough diversion to deter the Japanese from other adventures. In accordance with these objectives, U.S. support for China was mainly to offer China weapons and to help reorganize the Chinese troops to obtain effective united operations. Thus, the U.S. government placed aid to China after aid to England and the Soviet Union. Since the Allies concentrated their efforts in defeating Germany first, the European theater received first priority in the supply of strategic materials. Little was left to supply the China theater and few airplanes were available for the Chinese.

Americans officially and unofficially promised much to the Chinese government in 1942 but had delivered very little to increase the size of American air units in China or to strengthen the Chinese Air Force. The dollar value of Lend-Lease aircraft, parts, and accessories delivered to the British, Russian, and Chinese governments in 1941 and 1942 reflected the differences. Britain received $285.5 million in comparison to $2.7 million for China, a mere trickle of aircraft delivered to the Chinese.[67] By the end of 1942, Washington had exported more than 2,500 planes to Russia. Of these, about 1,550 were in Russia by the end of the year, far more than the Russian aircraft deployed at Stalingrad for their counteroffensive.[68] In the meantime, China had received only 19 A-29 bombers and 27 P-40, 41 P-43, and 82 P-66 pursuit planes.[69]

The U.S. government not only assigned low priority to China in the allocation of Lend-Lease materials, but also gave low priority to the delivery of already allocated materials to China itself. After the closing of the Burma Road in the spring of 1942, the ATC operated only 25 planes from India, far below the 100 that Roosevelt had promised, and the monthly delivery to China fell below 100 tons for some months in 1943.[70]

The British government also diverted aircraft allotted to the Chinese to other theaters. As combined chiefs of staff in Washington, the British proceeded to cancel factory production of bombers assigned to the Chinese for the balance of 1942 and diverted them to other fields— Britain, Russia, and Australia.[71] The British justified the diversion of those aircraft to Britain, claiming that the Lockheed Corporation had built the planes originally for the Royal Air Force rather than for the Chinese Air Force. The British ambassador Sir Jamel Seymour Horace claimed that those aircraft were "equipped with British type guns."[72]

By June 1944, Great Britain had obtained approximately $3 billion in aviation aid, including over 18,500 aircraft. The Soviet government had received over $1 billion including 7,600 airplanes. China ranked third among Lend-Lease recipients in money and number of aircraft. It had received only $100 million and 1,012 airplanes.[73] From June 22, 1941, to September 20, 1945, a total of 14,018 American aircraft were delivered to the Soviet Union.[74] To the Chinese, Washington was blocking China's legitimate goal of building its air power in favor Britain and Russia, although Chongqing repeatedly raised its concern about air aid to China.

The Chinese were not happy with the U.S. allocation policy because the Chinese developed their own military strategies after the eruption of the Pacific War. Believing that war in China should be carried on largely from the air, Chiang formulated his military strategy which was totally opposite to the American one. After China formally declared war against Japan on December 9, 1941, the Chinese government drafted a plan on December 20 for joint operations which were to be carried out by the U.S. Navy and Air Force, the Chinese Army, the British Navy and Air Force, and the Russian Army and Air Force. Allied forces were to destroy the Japanese Air Force first to gain air superiority in July 1942. Then, these air forces were to initiate air attacks on the Japanese Army and use aerial bombardment to cut the Japanese Army's supplies. Finally, the Chinese Army would annihilate the Japanese Army, which would end the Pacific War in December 1942.[75]

In this regard, Chiang favored the Chinese Air Force over the Chinese Army and his greatest concerns were the air force and air transport. The Chinese defeat in Burma in early 1942 further convinced him that air power was crucial in battles. For this reason, he persistently requested air support from Americans, clamoring for American air assistance. On the other hand, Chiang strongly urged Washington to send more transport planes and pressed Stilwell to draft a complete air transport plan. When few aircraft were made available to China, the Chinese believed that the United States and Great Britain were acting together to prevent the Chinese from obtaining aircraft. Therefore, dispute on air aid to China was intensified between the Chinese and U.S. governments. The different military strategies and goals between the United States and China were the principal cause of arguments over air aid and air supply to China.

It is understandable that the United States was not able to support offensives in Europe and the Pacific simultaneously, giving first priority to Europe in allotting Lend-Lease materials. But the question is: How much war material should the China theater receive rationally? When American warplanes originally delivered to China were reallocated by the War Department, Stilwell complained: "What can I say to the G-mo [Chiang Kai-shek]? We fail in all our commitments and blithely tell him to just carry on, old top."[76] Even Clarence Gauss, American ambassador to China at that time, criticized American allocation policy toward China and defended China's irritation. He stated:

> Planes or other military materials or equipment intended for China, have had to be diverted to other fields—to Britain, to Russia. . . . [I]t is probably difficult for Generalissimo to bring himself to the point of being persuaded that the need there in other areas is always so acute that aid must be diverted there and none come to the Chinese. His irritation is understandable.[77]

As director of the Office of Far Eastern Affairs in the State Department, Stanley Hornbeck, one of the most prominent, scholarly, and influential China experts within the Roosevelt administration, was also questioning American Lend-Lease policy toward China. After examining Lend-Lease materials given to Great Britain, the Soviet Union, and China, Hornbeck held that the Chinese justifiably thought that Americans discriminated against them. Defending the Chinese critics of the U.S. aid policy toward China, Hornbeck wrote:

> It is believed that we should make every effort to work out with them [the Chinese], at this point, something which constitutes new and clear evidence of our confidence in them and of our willingness and intention to confer, to collaborate and to cooperate with them in the same spirit and in the same manner and liberality in which we do with the British and the Russians.[78]

These statements suggest that the Chinese complaints of the U.S. allocation policy were rational and cogent, and their demand for more U.S. air aid and air supply was reasonable.

The Issue of Chinese Army Reform

Stilwell, an orthodox soldier and admirer of infantry, dismissed guerrilla warfare and sabotage as illegal action. Frustrated with the Chinese Army in the first Burma campaign in the spring of 1942, Stilwell was determined to increase its strength. On May 26, 1942, Stilwell presented Chiang Kai-shek with a plan to reorganize the Chinese Army by a sharp reduction in the number of divisions and a sweeping purge of inefficient high commanders. His major purpose was to retrain and re-equip 30 picked Chinese divisions as the nucleus of a new Chinese Army. Stilwell hoped the Chinese Army reform program would contribute to the establishment of an efficient army and to the political reform in China.[79] In Washington, General Marshall and Secretary Stimson supported Stilwell's program because they were concerned mainly with military campaigns against Japan.

To carry out his army training program, Stilwell tried to stockpile military stores while Chennault began an ambitious program of airfield building. When Chennault tried to take the Hump supplies to strengthen his air force, Stilwell had difficulty in building the Chinese Army for the Burma counterattack because he had to allocate so much war material to Chennault. Stilwell's and Chennault's programs were not compatible; each wanted five-eighths of the available Hump aircraft supplies.

When Stilwell tried to allot the lion's share of the Lend-Lease supplies to support the Chinese Army, Chennault and Chiang, on the other hand, sought to improve the logistical position of American air units in China and to obtain a small proportion for the CAF. Eventually, disputes arose over the allocations of Hump tonnage and anti-Stilwell sentiments began to develop within the Chongqing government. Stilwell's reform program became one of the sources of arguments over air supply to China.

From the military point of view, American generals and leaders were right in identifying such problems as ineffectiveness, increasing corruption and demoralization, and incompetence in the Chinese Army, but they did not appreciate that at that time many Chinese military forces were still provincial in nature. The Chinese Army under Chiang's control was still a loose coalition of semi-independent military forces bound together by personal loyalties and material rewards. By 1942, the Chinese Army had about 4 million men in 380 divisions, but Chiang actually controlled only about 30 divisions. The rest were divided among 12 zonal commanders, who were not primarily and fully loyal to Chiang. Stilwell's reform requiring Chiang to surrender his full powers of command over the Chinese Army would threaten the delicate political relationship and balance of power that ensured Chiang's paramount position in the Chinese government. Reduction in the number of troops in time of peace had caused disturbances; these radical measures were virtually impossible politically during wartime.

After establishing the National government in Nanjing, Chiang attempted to carry out a military reform at the beginning of 1929 to reduce the power enjoyed by the local warlords. On January 1, Chiang invited regional military leaders to attend a National Army Disbandment Conference in Nanjing. The aim of this meeting was to reduce the size of China's military forces of about 2.3 million soldiers, and Chiang attempted to use this opportunity to weaken his rivals' power. The conference called for the reorganization of all troops into one national army. The warlords demanded that the National government take the first step in military cuts while Chiang requested that any cuts start with the provincial forces. After one month of discussion, no agreement was reached between Chiang and other militarists who had perceived Chiang's purpose. Immediately after returning to their provinces, these warlords prepared for a military confrontation with Chiang. In March Li Zongren of Guangxi province started a revolt against the Nanjing government, claiming that Chiang's army reduction plan favored his own First Army Group at the expense of others. In May Feng Yuxiang rebelled against Chiang. These rebellions initiated a period of continuous civil war in China between 1929 and 1930. Chiang's first military reform failed, and therefore he never tried again and was unable to solve the problem of regional militarism in the following years.[80]

Stilwell seemed to have overestimated the potentiality of the army reform program and underestimated its difficulties. As Arthur Young, financial adviser to China during the war, commented, "Stilwell went so far as to say that, with army reform, China will be able to do her part and refute her critics, and will emerge at the end of the war with the means of assuring her stability."[81]

Stilwell spoke the Chinese language fluently, but it was not important, because the English language was the official language between the Chinese and Americans. In spite of his presumed familiarity with Chinese politics and military, he knew very few Chinese leaders personally. He seemed not to understand the immense complexity of modern Chinese military politics when he attempted to use the Chinese Communist forces in the prosecution of the war.

During the war, the U.S. government recognized China as a "Great Power" and gave aircraft, loans, arms, ammunition, and military equipment to the Chongqing government, China's legally recognized government. The Chinese Communists manufactured their crude weapons, developed a number of bases behind Japanese lines, and initiated guerrilla warfare against the Japanese Army. Since Japan controlled only major cities and railways in East and North China, the vast rural areas became a political vacuum after 1939, which created a golden opportunity for the CCP to extend its influence there. With the effective use of guerrilla warfare, the implementation of economic and political reforms, and an appeal to anti-Japanese nationalism, the Communists built a number of anti-Japan base areas, known as the "border-region" governments, supported by the peasants. The expansion of the Communist forces in North and East China frightened the Nationalists. As the Japanese advances slowed down, the old animosities between the CCP and KMT resurfaced. Beginning in 1941, Chiang imposed a blockade on the CCP central base at Yanan. The Communist troops and Nationalist Army fought separately against the Japanese, and occasionally against each other.

It is true that Chiang was reluctant to engage the Japanese enemy after 1939 and used some 200,000 of his best troops to blockade the Communists in North China. But Stilwell did not understand why Chiang did not want to send his troops to fight the Japanese Army because he had another enemy—the Communists— and he aimed at maintaining Kuomintang domination of China, particularly in the postwar period. In Chiang's mind, the Japanese were a disease of the skin, but the Communists were the disease of the heart, although the Nationalist government renamed the Communist Army the Eighth Route Army in 1937. After the war started, Mao Zedong's policy was 70 percent expansion, 20 percent dealing with the Guomindang, and 10 percent resisting Japan. Chiang realized how dangerous the Communist threat was during the war and believed that the Communists were a cancer that had to be removed before China was able to survive to fight the Japanese. The Communist influence in rural China posed a greater threat to his government and the country than did the Japanese. The chief American policy makers seemed not to understand the nature and tactics of the Communists.

In addition, Stilwell's methods and approaches to cooperation with Chiang were questionable. As President Roosevelt commented:

> Stilwell has exactly the wrong approach in dealing with Generalissimo Chiang who, after all, cannot be expected, as a Chinese, to use the same methods that we do.
>
> All of us must remember that the Generalissimo came up the hard way to become the undisputed leader of four hundred million people—an enormously difficult job to attain any kind of unity from diverse groups of all kinds of leaders. . . .
>
> Besides that the Generalissimo finds it necessary to maintain his position of supremacy. You and I would do the same thing under the circumstances.[82]

Roosevelt seemed to grasp the Chinese politics and military better than Stilwell.

Like Roosevelt, Wedemeyer seemed to understand China's politics better. Instead of Stilwell's rougher ways, Wedemeyer's approach to Chiang was sympathetic and friendly. He relied on advice and persuasion and established good relations with the Chinese, which enabled him to carry out the army reform program that Stilwell had attempted to implement. After replacing Stilwell, Wedemeyer devised a plan for training and equipping 39 divisions as a modern strike force and created a joint Chinese-American staff with frequent meetings in 1945. His reform was successful. As historians Romanus and Sunderland commented, "Many years have passed before rebuilding the Chinese forces within China received first priority from the Chinese and American Governments; the lost years were now exacting their price in the unreadiness of the Central Government to re-establish its authority in its own country."[83]

Yet Stilwell and other American military and civilian leaders cannot be criticized for the implementation of their unrealistic Chinese Army reform program because their focus was on military considerations. Arguments over the Chinese Army reform resulted from some American leaders' lack of understanding of and consideration for the complicated Chinese society in general and China's complex military politics in particular.

The Issue of Air Power

Stilwell held that the way to defeat Japan was not through air war but through ground assault because the Chinese Air Force was totally incapable of effectively opposing Japan. China was the logical base of such a counteroffensive; the most essential form of combat in China was on the ground, not in the air. Stilwell advocated concentration on the land war while Chennault favored an intensified air war against Japan. As Chennault commented:

> Stilwell was an even more violent partisan of the infantry than I was of air power. His acceptance of the airplane was limited to using it as a method of personal transport and later as a means of emergency supply to rescue his force from the results of too casual logistical planning.[84]

Stilwell's romantic view of the foot soldier probably contributed to the Chinese defeat in Burma in early 1942. His neglect of air power and refusal to strengthen the air forces in the China theater heightened the growing tension between him and Chiang because Chiang agreed with Chennault, preferring to keep his soldiers in reserve for what he believed was the impending civil war against the Communists. It is obvious that different perspectives on the role of air power in the China theater were one of the sources of arguments about air aid to China. The different beliefs on how best to fight the war often embroiled Stilwell and Chennault in the internal affairs of China.

Some historians seek to justify Stilwell's overestimation of the role of ground force and underestimation of air power in the war. Stilwell cannot be criticized casually for his low estimate of air power. He was accurate that, without the protection of ground force, air attacks on Japan would provoke retaliating Japanese offensives, and air assaults would not determine the outcome of war. The Japanese proved Stilwell correct in their 1944 Operation Ichi-Go, aimed at the capture of strategic points along the Hunan-Guangxi, Canton-Hankou, and southern Beiping-Hankou railways, with the initial objective being the destruction of Guilin and Liuzhou airfields in order to alleviate the threat to Japan. This operation overran the East China airfields. But the questions are: Could the Chinese Army have played a much more important role in fighting the Japanese if it had more air support? Could the Chinese Army have cooperated well with the air force during military campaigns?

The Chinese Army might have turned back Japanese advance if it had been given more support. During the campaign, the Chinese Air Force and the American Air Force in China dealt a blow to the Japanese. Chinese air attacks against Japanese vessels, vehicles, and railways were "so severe that daytime transportation of supplies to the front became increasingly difficult." The number of the Japanese pilots and planes "gradually decreased and, in spite of every effort, it was found impossible to replace them."[85] As a result, the fighting capacity of the Japanese Air Force suffered. As a Japanese historian admitted:

> During the Ichi-Go Operation, the main mission of the Fifth Air Army was cooperation with the ground forces. At times, however, it was compelled to engage the enemy air force in the air and to raid enemy air bases at night. The Fifth Air Army sustained heavy losses during this operation and received little replenishment. It was compelled, therefore, to avoid air combat with the superior enemy air force whenever possible.[86]

Dr. Joe G. Taylor argues that if the Chinese Air Force and the Fourteenth Air Force had provided more air support to the Chinese Army during the campaign, the Japanese would not have been so successful:

> During the Ichi-Go campaign, shortage of fuel was the chief factor limiting Fourteenth Air Force operation. Throughout its combat life, this organization had more aircraft and crews than it could use to fullest effect. Fuel conservation was partly responsible for the substitution of night for daylight sea sweeps, and it was almost wholly responsible for the use of fighter-bombers rather than the more accurate B-25's on bridge targets. Fuel shortage made it necessary to ease the pressure on Japanese-operated railroads after March 1945. Probably the Japanese offensive of 1944 could not have been halted by air action alone under any circumstances, but the advance could have been slowed and losses inflicted to such a degree as to make the Japanese victory Pyrrhic if only more fuel had been available.[87]

Even the Chinese Communist historians make valuable comments on the achievements of the Chinese Air Force and the American air force in China between 1943 and 1944:

> In the second half of 1943, the Chinese Air Force was strengthened. The Chinese Air Force and the American air force in the China Theater aggressively attacked the Japanese Air Force, bombed and strafed extensively Japanese airfields and military centers, and cut off the Japanese Air Force's transportation line. . . . In the fall of 1943 the Fourteenth Air Force began to use the latest model of P-40 fighters which made the American air force in China more powerful. In the latter of 1943, the Chinese Air Force and the American air force in China no longer only defended themselves, and their airplanes in large groups began to strafe and bomb intensively and frequently the enemy's important air bases, harbors, railroad stations, and army centers. Their reconnaissance planes flew over Japan's occupied areas in China, including Manchuria; their bombers interrupted often the enemy's land and water transportation lines, their B-25 bombers from Yunnan dealt a great blow to Japan's harbors in North Vietnam and Hainan island of China. Therefore, the Chinese Air Force and the American air force in China posed a threat not only to the Japanese Army in the China theater but to Japan's homeland itself which was strafed by American bombers stationed in China. . . . At the beginning of 1944 the Chinese and Americans became more aggressive in air raids on Japan's airports, harbors, and transportation ships, inflicting a great blow upon the enemy.
>
> In the spring of 1944, Japan started a general offensive—Operation Ichigo, which aimed at clearing the "continental corridor" from North China to Indochina and destroying the Chinese and American airbases. During the campaign, the Chinese Air Force and the Fourteenth Air Force enjoying air advantages dealt the enemy hard blows in support of ground forces, but China's military mistakes and errors, low morale of the weak Chinese Army, poor leadership of the Chinese military leaders who adopted the policy of passive resistance against Japan, and active anti-Communism contributed to Japanese occupation of several important cities in Central and South China. Since seven airbases and 36 airports fell to the hands of the Japanese, the Chinese and Americans had to move their airbases to Southwest China.
>
> Although Japan occupied the airbases and airports in Central and South China and the Chinese Air Force and the Fourteenth Air Force were forced to relocate in Southwest China, the Chinese and Americans continued the air attack over long distances on the Japanese and dominated the air in China. By the end of 1944 the Chinese Air Force and

the American air force had gained again air superiority in China while the Japanese Air Force had declined.[88]

Wedemeyer's strategies and use of air power in China also proved the importance of air support for ground forces. In mid-October 1944, when Major General Wedemeyer assumed responsibility for aid to China, U.S. Air Force officers in both China and India were extremely pleased. American air commanders were "delighted" with the choice of Wedemeyer because he was a "thoroughly air-minded" officer who would "support Chennault to the limit."[89]

Unlike Stilwell and other American leaders, Wedemeyer believed that China had tried its best to fight Japan. He wrote: "I began to understand that the Nationalist government of China, far from being reluctant to fight as pictured by Stilwell and some of his friends among the American correspondents, had shown amazing tenacity and endurance in resisting Japan."[90]

Wedemeyer stressed the importance of air power in the China theater. In a letter to Marshall on November 10, 1944, he claimed that air strikes "could practically destroy and certainly seriously disrupt the focal point of Japanese communications in Hankou. A similar strike might very profitably be directed against Shanghai and yet another feasible objective would be Canton."[91] To employ air power against Japan efficiently, Wedemeyer sought the services of competent air officers. Since there was no air specialist on Stilwell's staff, after arriving in Chongqing, Wedemeyer urgently requested Washington to send a group of special air advisers to China.[92]

After arriving in China, Wedemeyer started investigating the status of the CAF and found that it had not received any of the aircraft which the Munitions Allocations Board had authorized.[93] Wedemeyer promised the Chinese that Washington would support the building of an effective postwar Chinese Air Force in 1945. Accordingly, he proposed to the War Department to augment the strength of the CAF by allocating 415 aircraft for an air service of 17 squadrons.[94] He was successful in securing the approval of Washington for more planes in the China theater. He also increased the allocation of tonnage to the Fourteenth Air Force which received 14,170 tons of materials in November, in comparison with 10,456 tons in August, 8,426 tons in September, and 11,876 tons in October.[95]

Wedemeyer was an infantry officer but made great use of the air force. When he launched a military campaign to repel the Japanese offensive against Guiyang, he quickly ordered the transport of Chinese troops by air from the northwest to the critical spot. He directed the U.S. Fourteenth Air Force under Chennault's command against the Japanese forces; powerful American air assaults succeeded in slowing Japan's advance. Fresh reinforcements and air support enabled the Chinese to force the Japanese to retreat from Guizhou, and the military situation became favorable to the Chinese. Wedemeyer succeeded in realizing his initial goal—to hold American supply bases in China.[96] His boldly effective strategies and efficient use of air power earned the admiration of the Chinese as well as Generals Stratemeyer and Marshall, who regarded him as a "strategist of the whole war."[97]

A Japanese military officer later admitted the successful achievements of the Chinese Air Force and the American air force in China after Wedemeyer took command in China:

> At the beginning of November [1944], as the Eleventh and Twenty-third Armies closed in on Kweilin and Liuchowhsien, the enemy air force from Chihkiang made fierce attacks against the Japanese rear area transport facilities and front line units and kept the lines of communication under constant pressure. Using the airfields in Suichuan and Kanhsien districts as staging bases, it made surprise attacks on vessels in the Yangtze River and the rear air bases at Canton and Nanching. With the progress of the operation to destroy the air bases in the Shichuar and Kanhsien districts, the enemy, after destroying these air based, retreated. Thus, enemy air bases in southeast China were completely destroyed. The enemy, however, continued night bombing attacks on Wuchang and Hanhou and aerial mine-laying in the Yangtze River. They also began to make daylight attacks, mainly by fighter planes, on strategic points along the Tienching-Pukou railway.
>
> Furthermore, the enemy air force made furious daylight attacks on the Hunan-Kwangsi area, the focal point of the operation. It was reinforced and began attacks on such important air bases in the rear areas as Hankou, Nanching, Shanghai, Licheng and Peiping as well as continuing its attacks against the district along the Tienching-Pukou railway and the area to the west of the line.
>
> . . . strong enemy air forces based in Chihkiang in central China and Laohokou in northern China continued to harass the Japanese forces and to strike at the railroads and motor highways in north and central China and the occupied areas along Yangtze River and Hsiang Chiang in central China . . . the enemy's main objective in air operations at this time was to cut the Japanese land and sea communications in order to isolate the China Expeditionary Army. B-24's and other enemy planes conducted round-the-clock patrols over the sea and attacked even the smallest Japanese vessel in order to complete the blockade. As a result, after March sea communications over the East China Sea were at an almost complete standstill.
>
> On land, attacks were centered in north and central China where the P-51 and P-47 fighters concentrated on attacking rolling stock and other vital transportation targets. Occasionally, these planes attacked lines of communication in the real areas. They seriously damaged the railway system and made it almost impossible to operate it during the day.
>
> . . . the air over China was dominated by the enemy air force during the day, not only completely disrupting land transportation but also preventing air movements and air training.[98]

Under Wedemeyer's instruction, Stratemeyer[99] advanced the delivery of aircraft by releasing P-51s and B-25s to the CAF between June and July in 1945.[100] The Chinese Air Force played a more important role in the last year of war. The success of American air support and air transport indicated the significance of air power in the China theater, proving that the Chinese ground army could play an important role with the assistance of air power.

The Issue of the Ledo Road

An increase of U.S. air supply to China was one of central arguments between Chiang and the United States. The Chinese had been complaining about the small amount of cargo delivered over the Hump to China and pressed the U.S. government to increase the air transport volume from India to China after 1942. The amount of cargo transported by air to China became a source of conflict between the Chinese and American leaders. It was difficult for the War Department to increase the amount of cargo over the Hump to China because the European theater was in urgent need of a large number of transport aircraft during the critical years of the war. In addition, construction of the Ledo Road absorbed a large volume of cargo transported over the Hump, and reduced the amount of cargo to China.

To increase the flow of materials into China, Stilwell decided to build the Ledo Road. The construction of the road was a huge engineering task because most of the route lay in difficult mountain and jungle country. After Japan occupied Burma in May 1942, the China-Burma-India theater was cut into two separate strategic packages. To hold the two pieces together, the airlift, the Hump Operation, provided the Chinese with ammunition, bombs, trucks, gasoline, and other war materials. All instruments of modern warfare were supplied by air.

To end the Hump route's drain on transport aircraft needed in Europe, Stilwell decided on October 29, 1942, to build and to maintain a road forward from Ledo to the Chinese lines. Stilwell's proposal won the support of both Stimson and Marshall, who held that opening a land route to China should receive priority over a substantial increase in Hump tonnage to the air forces in the China theater. On December 8, Roosevelt approved the project of Ledo Road and instructed the War Department to give it a priority second only to that enjoyed by the North African campaign, which then was under way. Marshall promised Stilwell that 6,000 more troops and 63,000 tons of supplies for early shipment to India would be allotted for the road's construction. The troops, scheduled to sail from the west coast in January 1943, would reach Karachi in March or April, with the supplies following in about a month.[101] Therefore, a large portion of tonnage carried across the Hump went to support this road in the following years.

Doubting the efficiency of the Ledo Road and stressing U.S. air power in China, the British chiefs of staff suggested canceling this project during the Trident Conference of May 1943. They proposed that the first priority for building up the Hump be to support the Fourteenth Air Force because, they argued, the Hump would bring more cargo into China than the Ledo Road.[102] But no one paid attention to the British suggestion. Later, at the Quebec Conference of August 1943, those who supported the road plan claimed that when the two-lane all-weather road was completed in early 1945, it would be able to carry 36,000 tons monthly in the beginning and work up rapidly to 65,000 tons. The Chinese forces by then would receive enough war materials to initiate a Sino-American counteroffensive from East China in the spring of 1945. But some British and American engineers were not optimistic about the road at that time, predicting that a one-way gravel road into China would not be completed until January 1946. The road would come too late to assist the Allies in defeating Japan. The U.S. War Department's Operation Division and the Air Force suggested cancellation of the new road, arguing that air delivery could do the job. Chennault had no faith in the Ledo Road as a means of saving China, arguing that available resources should be concentrated on the buildup of the Fourteenth Air Force.[103] The Chinese were also more interested in the Hump than in the Ledo Road. In spite of the opposition to the construction of this road, Stilwell was the strongest proponent of this land route. As General Wedemeyer said, "He conceived the plan and fought it through the council rooms."[104]

When the road was completed in January 1945, it had absorbed a huge amount of manpower and material resources. The price of construction of this road was very high. The total cost of the road to the Allies was about $148,910,000. About 17,000 American engineers had worked on the road and there was a total of 1,133 American fatalities. Ironically, when the road was completed, delivery of war materials to China was still more efficient by air than by land. The ATC achieved a monthly delivery capacity of 35,000 tons in December 1944. The figure continued to climb until it reached 71,000 tons in July 1945 while the Ledo Road, the so-called Stilwell Road, facilitated delivery of only 6,000 tons of cargo in the same month. The ATC delivered 555,000 tons to China in 1945 while the Ledo delivered only 147,000—four times as much material as the road and pipelines combined.[105] ATC officers claimed that they could deliver 80,000 tons a month by air if the ACT was reinforced by about 150 more C-54s and an additional 968 officers and 4,326 enlisted men.[106]

It is obvious that the decision to build the Ledo Road was questionable. More war materials would have been delivered to China if the road had not been built. Major General William Turner argued that the Ledo Road was a waste because the air delivery system would have been more efficient if the U.S. Army Air Force had invested a considerable amount of capital and manpower in the ATC's Hump operations after 1942.[107] In other words, without construction of this road, the conflict over air support for China might have been significantly reduced.

During the Sino-Japanese war, the U.S. government provided Lend-Lease aircraft and airlifted war supplies to the Chongqing government in an effort to boost the morale of the Chinese, mire hundreds of thousands of Japanese troops in China, and build air bases for America's final air strikes on the Japanese. The Chinese Air Force assisted by Americans and the American air force in China played an important role in the war. However, disagreements over U.S. air aid and air supply between Washington and Chongqing existed, which contributed to a Sino-U.S. diplomatic and military cooperation crisis. The Sino-U.S. arguments on air aid and air supply came mainly from the different military strategies of China and the United States, Stilwell's attempted reform of the Chinese Army and neglect of air power, and construction of the Ledo Road.

A man of principle, Stilwell, as the American representative, tried to help China resist and defeat the Japanese. He was determined that the Chinese Army, trained and equipped by Americans, would win the war, while Chiang was not interested in reforming his army and in using his best troops to fight the Japanese. Chiang was more interested in using American air aid to strengthen his power and to drive the Japanese out of China.

Stilwell did not see himself as Chiang's agent to satisfy his desires and needs and attempted to refrain from politics and politicians. Yet it is unrealistic to expect that military policy at the high level can be divorced from political objectives. Since China rose to the status of one of the "Big Four" and Chiang was supported by Washington as China's heroic leader throughout the war, Stilwell's military objectives conflicted not only with Chiang's war strategies, but also with Washington's China policy during wartime. As a result, the failure of Stilwell's mission in China was unavoidable. If Stilwell, like his successor Wedemeyer, had been more adept at operating in a political environment, he might have been more successful in realizing his military goals. Even mainland China's historians pointed out Stilwell's weakness during that time:

> Stilwell was a professional soldier, rather than a politician or a diplomatic agent. He was educated that the military was divorced from politics, and soldiers should not be involved in political affairs. This idea was in conflict with China's reality at that time. Stilwell's consideration was very simple. He tried to mobilize China's military resources to attack the Japanese Army efficiently without considering any other political or personnel factors. To Chiang Kai-shek, if Stilwell wanted to establish military cooperation with the Chinese, he had to understand China's politics first. However, Stilwell, who did not consider any political factors, often got into trouble in China. This was inevitable.[108]

Stilwell sought energetically to carry out his mission exactly as it had been stated. He stood as the middle man between two governments with few resources at his command and many problems. Some conflicts between Chongqing and Washington could have been avoided or reduced if Stilwell and other American military leaders had understood more about Chinese culture and politics. Wedemeyer's successful mission proved this when he not only repaired the break in the Sino-U.S. entente that Stilwell had created but also built the Chinese Air Force to a new high.

Table 6.1
American Aircraft to China, 1937–1945

U.S. Aircraft to China, 1937–1945	
Bombers	244
Fighters	1,038
Scouts	15
Transport	97
Total	1,394

Source: Yingqin He, *Rijun jinhua banian kangzhanshi,* table 21.

NOTES

1. Major E. H. Alexander to Maj. Gen. Arnold, December 31, 1941, 839.01, Record on the Chinese Air Force, Army Air Force, AFHRA.

2. Interview with George Paxton, AVG, August 25, 1942, conducted by members of G–2, War Department, 863.549-2, ibid.

3. Chennault to War Department, April 2, 1942, 145.95, War Plane Files, Army Air Force, ibid.

4. General Marshall to Stilwell, April 23, 1942, File, Combined Chiefs of Staff, 091.711, China, Records of the Combined Chiefs of Staff, RG 407, NA.
5. For more details of the American Volunteer Group, see Charles Bond, Jr., and Terry Anderson, *A Flying Tiger's Diary* (College Station, Texas: Texas A & M University Press, 1984); and Daniel Ford, *Flying Tigers: Claire Chennault and the American Volunteer Group*.
6. Report on Conditions in China, Office of Strategic Command, August 27, 1942, 170.2276-1, AFHRA. There is considerable controversy over the AVG's claim because some figures could not be confirmed officially or by estimate.
7. Generalissimo Chiang Kai-shek, *Resistance and Reconstruction: Messages During China's Six Years of War, 1937–1943* (New York: Harper & Brothers, 1943), p. 281.
8. See Herbert Weaver, "The Tenth Air Force," in *The Army Air Forces in the World War II*, vol. 4 edited by Wesley Craven (Chicago: University of Chicago Press, 1950).
9. Brig. Gen. Chennault to Brig. Gen. Bissell, September 16, 1942, 864.311, Operational Correspondence, Fourteenth Air Force, AFHRA.
10. Brig. Gen. Clayton Bissell to Brig. Gen. Claire L. Chennault, December 25, 1942, ibid.
11. Bomber Log, China Air Task Force, 1942, 864.306A; Army Air Force Evaluation Board, "Report on China-Burma-India Theater," 138. 7–9, ibid.
12. Carl Molesworth, *Wing to Wing: Air Combat in China, 1943–1945* (New York: Orion Books, 1990), pp. 3–10.
13. Shipping and Enemy Aircraft Losses, Chinese-American Composite Wing, November 1, 1943–November 20, 1944, 865.375, ibid.
14. Major Collums to Chennault, January 8, 1945, Correspondence File, Chinese-American Composite Wing, 865.310, ibid.
15. For a detailed study of the CACW, see Molesworth, *Wing to Wing*.
16. Joe G. Taylor, "Air Interdiction in China in World War II," USAF Historical Studies No. 132 (USAF Historical Division, Air University, 1956).
17. Lt. Gen. Albert C. Wedemeyer to Chennault, May 7, 1945, Correspondence File, Chinese-American Composite Wing, 865.310, AFHRA.
18. Strength Report, July 1945, Fourteenth A.F., 862.052-1, ibid.
19. *The China White Paper* (Stanford, Calif.: Stanford University Press, 1967), pp. 28–29.
20. *Hongjun kangri zhanshi* [The combat history of Chinese Air Force fighting Japan] (Taipei: Intelligence Department, Air Force Headquarters, 1963), 8: 35–36.
21. Hawthorne Cheng, *China: After Seven Years of War* (New York: Macmillan, 1945), pp. 184–203.
22. Brigadier General L. S. Smith to Commanding General, Air Force Flying Training Command, November 13, 1 942, 200.7276-1, vol. 1, 1942, Chinese Training, AFHRA.
23. General Arnold to Command General (CG), Army Air Force Flying Training Command, June 16, 1943, Chinese Training, 220.7276-1, vol. 2, ibid.
24. Report on Interpreters, Foreign Training Program, March 1, 1945; Statistical Control Section, Hq., AAFTC, "Summary of Training Activities," AAFTC, 1944, vol. 3, ibid.
25. See Frank H. Heck, "Airline to China," in *The Army Air Forces in the World War II*, vol. 7 edited by Wesley Craven (Chicago: University of Chicago Press, 1958).
26. Ibid.
27. Ibid.
28. Interview with Madame Chiang Kai-shek, Washington, D.C., February 12, 1946, conducted by members of the Historical Division, U.S. Air Force, 105.5; Report on

Conditions in China, Office of Strategic Services, August 27, 1942, Intelligence Summaries File, 1938-1942, 170.2276-1, AFHRA.

29. Lauchlin Currie to the Secretaries of War and Navy, September 18, 1941, War Plans IV, C-1, Book 1, ibid.

30. Secretary Stimson to T. V. Soong, January 29, 1942, Adjutant General File, RG 407, NA.

31. Madame Chiang Kai-shek to T. V. Soong, January 17, 1942, Operational Correspondence, Fourteenth Air Force, 864.311, AFHRA.

32. Riley Sunderland and Charles F. Romanus, *Stilwell's Personal File: China-Burma-India, 1942–1944* (Wilmington, Delaware: Scholarly Resources, Inc., 1976), p. 208; Charles F. Romanus and Riley Sunderland, *Stilwell's Command Problems* (Washington, D.C.: Office of the Chief of Military History, Department of the Army, 1956), pp. 3–4.

33. Chennault to Arnold, July 16, 1942, War Plans Files, Army Air Force, 145.95, AFHRA; Chennault to Stilwell, August 13, 1942, Operational Correspondence, Fourteenth Air Force, 864.311; Chennault to President Roosevelt, October 8, 1942, ibid.

34. U.S. Department of State, *FRUS, 1942, China,* pp. 105, 124–25, 131–32.

35. President Roosevelt to General Stilwell, October 10, 1942, 091.711, Records of the U.S. Joint Chiefs of Staff, Record Group 218, National Archives, Washington, D.C. Hereafter cited as RG 218, NA.

36. Robert E. Sherwood, *Roosevelt and Hopkins: An Intimate History* (New York: Harper & Brothers, 1948), p. 463.

37. President Roosevelt to General Arnold, May 19, 1942, War Plans Files, Army Air Force, 145.95, AFHRA.

38. Romanus and Sunderland, *Stilwell's Mission to China,* pp.157–170.

39. Ibid., pp. 23–24.

40. Ibid., p. 167.

41. Ibid., p. 171.

42. Ibid., pp. 133, 167, 171. See also Soong to Chiang, June 26, 1942, *Zhonghua minguo zhongyao shiliao chupian: Duiri kangzhan shiqi*, 3: 165–66.

43. Chin-tung Liang, *General Stilwell in China, 1942–1944: The Full Story*, p. 58. See also Roosevelt to Chiang, June 27, 1942, *FRUS, 1942, China*, p. 89.

44. Ibid., p. 62.

45. Sunderland and Romanus, *Stilwell's Personal File,* p. 236.

46. Romanus and Sunderland, *Stilwell's Mission to China*, p. 180.

47. General Joseph W. Stilwell, *The Stilwell Papers* (New York: William Sloane Associates, 1948), edited by Theodore White, p. 128.

48. Memorandum, October 23, 1942, China, 400.3295, RG 218, NA.

49. Stilwell to Marshall, July 4, 1942, China, 091.711, RG 218, NA; Marshall to Stilwell, August 19, 1942, Policy File, 1940-1945, 862.164, Fourteenth Air Force, AFHRA; Stilwell to Marshall, October 11, 1942, 145.95, War Planes IV, D-1, ibid.

50. Record of conference between General Arnold and Generalissimo Chiang, Chungking, China, February 7, 1943, Policy File, Fourteenth Air Force, 862.164, 1940–1945, ibid.

51. Roosevelt to Generalissimo Chiang, March 6, 1943, 091.711, RG 218, NA.

52. Minutes of the Pacific War Council, August 11, 1943, Records of the U.S. Joint Chiefs of Staff, Record Group 218, National Archives, Washington, D.C. Hereafter cited as RG 218, NA.

53. Sunderland and Romanus, *Stilwell's Personal File*, p. 578.

54. Chennault to Wendell Willkie, October 8, 1942, in Anna Chennault, *Chennault and the Flying Tigers* (New York: P. S. Eriksson, 1963), p. 203; transcript of conference

of Stilwell, Chennault, Marshall, Arnold, and others, April 30,1942, in OPD File ABC 384 China (15 Dec. 43) Sec.1-A, RG 165, NA.

55. Stilwell, *The Stilwell Papers*, p. 205.

56. Interview with General Claire L. Chennault, Washington, D.C., April 5, 1948, conducted by members of the Historical Division, U.S. Air Force, 105.5-11, AFHRA.

57. President Roosevelt to Generalissimo Chiang, July 17, 1943, Chennualt's 201 File, 862.289A, 1942–1945, ibid.

58. Chennault to Arnold, March 12, 1944, Records on Operations in China, Army Air Force, 145.81-171, ibid.

59. T. V. Soong to President Roosevelt, September 27, 1943; Roosevelt to Marshall, October 15, 1943; Marshall to Roosevelt, November 17, 1943, War Plans Files, 1943–1945, Army Air Force, 145.95, ibid.

60. Sunderland and Romanus, *Stilwell's Personal File*, p. 1291.

61. *FRUS, 1944*, VI, *China*, pp. 58–59.

62. Chennault to Arnold, March 12, 1944, Records on Operations in China, Army Air Force, 145.81-171, AFHRA; Brig. Gen. L. S. Kuter to Arnold, April 18, 1944, ibid.

63. Lt. Gen. Daniel I. Sultan to Marshall, August 25, 1944, ibid.; Chennault to General C. J. Chow, August 16, 1944, and September 6, 1944, 862.161A, Chennault's Personal Correspondence, Ibid; Arnold to Harry Hopkins, September 3, 1944, Chinese Training, 350. 2B, RG 18, NA.

64. Arnold to President Roosevelt, April 20, 1944; Kuter to Arnold, April 16, 1944; Chennault to Arnold, May 12, 1944, Records on Operations in China, Army Air Force, 145.81-171, AFHRA.

65. General C. J. Chow to Chennault, September 21, 1944, Chinese-American Composite Wing Correspondence File, 865.310; Captain Kibler to AC/AS, Plans, September 21, 1944, Records on Operations in China, 145.81-171, AFHRA.

66. Chin-tung Liang, *General Stilwell in China, 1942–1944*, pp. 260–262.

67. Lend-Lease Statistics and Report Branch of the Overseas Section, Lend-Lease Administration, "Supplying the Air Forces of the United Nations," January 1, 1943, 201.38, AFHRA.

68. Richard C. Lukas, *Eagles East: The Army Air Force and the Soviet Union, 1941–1945* (Tallahassee: Florida State University Press, 1970), p. 232.

69. Long-hsuen Hsu, *History of the Sino-Japanese War, 1937–1945*, p. 510.

70. Barbara W. Tuchman, *Stilwell and the American Experience in China, 1911–1945* (New York: Macmillan, 1971), p. 359.

71. Memorandum, August 28, 1942, China, 400.3295, RG 218, NA.

72. *FRUS, 1942, China*, pp. 149–52.

73. Memorandum for Arnold, June 22, 1944, Records on Operations in China, Army Air Force 145.81-171, AFHRA.

74. Office of Foreign Liquidation, Department of State, *Report on War Air Furnished by the United States to the USSR* (Washington, D.C.: Department of State, 1945).

75. Chin-tung Liang, *General Stilwell in China, 1942–1944*, pp. 16–7.

76. *Stilwell Papers, 1948*, p. 119.

77. *FRUS, China, 1942*, p. 151.

78. *FRUS, 1944*, VI, *China*, pp. 846–47.

79. Romanus and Sunderland, *Stilwell's Mission to China*, p. 353.

80. For a study of the Disbandment Conference in 1929, see Robert E. Bedeski, *State-Building in Modern China: The Kuomintang in the Prewar Period* (Berkeley: Center for Chinese Studies Publications, University of California, 1981).

81. Young, *China and the Helping Hand, 1937–1945*, p. 275.

82. Ibid., p. 279.

83. Ibid., p. 362.

84. Claire Lee Chennault, *Way of a Fighter*, pp. 310–311.

85. *Air Operations in China, 1937–1945,* pp, 152, 157.

86. Ibid., p. 159.

87. Joe G. Taylor, "Air Interdiction in China in World War II," USAF Historical Studies No 132 (USAF Historical Division, Air University, 1956).

88. *A History of China's Aviation,* pp. 111–113.

89. Stratemeyer to Brig. Gen. William W. Welsh, Assistant Chief of Air Staff, Training, November 10, 1944, General George E. Stratemeyer Personal File, 825.161-1, AFHRA.

90. Albert Wedemeyer, *Wedemeyer Reports!* (New York: Henry Holt, 1958), p. 279.

91. *FRUS, 1944*, VI, *The Far East, China,* pp. 191–192.

92. Wedemeyer to Kuter, November 20, 1944, Records on Operations in China, Army Air Force, 145.81-171, AFHRA.

93. Wedemeyer later asked Stratemeyer who indicated that almost all of the aircraft which had been consigned to the Chinese during 1944 had been diverted to the Fourteenth A.F. See Stratemeyer to Wedemeyer, December 14, 1944, General George E. Stratemeyer Personal File, 825.161-1, ibid.

94. Wedemeyer to War Department, December 9, 1944, Records on Operations in China, Army Air Force, 145.81-171, Ibid.

95. *FRUS, China,1944,* p. 192.

96. F. F. Liu, *A Military History of Modern China, 1929–1949,* pp. 192–3.

97. Freda Utley, *Last Chance in China* (Indianapolis, Ind.: Bobbs-Merrill, 1947), p. 278.

98. *Air Operations in China 1937–1945,* pp. 179–187.

99. In March 1945, Stratemeyer replaced Chennault not only as an air adviser to Wedemeyer but also as senior air officer in China, who became the commanding general, Army Air Forces, China theater and directed both the Tenth and Fourteenth Air Forces from Chongqing. Arnold replaced Chennault with Stratemeyer on the basis that the latter officer was more capable of directing aerial logistics, the most likely type of air operations to take place during the rest of war. See Stratemeyer to Arnold, June 8, 1945, General George E. Stratemeyer Personal File, 825.161-1, AFHRA.

100. 24th Statistical Control Unit, Chinese Air Force Activities, Fourteenth Air Force, 867.308-2, June–July 1945, ibid.

101. Leslie Anders, *The Ledo Road: General Joseph W. Stilwell's Highway to China* (Norman: University of Oklahoma Press, 1965), pp. 15–22.

102. Romanus and Sunderland, *Stilwell's Mission to China*, p. 330.

103. Anders, *The Ledo Road*, pp. 143–9; Wesley Frank Craven and James Lea Cate, *The Army Air Force in World War II* (Chicago: University of Chicago Press, 1953), p. 201.

104. A. C. Wedemeyer, "Broadcast on January 28, 1945 on the Opening of the Stilwell Road," *China Handbook, 1937–1945*, p. 224.

105. Leslie Anders, *The Ledo Road*, p. 235.

106. Charles F. Romanus and Riley Sunderland, *Time Runs Out in CBI* (Washington, D.C.: Office of the Chief of Military History, Department of the Army, 1959), p. 319.

107. William Turner, "Technology or Manpower," *Air University Quarterly Review,* vol. V. (Fall 1952), pp. 3–322.

108. Wenzhao Tao, *Kangri zhanzheng shiqi zhongguo duiwai guanxi* [China's foreign relations during the war against Japan] (Beijing: Zhonggong dangshi chubanshe, 1991), pp. 325–326.

Chapter 7

U.S. Air Aid and the CCP's Anti-American Campaign, 1945–1949

The Japanese surrender came so quickly that it created an extremely critical military situation for the Nationalist government, which was unprepared for the sudden end of the war. The Communists, enjoying a distinct geographical advantage, began to move on a large scale to Manchuria, to the Yellow, Yangzi, and Pearl River valleys. The Communist Army started to seize cities under Japanese occupation, receiving Japanese surrender and military supplies. On August 11, 1945, 100,000 Communist troops began to penetrate Manchuria. To overcome the Communist geographical advantage, Chiang requested that the U.S. government airlift his troops from Southwest China to the Japanese-occupied areas.

The Nationalist government received many more Lend-Lease supplies during the last three months of the war than ever before because military operations in the European theater ended in May. By September, the U.S. government had granted to China military aid of $846 million under the Lend-Lease Act. Of the total amount, $187 million was for aircraft and other aviation material, including 1,378 aircraft of all types, 551 engines, and 619 propellers.[1]

The Lend-Lease program required clarification after the Japanese surrendered in August 1945. When the U.S. forces in China began to end their military role and the Lend-Lease was pending, the War Department instructed American commanders in China and India to continue military assistance to China in order to complete expeditiously the surrender of all Japanese forces in China. The War Department's policy thereafter enabled the Chinese Air Force to obtain a large number of American aircraft and large amounts of aircraft supplies. From August to September 1945, Albert C. Wedemeyer, commanding general of American forces in China, ordered the transfer to China of stocks valued at approximately $4.5 million. Some material, such as B-24 equipment and spare parts, was transported by the USAAF from India to Shanghai.[2] The War Department also sold to the Nationalist Air Force 70 C-47 cargo transports and some utility aircraft.[3] Many of the fighter aircraft were worth only the price they would bring as scrap. Although the GMD government agreed to pay $21.5 million for this property, which the USAAF estimated to be worth more than $50 million, the U.S. government did not receive a penny from the Nationalists.[4]

U.S. CONTINUATION OF AIR AID TO THE NATIONALISTS

To clarify U.S. Lend-Lease policy, President Truman announced on September 5, 1945, that all U.S. aid was given only to those countries engaged against Japanese military forces that had not surrendered. To enforce the new policy, Wedemeyer decided to continue U.S. aid to Chiang Kai-shek in an effort to complete the surrender of all Japanese forces in China. He was authorized to transport Nationalist forces by air and sea, to transfer the Lend-Lease equipment and supplies that already were in China to the Nationalist government, to turn over many airplanes, particularly transport planes, to the Chinese Air Force, and to deliver to the Nationalist forces huge stocks of aviation equipment and a large quantities of ordnance located in India and West China.[5]

While transferring and selling their airplanes and equipment to China, American policy makers, at Chiang's request, were discussing the size and composition of the postwar Chinese Air Force. Prior to V-J Day, the American government had embarked on programs to equip an air force commensurate with the Nationalist government's needs. For this reason, Wedemeyer drew up a plan, which was approved by Chiang Kai-shek in early August and by the USAAF commander, Henry H. Arnold, in mid-August. At that time, in Washington high-ranking officers of the War, Navy, and State Departments debated Wedemeyer's program for the revitalization and reorganization of the Chinese Air Force. Some officers suggested that the United States would support a smaller air force in China than that suggested by Wedemeyer and Arnold, because a relatively small air force was adequate for China after the war.

Both the War Department and the Joint Chiefs of Staff finally approved Wedemeyer's proposal to build for the Nationalists an 8 1/3 group program composed of four fighter groups, two transport groups, one medium bombardment group, and one heavy bombardment group, together with one photo reconnaissance unit. So equipped, the Chinese Air Force would be "of sufficient size to support a 39 division army."[6] On September 14, 1945, Truman agreed to expand Chiang's military power through completion of the wartime program for training and equipping 39 divisions and 8 1/3 air groups with a total of 936 planes.

In October, the War Department approved Wedemeyer's request to transfer to the Chinese Air Force all surplus property in China. As a result, China gained aviation stores when the Americans started transferring or abandoning the surplus property. In late October, General George E. Stratemeyer, commanding general of the USAAF in Burma and India, assigned a general officer with 400 servicemen to prepare for implementation of Wedemeyer's program of reorganization of China's air force. Stratemeyer then started to examine the Chinese Air Force's equipment and facilities to determine how many aircraft and what equipment the War Department would transfer to the Chinese. By that time, under the Lend-Lease program, the Chinese Air Force had acquired 42 B-25 medium bombers and 67 P-51 fighters as well as parts and equipment.[7] In November 1945, the Chinese Air Force possessed a total of 855 modern combat and transport aircraft.[8] It is obvious that airplanes were the most valuable equipment and surplus given to the Nationalists during these immediate postwar months.

In the meantime, the USAAF helped to transfer the Nationalist ground forces to North China and Manchuria in 1945 to prevent the Communist forces from taking over the big cities. Under the Lend-Lease program, mass transportation for the Nationalist Army from West China to East and North China started immediately because the war ended so abruptly that the Nationalist forces were not adequate to secure the surrender of the Japanese forces. In early September, the U.S. Tenth Air Force rushed three Nationalist armies to big cities, such as Shanghai, Beiping, and Nanjing.[9] In just November, 235 C-47 and DC-3 cargo planes flew 35,000 flying hours in four major airlift operations to transport 100,000 Nationalist Army troops to North China. These elite troops, equipped and trained by Americans, extended Chiang's military influence to Manchuria.[10]

The Nationalist government needed not only American aircraft but also American advisers to teach the Chinese to fly and maintain these airplanes. When the war was over, the Chinese-American Composite Wing began to curtail its activities, and the USAAF cadres were replaced with 13 Chinese Air Force units. More than 60,000 American Air Force personnel in China began to withdraw, and only a few of their air units had tasks in a continuing military mission to China.[11] To maintain American influence in the Nationalist Air Force during the post-war period, Wedemeyer suggested to the War Department that Americans help China develop its "military and commercial air potential" because the USAAF had been involved in Chinese military aviation during the war and should not lose this "favorable opportunity."[12] In the meantime, Chiang Kai-shek requested the Truman administration to send to China a U.S. military advisory group, headed by Wedemeyer, to teach Chinese to employ American tactics and weapons.

War Department officials were interested in Wedemeyer's suggestion and Chiang's request because a U.S. military advisory group could replace the U.S. military personnel in China, who were to return home in the near future. More important, U.S. military advisory groups would maintain American military influence in China.[13]

In October, the leaders of the Joint Chiefs of Staff began to discuss the program of sending a military mission to China after the Chongjing government had promised that no foreign advisory group was to be invited to China in five years.[14] Emphasizing the strategic significance of a military mission for the U.S. Army Air Force, Wedemeyer claimed during the meetings that if a military mission was sent to China, the U.S. Army Air Force would use the Chinese airfields and facilities to fight the Soviet Union in case of war, and the Chinese government might hire some former American pilots to fly C-54s when necessary. After a long discussion, the Joint Chiefs of Staff members finally agreed to send a military advising group in part to prevent Chiang Kai-shek from seeking Russian military assistance, especially air assistance such as the Nationalists had received during the Sino-Japanese war. Two advisory groups were to be established. One would be composed of air and ground officers arranged into two separate divisions. The other would be naval officers. The combination of both groups was officially called the Joint United States Military Advisory Group (JUSMAG). As a result, the State-War-Navy Coordinating Committee (SWNCC), representatives of the respective departments, helped to organize the JUSMAG. Although the officials of the SWNCC advocated sending a U.S. military advisory group to China, they admitted that the Soviet Union might interpret "the U.S. government support of China" as threatening because the Soviet Union was to "withdraw all Soviet forces from Manchuria."[15]

On February 25, 1946, President Truman authorized the establishment of the Joint United States Military Advisory Group for China in an effort to assist in the reorganization and consolidation of the various Chinese armies under the terms of general principles to which both the Nationalists and Communists had already agreed.[16] Two groups were therefore established: an Army Advisory Group which included army, air, and supply advisers, and a Naval Advisory Group. The Air Division of the Army Advisory Group was established to assist in the modernization of the Chinese Air Force and to give the Chinese technical advice on maintenance and operation.

A few days later, Brigadier General John P. McConnell, who was appointed as the chief of a liaison advisory group, established his headquarters in Nanjing. The formal activation of the air advisory organization, headed by McConnell, took place on April 1, 1946. The Air Division, composed of 83 American officers and 166 enlisted men, was to organize, train, and equip all Chinese Air Force units down to squadrons; establish American-style aircraft scheduling, maintenance, and logistical support systems; and develop the curriculum for the Chinese Air Force schools. In a word, the mission of the Air Advisory Group was to modernize the Chinese Air Force under the 8 1/3 group program.[17]

Because of the close wartime cooperation between American and Chinese Air Force personnel, the establishment of an air force unit in the Advisory Group represented a continuation of certain wartime assistance. Though the ultimate objective of the Air Division was to assist in the development of a modern air force, much of its early activities consisted of giving the Chinese Air Force advice on receiving, storing, and maintaining the vast amounts of U.S. surplus air equipment. Simultaneously, steps were taken to lay the groundwork for an air force organization fitted to the Chinese needs and capabilities. Work in this regard continued throughout the existence of the Air Division, and as experience dictated, changes in organization were recommended. The Air Division teams were also established at several Chinese Air Force bases, such as Beiping, Chengdu, and Hankou.[18]

While undertaking assignments involving the transfer to the Chinese Air Force of about 500 of the surplus aircraft which were operable and adaptable to the 8 1/3 group program, the Air Division helped the air transport groups of the Chinese Air Force to airlift the Nationalist Army from West China to East and North China before the civil war erupted. With American assistance, for example, the Second Air Transport Group used 14 C-46s to fly approximately

708 hours within 12 days, airlifting 35,000 troops and 900 tons of cargo from Suzhou to Jinan without any accidents.[19]

Although the U.S. government continued military aid to the Nationalists, the U.S.-China policy after the war was to encourage the Communists and Nationalists to establish a coalition government to end the strife. Aiming to occupy a dominant position in East Asia, the U.S. government required a relatively strong and friendly China capable of counter-balancing the Soviet Union and open to the penetration of American capital. As early as November 1944, Ambassador Patrick J. Hurley began to mediate between the GMD and the CCP for fear a civil war would expand the Soviet influence in China.

During the Sino-Japanese war, the Communists unceasingly built a mass peasant army and expanded their bases in North China. In 1937, they commanded only a few thousand square miles, a million people, and an army of 80,000, but by 1945, they controlled one-fourth of China, 100 million people, and boasted one million troops. Immediately after the surrender of Japan, Mao Zedong ordered his troops to seize the Japanese-occupied territory. Within two weeks of the collapse of Japan, the Communists expanded their territories from 116 to 175 counties. With American air aid and Japanese cooperation, however, Chiang Kai-shek regained control of all the major cities in South, East, and Central China while the Communists controlled only a few cities but much of the vast countryside in North China. In Manchuria, the Nationalist Army conflicted with the Communist Army, to whom the Soviets had turned over considerable quantities of surrendered Japanese arms. To avoid an enlarged confrontation, Chiang invited Mao to conferences in Chongqing, and American envoy Patrick Hurley himself went to Yan'an and encouraged Mao to talk with Chiang. Mao faced a difficult choice. If he refused the invitation, an immediate civil war would erupt, which would be favorable to Chiang. If he accepted it, he would be in the hands of Chiang, but a war would be delayed and he would have more time to mobilize and expand his military forces. Mao decided to attend the peace conferences after Hurley promised his safety. Mao arrived in Chongqing on August 28, 1945. After one month of negotiation, both parties reached an agreement on October 10, stressing the importance of peaceful reconstruction. But the conference did not produce any concrete results.

In November 1945, America's early effort to bring peace to China failed. The Communist representative, Zhou Enlai, left Chongqing on November 25, and American ambassador Patrick Hurley resigned as mediator with the failure of his peace-making effort. After Hurley's resignation, Truman immediately appointed General George Marshall to go to China as his personal ambassador to prevent fighting between these two forces. The official goal of Marshall's mission was to secure an immediate cease-fire to broaden the basis of the one-party GMD government, and to help create a strong, united, and democratic China. Before Marshall left, President Truman declared that the U.S. government would not use military forces "to influence the course of any Chinese internal strife."[20] However, American military leaders, who sympathized with the cause of the Nationalists, advocated support of Chiang Kai-shek. In December, Arnold told General Marshall that the USAAF strongly advocated equipping China with 3,867 aircraft in the next three years to establish a powerful Chinese Air Force composed of 18 groups.[21]

Marshall's high reputation, the great power of the United States, and the anti–civil war sentiment in China contributed to Marshall's early success. Encouraged by Marshall, the GMD and the CCP agreed to a political consultative conference in January 1946 and to stop fighting immediately. The January conference produced two major agreements. One was a political agreement to establish constitutional government; the other was a military reorganization agreement that demanded massive troop reduction on both sides and the integration of the Communist forces into a united national army. The GMD, however, did not grant the CCP equal political power, whereas the CCP resisted the integration of their armies. Large-scale fighting occurred in April 1946, and the resolutions of the Political Consultative Conference remained dreams on paper. In June, Marshall's last cease-fire arrangement expired, and the CCP demanded the withdrawal of American troops from China.

As the civil war renewed in June 1946, General Marshall, who was in charge of reaching a lasting truce between Chiang Kai-shek and Mao Zedong, felt sure that any direct American involvement would further bruise strong Chinese sensibilities regarding infringement of China's sovereignty. He also understood that if the United States assumed a direct responsibility for the conduct of civil war in China, it would have to commit sizable forces and resources over an indefinite period and thus would face grave consequences by making China an arena of international conflict.[22] Assuming that the U.S. policy of helping Chiang Kai-shek was incongruous with his mission to stop fighting between the Nationalists and Communists, Marshall asked the Truman administration to impose an embargo on all military facilities and equipment to the Nationalists in an effort to convince the Communists that his mission was mediating in good faith and not showing partiality to the other side. Because of the embargo, the Nationalists sought to purchase in July from the Foreign Liquidation Commission a number of planes for the Chinese Air Force, and the State Department refused to approve the purchase. Although these planes were intended to complete the earlier commitment to assist China in establishing an air force of 8 1/3 groups, the completion of that program was placed under temporary suspension.[23]

In addition, Marshall ordered the American advisory teams to withdraw from the Chinese military units engaged in fighting.[24] On August 11, Marshall instructed Air Director McConnell to order his group to cut all advisory relations with the Chinese air units which were taking part in combat operations against the Communists. The Air Division teams, which were advising the Chinese fighter groups actively engaged in combating the Communists, started to withdraw on order of General Marshall to prevent further involvement in the Chinese conflict. McConnell, who sympathized with the Nationalist government, continued to advise the Chinese Air Force on operational matters, however. Other Air Division members also continued to teach the Chinese aviators to employ or deploy their air units at the beginning of the war. Advice and guidance continued on such problems as planning an adequate training system for the Chinese Air Force, establishing an adequate personnel management system, instituting coordinated supply procedures, compiling suitable tables of organization and equipment for the Chinese Air Force, and developing proper operational procedures.[25]

As the civil war escalated, aerial transport became increasingly important because the Sino-Japanese war had already destroyed most of China's transportation network, and transport aircraft provided mobility. The Nationalist Air Force was fighting long distances from the center of its support during the war. Although American air advisers were not present with the Nationalist air units fighting against the Communists, they started projects that helped to improve the Nationalist Air Force. They helped the Nationalist air units to receive, maintain, and resupply their organizations with airplanes, parts, and equipment. A significant role the Air Division played was to provide assistance to the various support organizations of the Chinese Air Force. To improve the support organizations' ability no doubt would increase the power of the combat units. During the fighting, the Air Division sent three teams to the field, advising several Chinese service groups in many aspects of logistics and maintenance.

General Charles E. Thomas, Jr., who replaced General McConnell in late 1947, was authorized to increase the number of the personnel of the Air Division to 294. Developing the concept of area team, Thomas undertook to appoint teams to serve all of the air units in an area rather than providing advice to only one specific unit. A team in Beiping, for instance, was to advise not only the Second Air Region but also fighter squadrons and a squadron of B-25s stationed there. This team also made the technical specialists available to a number of units in that area. This approach raised the efficiency of the Chinese Air Force.[26] Nonetheless, due to the desperate military situation, American air advisers, together with other advisers, began to withdraw from China in January 1949, and the mission of the JUSMAG was over.

American air advisers played a role in increasing the efficiency of the Chinese Air Force, but Americans were more successful in training the Chinese pilots. Aviation training represented an important part of U.S. air aid to the Nationalists. During the Sino-Japanese war, the USAAF trained many Chinese aviators, which helped to strengthen the Chinese Air Force. When Japan surrendered in August, the USAAF planned to cancel some training programs for the Chinese. The suspension of such programs would seriously affect the expansion and modernization of the Chinese Air Force because the Nationalist government did not have adequate facilities and equipment to train its air crew members and technicians.[27] For this reason, Chiang Kai-shek encouraged the War Department to continue the training programs for the Chinese Air Force. To comply with Chiang's request, Truman authorized the USAAF to continue to train the Chinese crewmen and technicians who were in the United States or in China under the Lend-Lease program, although the War Department was concerned about the legality of a continuation of training the Chinese under this program.[28] As a result, the U.S. training programs of the Chinese aviators were not reduced in number. On the contrary, they were increased because of suspension of other countries' training programs and the American determination to modernize the Chinese Air Force.

In late September 1945, only 84 officers and 194 enlisted mechanics of the Chinese Air Force were receiving American training in China. But in November 1945, 2,259 Chinese Air force personnel were undergoing training in the United States and 192 were en route, while 1,285 were in China awaiting transportation for such training. By December of 1945, 2,167 Chinese had finished their training programs here.[29] The most important program in 1946 was the training of about 500 Chinese pilots and technicians in the C-46 cargo transport. The USAAF at Bergstrom Field, Austin, Texas, provided 60 C-46s; each Chinese aviator received more than 150 flying hours. The success of this extensive ground and transition flying program enabled the Chinese to fly the C-46 reliably when they returned home in late 1946.[30]

In 1947, the Air Advisory Group took charge of the Chinese aviators' training program, assigning eight officers and four enlisted specialists to provide appropriate advice to such units as the Chinese Flying School at Hangzhou and the Air Technical Service Command at Shanghai because the Chinese Air Force lacked training facilities and equipment. When the Cadet Aviation School reopened in 1947, it had only 60 trainers but 421 students. Due to a critical shortage of training aircraft and parts to keep the Chinese students safely flying, only 25 students had graduated from the school by mid-1947. Training conditions were improved when American advisers arrived at the school in June and started to train Chinese instructors and students.[31]

In the Fall of 1948, Americans advised the Chinese pilots on airlift procedures and parachute drops when they were using 150 C-46 transports to provide supplies to such beleaguered cities as Xi'an in West China and Shenyang in Manchuria. With American assistance, the Nationalist transport groups successfully flew 7,700 hours or 4,000 missions. The advisers with the Nanjing area team developed a new course in fighter tactics for the pilots of the fifth Fighter Group prior to their flying P-51s to attack the Communists.[32] From September 1945 to January 1948, about 3,300 Chinese personnel received American aviation training at a cost of $25 million, most of which was covered by Lend-Lease.[33] Many of them became the effective nucleus of the Chinese Air Force; the postwar Nationalist Air Force was for the most part American-trained.

The smashing military success of the Communists in late 1947 and early 1948 shocked many Americans, and some of them tried to use air power to save the Nationalists. Claire Chennault was one. Chennault testified in March 1948 before the House Committee on International Relations that the U.S. government had to increase military aid to China. Otherwise, there would be "the greatest failure of American policy in all of our history."[34] In April 1948, Whiting Willauer, vice president of Civil Air Transport (CAT),[35] told the American ambassador to China that Chennualt was planning to create another American Volunteer Group, but he intended first to get at least tacit agreement of the State Department for this move. Some CAT officials were convinced that a revival of American Volunteer Group backed by a sizable transport force would have great impact on the course of the civil war.[36] In November, Major General C. C. Liang of the Nationalist Army went to Beiping and obtained an interview with Fu Zuoyi, commander in chief of Bandit Suppression Forces in

North China. After learning that Fu had an interest in the establishment of an organization developed out of CAT, similar to the American Volunteer Group, Liang arranged a meeting with Fu for the CAT officials. During the conference, Fu was very favorable to this program and promised to encourage Chiang Kai-shek to invite Chennault to restore the "Flying Tigers."[37]

Later Chennault discussed with Chiang the possibility of the revival of AVG. Chennault boasted that a relatively small, balanced force of fighters and bombers under his control would halt the Communist drive, deal the Communists hard blows, and save the Nanjing government. Chennault also claimed that all pilots flying for CAT were ex–fighter pilots and over 100 ex–Fourteenth Air Force pilots had expressed willingness to serve in the new AVG. Although Chiang was interested in the reestablishment of AVG, he pointed out that the Nationalist government did not have funds available for the establishment of such a group. If such a group were formed, it would have to be at the expense of the U.S. government.[38]

When the rumors of a new AVG possible for China spread to Washington, Defense Secretary James Forrestal was very interested in this idea and encouraged Secretary of State Marshall to seize the opportunity.[39] The State Department, however, rejected the formation of an AVG, with or without U.S. government support or tacit approval. First, such a volunteer group would constitute a form of direct U.S. involvement in China's civil war, and the U.S. government had been opposing any direct American involvement in the CCP-GMD confrontation. Second, the participation of such an American air group would have possible serious repercussions in the treatment of U.S. nationals residing in areas that were then or might fall under the Communists' control. Finally, the establishment of such a group might incite the Soviet Union to establish a "Soviet Volunteer Group" to operate in support of the Chinese Communists, which would result in probable direct air combat between Soviet and American nationals over Chinese territory.[40] As a result, the American officials dismissed Chennault's suggestion and continued to prevent any Americans from participating in the civil war in China.[41]

As the military situation in China became critical, the Truman administration realized that the Nationalist government would collapse without American military and economic aid. In April, under the pressure of the Republican Congress, President Truman asked Congress for a grant to China of $463 million over a period of 15 months from February 1948 to June 1949. Of the total amount, $338 million was to be used for economic aid and $125 million for military assistance.[42] Of this $125 million, $87.5 million was for the army, $28 million for the air force, and $9.5 million for the navy. About one-fourth of these grants would be for the use of the Nationalist Air Force. The China Aid Act marked the return of large-scale U.S. aid to the Nationalists at a critical moment. From January 1948 to March 1949, the United States shipped to China aircraft valued at $4,454,105, four times the total value of the combination of ammunition of $569,071 and equipment of $282,987.[43] These aircraft, however, inadequate to deal with the deteriorated situation, came too late, and could not save the Nationalist government during the final stage of the civil war.

By early 1949, the Communists had occupied Manchuria and North China, and the Nationalist government seemed to break down rapidly. Chennault flew from Shanghai to Washington to encourage the Truman administration to adopt his plan to check the Chinese Communist expansion and to save the Nanjing government. On May 3, Chennault delivered a speech before the Senate's Armed Services Committee, and on June 30, he testified before the House Committee on Foreign Affairs, claiming that the Communists' victory in China would eventually contribute to the Communist triumphs in Indochina and other Southeast Asian countries. In this situation, Chennault argued, Truman should immediately send military advisers to serve in the Nationalist military force and increase military aid, particularly air aid, to the Nanjing government. Chennault boasted that air support would play a significant role in determining the outcome of the civil war.[44]

In the meantime, Chennault presented his plan to the high-ranking officials of the Department of State. On May 11, he met with Undersecretary James E. Webb and submitted his program to Dean Rusk, deputy undersecretary of state, but neither of them were interested in the Chennault plan. Seeing no hope for the corrupt Nanjing government, Secretary Dean Acheson and W. Walton Butterworth, director of the Division of Far East Affairs, expressed no enthusiasm for this plan. Leighton Stuart and Minister Counselor Lewis Clark also opposed it. They, asserting that American aid would not affect the course of the civil war, said, "It [the plan] is impractical and of doubtful value to furtherance of US national interest."[45] After considering all these comments, Acheson rejected Chennault's proposal and testified before the House Committee on Foreign Affairs in June, claiming, "I am not in a position to come to Congress and ask Congress for money at this time to do something which we do not believe can possibly be effective."[46] Chennault's final effort to use air power to save the Nationalists died.

Americans also were interested in civilian aviation in China during the postwar period. As the war ended, the Chinese commercial aviation corporations acquired American aircraft and equipment, too. The China National Aviation Corporation (CNAC) acquired about 50 C-47s and C-46s from the U.S. Army-Navy Liquidation Commission and the Surplus Property Administration after the war.[47] It also purchased six C-54s for $540,000 and had them converted to airline use at a total cost of $2.1 million. By May 1946, CNAC, owned 20 percent by Pan American Airways and 80 percent by the Chinese Nationalist government, had 25 aircraft in operation and employed 60 Americans.[48] Another airline was Central Air Transportation Corporation (CATC), which was established in the summer of 1941 and owned totally by the Nationalist government. CATC entered a new era when it purchased 12 C-47s from American Army surplus in India at the end of 1945 and gained 150 surplus C-47s and C-46s in 1946. It averaged 15.4 planes in service and employed 15 Americans in 1946.[49] Supported by the Nanjing government, both companies tried to expand their service after the war.

Meanwhile, a new aviation company was established. At the request of Colonel Ralph W. Olmstead, director of operations for the United Nations Relief and Rehabilitation Administration (UNRRA), Chennault and Whiting Willauer offered a proposal in February 1946 for the establishment of a new airline to transport UNRRA's materials from storage in warehouses on the coast to the people in inland areas. UNRRA would provided financial aid to the Chinese National Relief and Rehabilitation Administration (CNRRA) to purchase airplanes and facilities for Chennault's new airline. With the approval of the director-general of UNRRA, the Civil Air Transport was created in October 1946 when the Chinese government signed a contract with both Chennault and Willauer. Willauer had helped Chennault to organize the AVG and worked for China Defense Supplies during the war. Between 1944 and 1945, Willauer was the director of the Foreign Economic Administration's Far Eastern and Special Territories Branch, which was in charge of foreign aid, economic intelligence, and postwar planning. Learning that the Chinese government urgently needed to develop air transportation, Willauer returned to China in 1946 to work with Chennault and established a new airline.[50] CAT obtained 19 C-46s and C-47s from American government surpluses in Manila and Honolulu, and employed 39 American pilots and co-pilots. All CAT pilots were former USAAF airmen who could competently fly into battle areas.[51]

To increase American influence in the Chinese civil air transport industry, the U.S. government reached an air agreement with the Nationalist government. Before the war was over, Vice Foreign Minister Liu Jie led a Chinese delegation at the International Civil Aviation Convention in Chicago in December 1944. After the meeting, the GMD government was becoming more interested in reaching an air agreement with the U.S. government. In July 1946, Dr. Liu informed American Ambassador Leighton Stuart that Nanjing desired to negotiate a bilateral air agreement with Washington. After receiving this information, the State Department instructed Stuart to "conclude bilateral air transport agreement with China as soon as possible."[52] On December 20, 1946, the Sino-American Civil Air Transport Agreement between the United States and the GMD government was signed in Nanjing. The agreement included the following:

> The Government of the United States is desirous of obtaining the right for the United States carriers to serve other international traffic points in China beyond those mentioned in the present agreement. The Government of China does not wish to extend these points at present but will be ready to give prompt consideration thereto when conditions justify.
>
> The Government of the United States agrees that if at any time it should enter into an agreement with any other nation adopting formulae for the predetermination of frequencies or capacity, it will enter into a similar agreement with the Government of China.
>
> United States carriers will be authorized to serve additional traffic points in Chinese territory as soon as the carriers of any third country are so authorized, and on a basis of reciprocity Chinese carriers will also then be authorized to serve additional points in United States territory.[53]

This agreement, without doubt, would allow Americans to have more advantages than the other Western countries to increase their influence in the commercial aviation in China.

To help the Chinese to expand their commercial air service, the U.S. government also tried to send a civil air mission to China. In 1946, Marshall, special representative of President Truman in China, asked Major General John P. Lucas, chief of the Army Advising Group, to prepare a survey of civil aviation in China with recommendations for its rehabilitation. On January 15, 1947, Brigadier General J. P. McConnell, director of Air Division, prepared a memorandum, "Civil Aviation in China." This memorandum was directed toward the development of a first-class civil aviation establishment in China and suggested the appointment of an American adviser on Chinese civil aeronautics. On January 21, Stuart gave a copy to Chiang Kai-shek, who received it gratefully and read it carefully with great interest. Later, this document was sent to Minister of Communications Yu Dawei. On February 14, 1947, Yu told Stuart that Chiang had accepted the suggestion of the appointment by the U.S. government of a civil aviation adviser to China. American diplomatic officials in China were also interested in sending a civil aviation mission to China, believing that a well-organized, efficient, and modern civil aviation in China would contribute to the country's unification and stability under the rule of the Nationalist government. As a result, this memorandum was sent to Washington.[54] On January 27, 1948, the Smith-Mundt Bill was passed in Congress, and President Truman later affixed his signature to it. Under the provisions of this bill, the Civil Aeronautics Administration was authorized to send a civil air mission to China at the U.S. government's expense. This mission, consisting of a chief and five associates, was to be available to China in late summer or early fall of 1948. The Civil Aeronautics Administration promised that it would endeavor to pick a top-flight man with broad experience in both civil and military aviation as the chief of this proposed technical assistance mission to China.[55] With the military failures of the GMD in spring, Stuart urged the White House in April to send to China the civil air mission as early as possible because "the importance of the civil air transport industry not only to the general economic life of the country as a whole—especially in view of the lack of alternative modes of transport—but also as an important adjunct to the military forces has been recognized."[56] However, Washington failed to send this air mission due to the deteriorated situation unfavorable to the Nationalist government in 1948.

THE ROLE OF THE CHINESE AIR FORCE AND THE CHINESE AIRLINES

Chiang was determined to resolve the Communist problem by military means after mid-1946. Well trained and well equipped, the Nationalist Air Force had 160,000 personnel and had about 1,000 airplanes in June 1946, consisting of one B-24 bombardment squadron, one B-25 bombardment squadron, four P-40, P-47, and P-51 pursuit squadrons, one C-46 transport squadron, one C-47 transport squadron, and one reconnaissance squadron.[57] In the early phase of the civil war from July to December 1946, the Nationalist armies reaped victories at every encounter by capturing 165 towns and 174,000 square kilometers of territory from the Communists. The Chinese Air Force helped the Nationalist army fight against the Communists, flying 81 missions with 153 aircraft to attack the Communists from June to July in 1946. On July 5, four aircraft strafed the Communist Army near Kaifeng.[58] During the campaign in Rehe province from August to October, P-51 aircraft flew 69 missions and B-25 aircraft flew 8 missions, strafing the Communists.

In the spring of 1947, when the Nationalists managed to keep open four strategic corridors in North China, the Communists had controlled most of the North China countryside. Mid-1947 seemed to mark a turning point in the fighting. The victory-laden Nationalist military machine began to sputter when the Communists went on a general offensive in the second half of 1947, scoring victories in Henan and northern Hebei. In 1947, the Nationalist Air Force used P-51 and P-38 fighters to assist the ground forces in launching fierce offensives against the Communist Army, sent B-25 bombers to attack transportation lines, military centers, and military bases of the Communists, and employed C-46 and C-47 aircraft to transport war materials and armies by air to the cities occupied by the Nationalists. In May 1947, confronted by the Communists' attack, the Nationalists were forced to abandon the Liaodong Peninsula and concentrate their force around Shenyang. Because of a massive effort with unstinting use of air transport, the Nationalists succeeded in the resupply and reinforcement of their encircled garrisons, and finally managed to reestablish a precarious link between Shenyang and Changchun.

By the beginning of 1948, Lin Biao's army had inflicted losses of 150,000 Nationalist troops. The remainder were pressed into a small triangular area among Shenyang, Changchun, and Jinzhou, which represented less than one percent of Manchuria. In March 1948, the situation in Manchuria was desperate. The garrisons of Shenyang and Changchun were being resupplied only by air. Only the Nationalist Air Force was active, bombing cities and attacking trains, but to no great effect.[59] By mid-1948 Lin had so tightened the encirclement that he practically smothered the Nationalist defenders. When the Communist forces began to attack Jinzhou on September 12, the Chinese Air Force used four C-46 transports to airlift 5,000 troops from Shenyang to that city while P-47 fighters strafed the Communist troops to stop their offensives and to assist the defenders. Having destroyed 100,000 Nationalist troops, Lin conquered Jinzhou on October 15. The Nationalist Air Force lost 15 airplanes during this campaign. Changchun fell on October 18, and Shenyang on November 2. The Manchurian campaign cost Chiang 470,000 of his best troops. After Nationalist troops surrendered cities in Manchuria, the Chinese Air Force returned to bomb the surrendered cities at random, causing a loss of civilian lives and property but damaging no military objectives because of inaccurate bombing.

Operating simultaneously with the Manchurian battles, another Communist field army under Chen Yi conquered Shandung after fierce fighting at Jinan on September 26, 1948, although Chiang had ordered the transport of one division by air to the city and the Nationalist Air Force had flown 520 sorties during the campaign. The Jinan campaign cost Chiang 100,000 troops. After the fall of Jinan, Chiang ordered the GMD air force to bomb the city for three days, destroying all military warehouses, factories, and railroad stations there. After conquering Jinan, the Communist troops moved on to attack Xuzhou. Chiang deployed 400,000 of his mechanized troops, equipped with tanks, heavy artillery, and armored cars to defend this gateway to Nanjing. Although the Chinese Air Force flew 2,957 sorties to strafe the Communist troops during the Huaihai campaign, 100,000 Nationalist troops were annihilated by the Communists. Xuzhou fell on December 15. In November, when the Communist Army attacked the city of Suzhou, General Qiu Jingquan determined to deal a great blow to the Communists with the massive intervention of the Nationalist Air Force. The Chinese Air Force flew from 400 to 500 sorties per day from Nanjing and forced the Communists to retreat and momentarily loosen their grip around Suzhou.[60] However, by the beginning of 1949, the Nationalists had lost 1.5 million men and the Nationalist Air Force had only 401 aircraft, including 148 fighters, 91 bombers, and 162 transports. With such losses, the Nationalist forces collapsed.

In spite of American air assistance and aid, the Nationalist Air Force did not render valuable support in ground operations and its capabilities were not properly employed. It was mainly the transport operation that gained some measure of success by supplying encircled garrisons and cities until the arrival of relieving forces. Chinese Communist military historians have made comments on the role of the Chinese Air Force during the civil war:

> From June 1946 the Nationalist Government destroyed the peace agreement and launched the civil war by attacking the liberated areas in Central China. It employed about 1,000 aircraft provided with enormous American military aid during the civil war. During the three-year war, the Nationalist Air Forces flew 66,239 sorties to assist the ground forces in battles and to bomb heavily the liberated cities, bridges, warehouses, factories, and others, committing monstrous crimes. During the war, the People's Liberation Army dealt a blow to the Nationalist Air Force. It is estimated approximately that from 1946 to May 1950, the Nationalist Air Force lost 379 aircraft. From July 1946 to June 1947, three aircraft were captured and three were shot down. From July 1947 to June 1948, 10 captured and 81 shot down. From July 1948 to June 1949, 121 captured including those deserted. From July 1949 to June 1950, 55 captured including those deserted and 21 shot down.[61]

The military historian F. F. Liu commented, "The Nationalist Air Force, unopposed in the air, could have become a decisive factor in the entire outcome, but was hoarded by its incompetent air commanders instead of being employed effectively to support land operations."[62]

There were several reasons why the Chinese Air Force did not play an important role in the civil war. First, the Nationalist officers, mostly old and loyal contemporaries of Chiang Kai-shek, had little or no knowledge of the Chinese Air Force. Their lack of organization, skill, and honesty diminished the number of operationally ready aircraft on line. Despite the efforts of the Air Advisory Division, airfields and radio installations were inadequate because they were not properly maintained. As the Air Advisory Division official commented, "The CAF has questionable value as a military ally mainly because of their inability to conduct their administrative and logistical function satisfactorily."[63] The Nationalist military leaders were also indecisive in military campaigns. On April 20, 1949, when the Communists were crossing the Yangzi River, the Chinese Air Force, still able to fly 300 airplanes along the Yangzi River, might conceivably have turned the crossing into a holocaust. The Chinese Air Force, however, did not appear to give battle to the Communist armies. The Nationalist Air Force strafed Communist troops crossing the river, but under the heavy attack of antiaircraft fire, the Nationalist aircraft retreated, returning to their Shanghai base. As a result, the People's Liberation Army succeeded in crossing the Yangzi River without strong resistance and occupied Nanjing several days later.[64]

Second, despite American advisory assistance, the Chinese Air Force never attained satisfactory operating standards and refused to operate at low altitude in order not to lose aircraft or personnel. During the civil war, the CAF's bombing operations were usually conducted from such an altitude that accurate bombing of military targets was impossible; their strafing of the Communist troops was carried on from altitude of 1,500 to 2,000 feet; their supply drops to beleaguered garrisons and units were made from such altitudes that considerable quantities of the supplies dropped fell into the hands of the besieging Communists.[65] In June 1948, when the Nationalist Air Force was bombing the city of Kaifeng occupied by the Communists, the pilots flew too high and obtained no militarily significant results, only killing civilians. On June 29, 1949, the Chinese Air Force started a series of bombing raids over Shanghai and Nanjing. Fearful of Communist antiaircraft fire, the high-flying planes could claim only a few civilian victims, and the raids soon showed themselves to be ineffective. As General L. M. Chassin of the French Air Force in the Second World War wrote, "The Chinese Air Force, consisting of 8 1/3 groups, is far in excess of what a country bereft of gold credits can support. Although it has among its personnel over five thousand United States–trained pilots, it accomplished little, other than air-lifting troops and operating its transports for personal gains."[66]

Third, the morale of the Nationalist pilots was low. Japan's surrender gave the Chinese Air Force a sense of relief and a feeling of having accomplished a mission. The Nationalist pilots, hoping for a rest, disliked the civil war and nurtured an ingrained resentment against killing the Communists. In addition, the Communists began to penetrate the Nationalist Air Force by recruiting Communist members and incited pilots to desert with their planes to the Yan'an. On June 26, 1946, when Captain Liu Shanben and two other pilots were flying a B-24 bomber loaded with military equipment from Chendu to Kunming, they decided to go to Yan'an, the capital of the Communists. Born in 1915, Liu began to serve the Nationalist Air Force after graduating from the aviation school in 1938. He went to the United States in 1943 to learn how to fly the B-24 bomber and returned to China in 1945 after completing his studies. Under the influence of the Communists, Liu had planned to desert to the Communists for a long time, and this mission provided his opportunity. After arriving at Yan'an, they were received warmly by Mao and other Chinese Communist leaders. This was the first time that Nationalist pilots deserted with their aircraft to the Communists' occupied areas. From June 1946 to December 1949, 69 Nationalist pilots, seven of whom were Communists, flew 26 airplanes to

Yan'an, including three P-51s, one P-47, three B-24s, one B-25, six C-46s, four C-47s, and others.[67]

Finally, the arms embargo imposed by Washington from August 1946 to May 1947 also reduced the Chinese Air Force's efficiency. In August 1946, the U.S. War Department imposed an arms embargo on Nanjing, refusing to grant licenses for exportation of arms, weapons, and aircraft. The embargo and the delay in delivering supplies to China resulted in a shortage of spare aircraft parts. The attrition rate of combat aircraft was 3.2 percent per month, and that of transport aircraft was 2.7 percent per month. At that rate, it was estimated that the Nationalist Air Force had been reduced to one half of its normal operational strength, and by 1948 because of a lack of spare parts it became entirely ineffective other than a few transport aircraft.[68]

Due to the close relations of the civil and military aviation in China, the civil airlines made a substantial contribution to the Nationalist military forces in the movement of military and supplies into battle areas. The commercial airlines also assisted other agencies of the Nationalist government in the evacuation of government personnel and their families from besieged areas. When the Communists besieged Shenyang in May 1948, for example, the Chinese Air Force called upon CAT, CNAC, and CATC to undertake the evacuation of 7,000 Nationalist personnel and their families from the area and to provide food to the city. CAT made 22 flights daily; CNAC, CATC, and the Chinese Air Force made up to 50 flights each day. Flying food and other supplies into the city, these airlines and the Chinese Air Force effectively and successfully accomplished the task.[69] CAT had airlifted 17,208 tons of goods to Shenyang before the Communists occupied it in October 1948. In late 1948, when the Communists were attacking the city of Suzhou, CNAC, CATC, and the Chinese Air Force provided paramilitary support for the Nationalists. CNAC carried rice into the city and took wounded Nationalist soldiers from the city to Nanjing. The airline made 33 flights with 46 to 60 casualties per trip in a single day. CAT airlifted or airdropped 37,136 troops, 135 tons of ammunition, and 1,500 tons of food to the city. In November 1948, when the Communist forces were besieging the city of Taiyuan, Shanxi province, CAT, CNAC, and CATC airdropped military supplies and food to the city. CAT flew more than 45 trips a day, delivering one million pounds of food to the Taiyuan people within two weeks.[70] Between 1946 and 1949, CAT evacuated about 100,000 Nationalist personnel from besieged cities. Willauer later claimed that CAT had slowed up the Communist advance for several months.[71] The Chinese civil airlines played a role in helping the Chinese Air Force in the civil war. As historian William Leary points out: "During the battle, CAT, together with CNAC and CATC, operated as a paramilitary adjunct of the Nationalist Air Force."[72]

In sum, after the Sino-Japanese war, the U.S. government provided air aid to the Nationalists, but U.S. air aid was limited and trivial in the conflict between the Communists and Nationalists. Americans helped to transport the Nationalist troops by air, but in the end this airlift contributed to the Nationalists' defeat because it allowed Chiang to extend his military influence into North China and Manchuria beyond his ability to support his troops adequately or cheaply. The U.S. government equipped the Chinese Air Force, trained the Chinese pilots, and sent an air mission to China, but the Chinese Air Force played a limited role in helping the Nationalist Army fight the Communists.

THE CHINESE COMMUNIST ATTITUDE TOWARD U.S. AIR AID

After the surrender of Japan, the Communists continued to extend their influence in North China and to penetrate Manchuria. They took the surrender of 220,000 Japanese troops and killed another 10,000. They controlled about 200 towns, over 300,000 square miles of territory and nearly 20 million people. The Communists' successful achievement in such a short time period was partially due to Soviet Union assistance. The Russians provided assistance to the Chinese Communists until they completed their withdrawal from Manchuria on May 3, 1946. The Soviets assisted the Chinese Communists covertly and openly in their military actions against the Nationalists after the Japanese surrender, transferred Japanese arms to the Chinese Communist forces, directly participated in the expansion of Chinese Communist troops, provided training of Mao's army, and helped to restore the railroad in Manchuria in order to make possible large-scale regrouping and centralization of the units of the Chinese Communist Army.[73] As Chu De, commander-in-chief of the Chinese Communist Army, stated on July 16, 1946, "The victory of the People's Democratic Revolution had its roots in Soviet assistance and friendship."[74]

The Chinese Communists were receiving Russian aid, but they accused Americans of providing aid to the Nationalists. From the Communist leaders' perspective, U.S. military aid to the GMD government was a significant interference in China's internal affairs, delaying peace and aggravating the misery of the Chinese. As the U.S. military aid to the Nationalists continued after the war, the Communists were consistently critical of American air aid to Chiang. On November 8, 1945, the Communist newspaper *New China Daily* declared that General Chu De had written to General Wedemeyer to protest against American planes that had strafed the Communists in Hebei province and had dropped propaganda leaflets calling for Communists' withdrawal from Kuan in Hebei. Two days later, the *New China Daily* voiced further complaints against the alleged use of American aircraft to provide daily resupply for the Nationalist Army engaged against the Communist Army. After the Soviet Union confirmed their intention on December 12 to permit Nationalists entry into Manchuria by air, land, and sea, the next day some air-transported Nationalist troops were landed at Shenyang from which the Russians had not yet evacuated. The Communist propaganda lost no time attacking the U.S. airlift of the Nationalist Army.[75] On April 19, 1946, the Communists' New China News Agency in Yan'an complained against the American training program of the Nationalist pilots,

> According to reliable sources, United States is now busily helping Kuomintang to train two Bombing Squadrons for certain purposes. Every Squadron consists of 48 B-25 Bombers, totaling 96 Bombers in 2 Squadrons with 10 Training Planes in addition. All these planes are recently transported from America.[76]

In June 1946, Zhou Enlai, the Communist spokesman, criticized American air support of the Nationalist government, claiming:

> After victory over Japan, the United States still assisted the Kuomintang by providing supplies which were not for the purpose of fighting the Japanese any more. . . . In Manchuria, airplanes had been used extensively. . . . [T]hese planes undoubtedly were handed over to the Kuomintang through Lend Lease.[77]

To the Communist leaders, U.S. air aid was in part responsible for the renewal of the CCP-GMD conflicts. Strengthened by American aid, the Nationalist Air Force enabled the Nationalist Army to gain considerable success in forcing the Communist forces out of the big cities in Manchuria and North China and having military advantages in these areas. They claimed, "The U.S. Army Air Forces helped to transfer a large number of Nationalist ground forces to Manchuria and North China immediately after the Japanese surrendered. . . . The United States supplied the Nationalists with more than 1,000 airplanes, hundreds of tanks, dozens of battle ships."[78]

In spite of their protests, the Communist leaders remained too insecure in their own position to break completely with the United States at that time. They encouraged American mediators to force Chiang to withdraw the Nationalist Air Force from the war. The Communists, who had no airplanes and antiaircraft defense equipment, strongly opposed the participation of the Chinese Air Force in the civil war. On August 5, 1946, during the conference between General Marshall and General Zhou Enlai, head of the Communist delegation in Nanjing, Zhou complained:

> Now I have received two reports which are of a serious nature. The first one is at noon yesterday. Yenan was bombed as may also be testified by the American liaison officers there. 7 Nationalist planes of which 6 are P-47s and 1 B-24 came from the southern direction. The first strafed the outskirts of the city for over twenty minutes, shooting over 10,000 machine gun bullets. Subsequently they released 11 bombs. They had two places as targets. The first one is the airfield and one Nationalist B-24 plane which came over to the Communist area from Kunming the end of June was destroyed. The second target place is near Wang-chia-ping. That is the place where you met Chairman Mao Tse Tung and General Chu Teh and where Chairman Mao is living. I am still awaiting detailed report about the casualties.
>
> The American liaison officers also helped to examine the pieces of the bombs and bullet shells and testify that they are American made and are of the time bomb type.[79]

Two days later, Zhou lodged a stronger protest with the Nationalist government regarding the bombing of Yan'an by the Nationalist Air Force. Zhou's statement is as follows:

> First. Order be immediately issued for investigation into outrageous bombing; also officers held responsible be severely punished;
>
> Second. Entire Chinese Air Force be put under control and supervision of Executive Headquarters so as prevent Chinese Air Force from participating in civil war.
>
> . . . all lend-lease combat planes together with spare parts and ammunition previously handed over to Chinese Government be withdrawn immediately.[80]

On the whole, from August 1945 to August 1946, the Communists' propaganda was critical of American airlift of the Nationalist Army and American delivery of aircraft to the Nationalist Air Force, but its major target was the Chinese Air Force. Their attack had a substantial factual basis, but after August 1946, their attacks on the U.S. air aid became increasingly abusive and increasingly unrelated to the facts.

On August 30, 1946, Washington and Nanjing reached a surplus property agreement by which the United States agreed to sell to China civilian-type surplus properties and nonmilitary type-goods worth $900 million. Although Marshall insisted that this agreement would prohibit the sale of military goods, including aircraft, and that that surplus property would be important for the economic reconstruction of China after the Sino-Japanese war, Zhou alleged that 12 percent of the surplus property of this arrangement was air force surplus and equipment and that spare parts that the Nationalist Air Force urgently needed were not included in the military-type goods category.[81] As a result, after the conclusion of this treaty agreement, the anti-American propaganda campaign intensified. On September 30, 1946, Zhou sent a letter to Marshall, claiming the Nationalist Air Force had "made ruthless air raids into Liberated Areas, killing and wounding countless civilians."[82]

Zhou was misleading in his charge. The Nationalist Air Force assisted by Americans had cost the Chinese lives and had caused suffering to the civilians, but the Communists obviously exaggerated the casualty figures. Chassin had a different picture of the Nationalists' bombing in his book, *The Communist Conquest of China: A History of the Civil War, 1945–1949,* which is based on the intelligence reports sent by the mission of the French Second Bureau (Intelligence) from China to the General Staff for National Defense during the Second World War. Chassin wrote, "Bombing of villages by B-25's was seldom effective, especially since the bombing was at regular, set times . . . an excellent warning system existed in all the villages."[83] Condemning the Nationalist Air Force of killing "countless civilians," the Communist leaders intended to stir up anti-U.S. sentiment among the Chinese people because the Nationalist pilots trained by Americans flew American aircraft to drop American bombs on the Communist army and civilian population. Stories of American atrocities against the Chinese civilians were distributed widely at that time. As Mao Zedong claimed, "Aircraft, tanks, guns and ammunition for the civil war are shipped from the United States in great quantities."[84]

During the press conference in Shanghai on October 3, 1946, General Zhou Enlai made public the text of his letter of September 30 and accused the United States of having encouraged civil war in China by supplying war material to Chiang Kai-shek and demanded immediate withdrawal of all American military and naval forces in China. Zhou further stated that General Marshall's early efforts succeeded, but following his return to China, American attitude had changed. Instead of acting as mediators, Americans had sided with the Nanjing government. Claiming that evidence could be found in the American extension of the Lend-Lease agreement and the surplus property agreement, Zhou pointed out, "Present Chinese Air Force was able to operate only because of American bombs, fuel, and spare parts."[85] Realizing that he had failed in his mission,

Marshall returned to Washington on January 8, 1947, to become the secretary of state.

The Communist victories in 1948 made the Communist leaders confident that they would defeat the Nationalists in the near future. Their attitude toward the United States started to change in November. They began to abandon their policy of balancing Soviet and U.S. influence and to lean toward the Soviet camp.[86] A wave of attacks targeting the United States escalated in November 1948. One month later, Liu Shaoqi, the CCP leader, declared, "We have to encourage the Chinese people and other peoples in the world to fight U.S. imperialists and their accomplices all over the world."[87]

In aviation matters, the Communist propaganda organ used those rumors of the plan to create a new American Volunteer Air Force Group to attack Washington. Communist radio announced on November 26, 1948:

> Notorious American aggressor against China, Chennault, dares plan organization volunteer air force to fight the Chinese people. This plan has already been presented to Chiang Kai-shek and Fu Tso-yi. During period (existence) Chennault's air transport force (CAT) (INT) [has] already joined Kuomingtang air force and, moreover, already gave effect to plan of slaughtering the people. This makes clear American imperialistic government is just tolerating these actions of slaughtering Chinese people. The Chinese people and Chinese People's Liberation Army resolutely oppose [propose] to end such actions.[88]

The Sino-U.S. Air Transport Agreement also became a target of this attack. Washington encouraged Nanjing to sign this agreement in December 1946 because Americans were active in seeking rights for commercial airlines, and the USAAF tried to acquire landing fields in China during the postwar period. To Americans, this agreement was based completely on principles of equality and reciprocity, but the Chinese Communist propaganda violently attacked this agreement as proof of U.S. imperialistic aims, charging that Washington aimed at reducing China to a colonial dependency. The Sino-U.S. Air Transport Agreement as well as other treaties were condemned as the most unequal treaties. As Mao claimed, "the Sino-U.S. Air Transport Agreement," together with such treaties as the Sino-U.S. Treaty of Friendship, Commerce, and Navigation, and the Sino-U.S. Bilateral Agreement, were intended to make China into "a U.S. colony" and to "protect the independence and integrity of Chiang Kai-shek's reactionary state."[89] To the Communist leaders, the Nationalists were America's agents because these treaties were so commercially favorable to Americans. The Communists spoke of Chiang as America's instrument, comparing him to "traitors" like the Dowager Empress Cixi and Wang Jingwei. They claimed that Chiang had turned over the economic, political, and military control of China to the United States. Chiang was reduced to playing the role of the "running dog" of the U.S. imperialism. Clearly, the Communists exploited these treaties to promote the anti-U.S. nationalism in China.

In 1949, anti-American campaigns increased in intensity and in extremism. The Chinese Communist leaders' attitude toward the United States was unambiguous: America was becoming China's principal enemy in a sharply bipolar world, the enemy which was surrounding and threatening China just as Japan had done earlier. On January 14, in a statement on the current situation in China, Mao accused Chiang of inviting "the U.S. navy and air force to occupy China's territory and territorial sea and air" and of obtaining "from the U.S. government huge quantities of aircraft, tanks."[90] In August, Mao proclaimed,

> U.S. naval, ground and air force did participate in the war in China. . . . The U.S. air force controlled all of China's air space and took aerial photographs of all China's strategic areas for military maps. . . . Chennault's air fleet took an extensive part in the civil war. Besides transporting troops for Chiang Kai-shek, the U.S. air force bombed and sank the cruiser *Chungking* which had mutinied against the Kuomintang. All these were acts of direct participation in the war, although they fell short of an open declaration of war and were not large in scale, and although the principal method of U.S. aggression was the large-scale supply of money, munitions and advisers to help Chiang Kai-shek fight the civil war.[91]

In Mao's eyes, the United States had become the CCP's enemy because the U.S. military forces had involved themselves in the civil war against the Chinese Communists. This statement, conjured up by the Communist leaders, misled the Chinese people: the Chinese Air Force, not the USAAF, bombed and sank the Nationalists' cruiser *Chongqing* which deserted to the Communists;[92] the USAAF did not control all of China's air space; the USAAF had stopped transporting Nationalist troops after the civil war started.

By exploiting the Chinese fear of imperialism, the Communist leaders had succeeded in manipulating Chinese nationalism to discredit the United States through the attack on U.S. air aid to the Nationalists. Their attack on U.S. air aid had become an element in the anti-U.S. campaign with which the masses were mobilized during the civil war. The persistence of the Communists in attacking U.S. air aid gained wide public support and proved effective in increasing hatred of many Chinese for the U.S. government.[93]

In face of the Communists' expansion immediately after the end of the Sino-Japanese war, the Truman administration continued to provide Lend-Lease aircraft to the Nationalists and helped to airlift the Nationalist Army to Manchuria and North China to extend Chiang Kai-shek's influence there. At the same time, Americans were also interested in Chinese civilian aviation by transferring aircraft to the Chinese commercial airlines, creating a new aviation cooperation, and reaching an aviation agreement with the Nanjing government. When the civil war was renewed in June 1946, Washington imposed an arms embargo on Nanjing in an effort to remain neutral in a civil strife. Due to a shortage of aircraft spare parts and a delayed shipment of aircraft, plus the poor leadership of the CAF, low morale of the pilots, the Nationalist Air Force played a limited role in the civil war.

The U.S. air aid to the Nationalists was limited, but the CCP, which was receiving Russian assistance, condemned the U.S. government of providing air aid to the Nationalists. The CCP escalated their propaganda attack on Washington and promoted anti-American feeling among the ordinary people, which in part made the CCP-U.S. conflict unavoidable. The making of the CCP-U.S. conflict had several factors, but one of the factors was that Mao was inflexible in his view of the Truman administration after the Sino-Japanese war. The CCP leaders used the limited U.S. air aid to GMD for propaganda purpose to make the United States their enemy.

NOTES

1. Office of the Chief of Military History, Department of the Army, "The United States Army in World War II, Statistics: Lend-Lease," p. 33, AFHRA. But according to figures of the Tenth A.F. Statistical Control Unit, the U.S. government had provided China with 2,152 aircraft by September of 1945. See Memorandum, 39th Statistical control unit, Headquarters, India-Burma Theater to Ad hoc Lend-Lease committee, September 20, 1945. This figure probably includes the aircraft delivered to India too.

2. Memorandum for General Marshall, August 27, 1945, 145.81-182; Stratemeyer to Arnold, October 9, 1945, 860.161-1, Stratemeyer's Personal File, AFHRA.

3. Memorandum, Headquarters, U.S. Force, China Theater, December 4, 1945, AC/AS, A-4, International Branch, Flying and Navigation Equipment, Records of the United States Army Air Force, RA 18, NA.

4. Arthur Young, *China and Her Helping Hands, 1937–1945*, p. 351.

5. G-4, War Department to General Arnold, September 5, 1945, AC/AS, A-4, Lend-Lease, China, RG 18, NA.

6. Department of State, *FRUS, 1945, The Far East,* 7: 559.

7. Lieutenant General Ira C. Eaker to Stratemeyer, October 25, 1945, 860.161, Stratemeyer's Correspondence File, AFA; War department to Wedemeyer, October 19, 1945, AC/AS, A-4, International Branch, Country File, 1941–1947, China, RG18, NA.

8. Brigadier General McConnell to Commanding General, Army Air Force, April 8, 1947, 861.011-2, AFHRA.

9. *New York Times*, September 6, 1945, p. 3.

10. Tenth A.F. Monthly Statistical Review, 168.7017-5, General James F. Whisenand Collection, AFHRA.

11. U.S. Department of State, *United States Relations with China, with Special Reference to the Period, 1944–1949* (Washington, D.C., 1949), p. 338. Hereafter cited as *USRC, 1944–1949.*

12. Wedemeyer to War Department, September 3, 1945, Joint Chiefs of Staff File, 452, China, 4-3-45, Sec. 2, RG 218, NA.

13. Secretary of State Byrnes to Ambassador Patrick Hurley, September 11, 1945, Joint Chiefs of Staff File, 452, China, 4-3-45, Sec. 2; Memorandum by Robert Patterson, Assistant Secretary of War, September 12, 1945, RG 218, NA.

14. Joint Chiefs of Staff Paper 1330/4, October 5, 1945, Joint Chiefs of Staff File, 452, China, 4-3-45, Sec. 2, ibid.

15. *FRUS, 1945, The Far East,* 7: 591, 597, 583–89.

16. Department of State, *USRC, 1944–1949,* pp. 339–40.

17. Special Orders 27 and 28, Headquarters, Army Air Forces, China Theater, February 22 & 27, 1946, 861.04-1, AFA; General Order 42, Headquarters, Army Forces, China Theater, March 29, 1946, ibid.; JCS 1330/10, 861.011-1, AFHRA.

18. Department of State, *USRC, 1944–1949,* p. 343.

19. "History, Liaison Team, Second Air Transport Group, February 1, 1946–June 30, 1946," 861.01, February–June 1946, AFHRA.

20. President Harry S. Truman, "United States Policy toward China," *Department of State Bulletin* (December 16, 1945), pp. 945–6.

21. Arnold to Marshall, December 11, 1945, 145.81-182, AFHRA.

22. Ernest May, *The Truman Administration and China, 1945–1949* (Philadelphia: Lippincott, 1975), p. 82.

23. Acheson to the Foreign Liquidation Commission, August 13, 1946, *FRUS, The Far East: China*, 10: 783.

24. Dean Acheson, *Present at the Creation: My Years in the State Department* (New York: W. W. Norton, 1969), p. 207; John F. Melby, *The Mandate of Heaven* (Toronto: University of Toronto Press, 1969), p. 145.

25. "History of A-3 Section, August 1946," 861.01, August 1946, AFHRA.

26. Brigadier General Charles E. Thomas, Jr. to Lieutenant General C. J. Chow, October 30, 1947, "History of the Command Section, October 1947," 861.01, October 1947, AFHRA.

27. *FRUS, 1945*, vol. VII, *The Far East,* 7: 629.

28. Memorandum, Major General, C. C. Chauncey, Acting Chief of Staff to Admiral Frank Leahy, January 12, 1946, 145.81-182, AFHRA; Memorandum, Arnold to Admiral Leahy, December 5, 1945, 350.2, Training Chinese Student, 1945, RG 18, NA.

29. Department of State, *USRC, 1944–1949*, pp. 810–11.

30. Memorandum for Assistant Chief of Air Staff, A-4, February 18, 1946, AC/AS, A-4, International Branch, Subject File, 1941–1947, RG 18, NA. See also Herbert C. Prouty, "China in the Sky," *Army Air Force Review* (September 1946), pp. 28–31.

31. Report to Director, Air Division, June 4, 1947, 861.01-1, AFHRA.

32. "History of the Advisory Team, 10th Air Transport Group, August 1948," 861.01, October 1948, AFHRA; "History of the Advisory Team, 20th Air Transport Group, August 1948," ibid.; "History of Liaison Team, 5th Fighter Group, Nanking, August 1948," 861.01, August 1948, History of Liaison Team, 5th Fighter Group, Nanking, October 1948," 861.01, October 1948, ibid.

33. Statement of Training Charges, Chief, Control Section, Supply Division, Headquarters, Air Material Command, Wright-Patterson AFB, Ohio, January 27, 1948, AC/AS, A-4, International Branch, Classified Central File, 1948, RG 18, NA; Chinese Training, Statistical Worksheets, AC/AS, A-4, ibid.

34. *United States Policy in the Far East: Selected Executive Session Hearings of the Committee, 1943–1950*, vol. 7 (Washington D.C., 1976), pp. 241–68.

35. Whiting Willauer, who helped Claire Chennault to establish the American Volunteer Group in 1942, worked for Chennault's 14th Air Force during the Sino-Japanese war, mainly helping to solving the logistic problems. See Willauer Biography, Whiting Willauer Papers, Firestone Library, Princeton University.

36. Ambassador in China (Stuart) to Secretary of State, April 17, 1948, *FRUS, 1948*, 8: 289–290.

37. Secretary of state (Marshall) to ambassador in China (Stuart), May 25, 1948, *FRUS 1948*, 8: 289–290; Ambassador in China (Stuart) to the secretary of state, November 18, 1948, ibid., p. 292; Consul general at Tsingtao (Turner) to Secretary of State, November 24, 1948, ibid., p. 293.

38. Consul General at Shanghai (Cabot) to the secretary of state, 8 December 1948, ibid., p. 296.

39. Walter Millis, ed., *The Forrestal Diaries* (New York: Viking Press, 1951), pp. 533–34.

40. Memorandum by the Director of the Office of Far Eastern Affairs (Butterworth), November 29, 1948, *FRUS, 1948*, 8: 294–5.

41. For the details of the Chennault plan, see William M. Leary and William Stueck, "The Chennault Plan to Save China: U.S. Containment in Asia and the Origins of the CIA's Aerial Empire, 1949-1950," *Diplomatic History* 8 (1984): 349–64.

42. "China Aid Act of 1948," in *China: U. S. Policy Since 1945* (Congressional Quarterly, 1980).

43. Department of State, *USRC, 1944–1949*, pp. 945–46, 952–53.

44. *Congressional Record*, 81st Cong., 1st Sess., May 3, 1949, pp. 5480–84; U.S. Congress, House Committee on International Relations, *United States Policy in the Far East* vol. 8, pt. 2, pp. 285–326.

45. Webb to Butterworth and Butterworth to Webb, May 10, 1949; Webb to Stuart, May 25, 1949 and Stuart to Webb, May 30, 1949, *FRUS, 1949,* 9: 517–525

46. U.S. Congress, House Committee on International Relations, *United States Policy in the Far East* vol. 8, pt. 2, p. 260.

47. For a study of the Chinese civil aviation in the 1940s, see William Leary, *The Dragon's Wing*, pp. 193–224.

48. Young memorandum, "Cost of Aviation in China," May 14, 1946, Arthur Young Papers.

49. "Central Air Transport Corp. Grows Rapidly," *China Economist*, 2 (August, 1948), p. 203; *World Aviation Annual, 1948* (Washington, D.C., 1948), pp. 507–8.

50. For the details of Whiting Willauer's role in the establishment of CAT, see William Leary, "Portrait of a Cold War Warrior: Whiting Willauer and Civil Air Transport," *Modern Asia Studies*, 5 (1971): 373–88.

51. For the details of the creation of CAT, see Leary, *Perilous Missions,* pp. 3–22. See also *Hanghong shiliao* [Historical documents on aviation] (Taipei: Guoshiguan, 1991), pp. 598–612.

52. *FRUS, 1946*, 10: 1228–30.

53. Department of State, *Treaties and Other International Acts Series,* No. 1609 or 61 Stat. (pt. 3) 2799. For Department press release concerning the agreement, see Department of State, *Bulletin*, January 5, 1947, p. 30.

54. *FRUS, 1947*, 7: 1007–12.

55. *FRUS, 1948*, 8: 801–2.

56. *FRUS, 1948*, 8: 803–4.

57. Shushan Li, *Chungkuo hungchu zhanshih* [The fighting history of the Chinese Air Force] (Guangzhou: Guangdong renmin chubanshe, 1996), p. 49.

58. The Counselor of Embassy in China (Smyth) to the Secretary of State, July 14, 1946, *FRUS, 1946*, 9: 1355–6.

59. Lionel Max Chassin, *The Communist Conquest of China: A History of the Civil War, 1945–1949*, (Cambridge, Mass.: Harvard University Press, 1965), p. 118, p. 167.

60. The consul at Mukden (Ward) to the secretary of state, October 31, 1948, *FRUS 1948*, 7: 530.

61. Jin Yao, *Zhongguo hanghong shi,* pp. 130–133.

62. Liu, *A Military History of Modern China*, p. 262.

63. Department of State, *USRC, 1944–1949*, p. 344.

64. Jack Belden, *China Shakes the World* (New York: Harper & Brothers, 1949), p. 455.

65. Department of State, *USRC, 1944–1949*, p. 343.

66. Chassin, *The Communist Conquest of China,* p. 256.

67. *Dangdai zhongguo hongjun* [China today: Air Force] (Beijing: Zhongguo shehui kexue chubanshe, 1989), pp. 15–17.

68. Department of State, *USRC, 1944–1949*, p. 181.

69. Consul at Mukden (Ward) to Secretary of State, May 19, 1948, *FRUS, 1948*, 7: 246.

70. *Hanghong shiliao* pp. 658–683; *CAT Bulletin,* 2 (March 1, 1949).

71. Willauer Taped Memoir, December 1, 1960, Willauer Papers.

72. Leary, *Perilous Missions*, p. 51.

73. See O. B. Brisow & B. T. Koloskov, *Soviet-Chinese Relations, 1945–1970* (Bloomington: Indiana University Press, 1975), pp. 52–58.

74. See M. I. Sladkovsky, ed., *Leninism and Modern China's Problem* (Moscow: Progress, 1972), pp. 26–27.

75. Chassin, *The Communist Conquest of China*, pp. 64–66.

76. Mr. Walter S. Robertson to General Marshall, April 30, 1946, *FRUS, 1946*, 9: 805–6.

77. Minutes of meeting between General Marshall and General Chou En-lai, June 3, 1946, *FRUS, 1946*, 9: 952–53.

78. See the CCP Central Committee to Zhou Enlai, May 28, 1946, *Zhonggong zhongyang nanjingju: Zhonggong lishiziliao* [The Nanjing Bureau of the CCP Central Committee: The CCP history documents] (Beijing: CCP Historical Press, 1990).

79. Minutes of Meeting between General Marshall and General Chou En-lai at No. 5 Ning Hai Road, Nanking, 3 August 1946, *FRUS, 1946*, 9: 1443–8.

80. General Chou En-lai to General Marshall, August 5, 1946, *FRUS, 1946*, 9: 1450–1.

81. Department of State, *USRC, 1944–1949*, pp. 180, 654–56, 1048.

82. Memorandum by General Chou En-lai to General Marshall, September 30,1946, *FRUS, 1946*, 10: 258–9.

83. Chassin, *The Communist Conquest of China*, p. 162.

84. See "Manifesto of the Chinese People's Liberation Army," October, 1947, *Selected Works of Mao Tse-tung* (Peking: Foreign Languages Press, 1961), 4: 147–152.

85. Ambassador in China (Stuart) to the secretary of state, October 3, 1946, *FRUS, 1946*, 10: 278–9.

86. For the details of the CCP's new attitude toward the United States, see *Zhou Enlai Nianpu* [Zhou Enlai's chronicle] (Beijng: Renmin chubanshe, 1989), edited by Zhonggong zhongyang wenxien yanjiushi [The Department of Archives and Research, the Central Committee of the CCP], pp. 794–5. Historians differ on when the CCP leaders started to shape an anti-U.S. policy. Some believe that Mao did not adopt a radical policy toward the U.S. government until after an enlarged Politburo meeting in January 1949. For this statement, see Hao Yufan and Zhai Zhihai, "China's Decision to Enter the Korean War: History Revisited," *China Quarterly,* vol. 121 (March 1990). But other scholars argue that Mao started to develop a hostile policy toward Americans in early November 1948. For this argument, see Michael Sheng, "Chinese Communist Policy toward the United States and the Myth of the 'Lost Chance,' 1948–1950." The present author believes that the latter's argument is more persuasive.

87. Shaoqi Liu, "On Internationalism and Nationalism,"*Renmin ribao* [People's Daily], November 7, 1948, p. 1.

88. Consul general at Peiping (Clubb) to the secretary of state, 27 November 1948, *FRUS, 1948*, 8: 293–294.

89. *Selected Works of Mao Tse-tung* (Peking: Foreign Languages Press, 1961), 4: 310–313.

90. See "Statement on the Present Situation by Mao Tse-tung, Chairman of the Central Committee of the Communist Party of China, January14, 1949," *Selected Works of Mao Tse-tung* (Peking: Foreign Language Press, 1961), vol. 4.

91. "Farewell, Leighton Stuart!" August 18, 1949, ibid.

92. For the details of the bombing of the cruiser *Chongqing*, see Changying Jiang, *Zhongguo hanghong shi*, p. 112.

93. For the study of the anti-American demonstrations and protests from 1945 to 1949, see Suzanne Pepper, *The Civil War in China: The Political Struggle, 1945–1949* (Berkeley: University of California Press, 1980), pp. 52–58, 72–78, 115–116, and Jeffrey Wasserstrom, *Students Protest in Twentieth-Century China: The View from Shanghai* (Stanford, Calif.: Stanford University Press, 1991).

Conclusion

From 1929 to 1936, China went through a difficult period when local warlords were fighting against the Nanjing government. The Nationalist government was not stable since much of the huge country of China was still ruled by warlords who maintained their own bases and armies and established their own taxation systems. They tried several times to organize another "National government" to challenge Nanjing's authority. These challenges indicate a significant characteristic of this period—political disunity and instability. When Chiang attempted to expand the authority of his National government into their domains, they resisted by force. Yet these rebellions were not the only, or the most serious, confronting Nanjing. The Chinese Communist Party initiated a long struggle against the Nationalist rule, with guerrilla wars in the countryside and underground sabotage in the cities. Their bases, only a few hundred miles from Nanjing, continued to expand rapidly and attract many peasant converts. Chiang took the CCP as the major threat to his power.

In the face of these threats, the Nationalist government determined to suppress the rebellions by sending in the Nationalist Army. The prospects for the Central government to dispatch military forces to destroy the warlords seemed dim, in part because railways and roads were scarce and China lacked other modern means of transportation. It was also difficult for the Nationalist Army to eliminate the Communist bases because most of them were situated in inaccessible and easily defended mountainous areas. The advantages of having an extensive and efficient air force became increasingly apparent. A modern air force would help the Nationalist Army to destroy both the local warlords and the Communists and to extend the Central government's influence to more distant provinces.

Meanwhile, beginning with the Manchurian Incident of 1931, the Japanese continued to expand their territorial control into North China. The battles between the Chinese and Japanese air forces over Shanghai in February 1932 brought the Chinese government to appreciate, as it had not earlier, the importance of military aviation, not only in unifying China, but also in defending the country against external aggressors. Therefore, the Chinese government grew increasingly "air minded," and attempted to arouse the interest of the Chinese people in developing Nationalist military aviation. The Nationalist government was determined to establish a powerful air force in China and formulated a comprehensive scheme to that end. In the years that followed, Nanjing built military airdromes, purchased airplanes abroad, and negotiated with foreign businessmen to establish aviation schools and airplane factories in China. The United States became one of the main sources for military airplanes. The Chinese needed American assistance; American businessmen were highly attracted to business prospects in China.

American businessmen played a very active role in encouraging the Nationalist government to arrange for American assistance in the development of all of the elements of modern military aviation. Between 1932 and 1936, the Nationalist government stepped up activity on all fronts of military aviation development. In 1932, Colonel John Jouett led an air mission to China to organize a modern aviation school at Hangzhou to produce competent pilots for the Chinese Air Force. At the same time, William Pawley helped to establish the first modern aircraft factory in China for the Nationalist government. American influence was the strongest foreign influence in China's military aviation. Nationalist wings were almost completely American in equipment and training. Nationalist military aviation had reached a turning point, and China had come of age in military aviation.

The expansion of the Nationalist Air Force relative to local air forces changed the balance of power in China, which, in turn, greatly affected the political situation. With strength in the air, Chiang Kai-shek's military power gained a preponderance over his internal rivals. His strong air force assured his retention of uncontrolled or poorly held regions of China by helping the Nationalist Army in suppressing the Fujian uprising swiftly, quelling the Communists in South China, and putting down rebellion in Guangdong and Guangxi. For the first time since 1912, most of China was effectively unified. The Nationalist Air Force played a crucial part in ending the prolonged civil warfare and in unifying China into a single political organization. Without such an efficient air force, the civil wars would have lasted longer, and most of China would not have been unified before the outbreak of the full-scale Sino-Japanese war.

The expansion of the Nationalist Air Force also had an impact on U.S.-Japan relations. The Japanese government and military had been concerned over American assistance to the Chinese Air Force in the years prior to the reopening of the hostilities in 1937 because the Chinese Air Force assisted by Americans played an important part in helping the Nationalist Army unify China from 1932 to 1936; a unified and stable China could inhibit the expansion of Japan in China and East Asia. Furthermore, Japan's Monroe Doctrine claimed East Asia as Japan's sphere of influence and asserted that China should be in a state of tutelage to Japan. Export of American planes to China extended American military influence, which challenged Japan's New Order in East Asia. Japan protested to the Washington government, hoping to discourage American air assistance to China. American air aid to China constituted one point of the conflicts between Washington and Tokyo and continued to grow in the late 1930s. For Americans, the central issue was that Japan attempted to close the door of China to Americans as well as to other Westerners while the United States and Great Britain advocated the Open Door

Policy. Washington denied any involvement in air assistance to China, stating that American air assistance was for commercial interests only, but in face of the Japanese challenge, the U.S. government tried to appease Japan by limiting American assistance to the Chinese Air Force.

After the outbreak of the Sino-Japanese war in 1937, despite the Japanese numerical superiority in airplanes, the Chinese Air Force successfully resisted the Japanese air raids in the first phase of air operations from July through December 1937 because of the superiority of the American airplanes and the skill of the Chinese pilots trained by the American instructors. As a result, Japan was determined to discourage American air assistance to China. Under Japan's pressure as well as domestic isolationists' protest, the Roosevelt administration imposed a partial arms embargo against both China and Japan and prohibited American pilots from flying for China. The new U.S. policy toward East Asia weakened the Chinese war effort, especially the power of the Chinese Air Force. The new American policy not only appeased the Japanese but also benefited the Japanese military force, which helped the Japanese gain sufficient war materials from the United States. Because the U.S. government refused to provide air aid to China, the Chinese turned to the Russians, and the Soviet Union began to provide air assistance to the Nationalist government at the end of 1937. The Soviet government's air aid increased the CAF's ability to resist the Japanese in the air and boosted the morale of the Chinese between 1938 and 1940.

Although the Soviets dominated the Chinese Air Force during that time, Washington encouraged the sale of military airplanes to China. American instructors and engineers were also active in South China in helping the Chinese manage their aircraft factories and supervise their aviation schools. The attitudes and policies of the American government inspired and strengthened the Chinese in their resistance to Japan, and, as a consequence, contributed to the growing tension between the Japanese and American governments. The Japanese government regarded the large numbers of American airplanes exported to China as a threat because the Chinese pilots were flying superior American airplanes in their resistance to Japan's air assault and in their attack on Japan's homeland. Meanwhile, the State Department advocated a moral embargo on all types of aircraft to Japan, which resulted in a practically complete halt to American aircraft exports to Japan and escalated the tension between Japan and the United States.

When the situation in the European theater forced the Soviet Union to withdraw its aviators and to stop delivery of aircraft to China, Chongqing began to petition American officials for support. In Washington, some high-ranking officials, who sympathized with China, advocated air assistance to the Chinese. Therefore, Roosevelt decided in 1941 to provide large numbers of the latest airplanes to China as part of its Lend-Lease aid and to allow American pilots to serve in the Chinese Air Force. The Japanese, therefore, claimed that the American government's air support of China meant that the United States had involved itself in the war against Japan by the end of 1941. The Pearl Harbor attack had several causes. The oil embargo of July 26, 1941, was the final one, but the air assistance given by the American government openly after 1940 certainly helped influence the Japanese decision.

American air assistance to China not only had a great effect on U.S.-Japan relations, but also on the Anglo-American relationships before the Sino-Japanese war. A stable, strong China was the common interest of both the United States and Britain in the 1930s. To increase stability and to prevent civil strife in China, the American and British governments attempted to cooperate with the Nationalist government in its efforts to maintain effective control of importation to China of airplanes, arms, ammunition, weapons, and other war materials. To help the British businessmen to develop new markets in China, Britain, in 1926, was the first country to remove commercial airplanes from articles listed in the China arms embargo, followed by the United States, which lifted the arms embargo on commercial aircraft in 1928. The revocation of the ban on the export of commercial planes by both the American and British governments opened a loophole in the China arms embargo implemented by the major powers because Chinese military authorities converted many of the imported aircraft directly and immediately for military purposes and employed them in fighting. The lifting of the embargo on commercial aircraft, leaving both Britain and the United States open to charges of violating the embargo, seemed to wreck the China Arms Embargo Agreement and weaken the enforcement of the China arms embargo.

After the Chinese arms embargo was lifted in 1929, in an effort to prevent local military authorities from obtaining Western arms, especially airplanes, both the United States and Great Britain as well as other powers favored stricter control of arms sales to China in order to discourage civil war and to promote stability in China. The Anglo-American cooperation policy with regard to arms sales to China, one of a number of important instances of cooperation between those two governments in East Asia politics in the 1930s, was decisive in securing uniformity on the part of other powers to prohibit arms sales to China, except those approved by the Nanjing government from 1930 to 1936. American and British aircraft businessmen were active in South China because the war and conflicts between the Nanjing and Canton governments offered those businessmen welcome opportunities. Vying for the aircraft market in South China at the time of the Great Depression, American and British aircraft dealers, lured by lucrative profits and encouraged by the local warlords, were competing for aircraft orders there. Their competition inevitably forced their governments to adopt their own arms policies toward China.

American businessmen profited from sales of military airplanes to the Nanjing government, and the British businessmen did not want to lose that potential aircraft market. They tried to open the market by delivering airplanes through Hong Kong to the local governments in South China without the permission of the Nanjing government. The British government provided assistance to the British businessmen much as did the U.S. Commerce Department. As a result, American businessmen were dissatisfied with what the British were doing and complained about the rigid control by the State Department over the export of American airplanes to China. For fear of placing its nationals at a disadvantage in the aircraft sales to South China, neither the State Department nor the Foreign Office was willing to control tightly aircraft sales to warlords in South China when the other did not comply with Chinese requests. Anglo-American cooperation was difficult at best because both governments necessarily served their own commercial interests, especially the interests of the aircraft industry endangered by the ravages of the Great Depression. Their cooperation suffered when their divergent interests conflicted.

To solve the problem, the State Department discussed the issue with the British Foreign Office. The arguments between the State Department and the Foreign Office over aircraft sales to South China gave rise to suspicions between Washington and London and disclosed the accumulated grievances that existed between these two countries.

By 1936 a large number of airplanes, mainly American and British, had been delivered to South China without Nanjing's permission. With strength in the air, the militarists controlling the southern provinces revolted, and the civil war started again in 1936. The major powers' control of arms sales to China proved unworkable. Anglo-American aircraft competition in China contributed to the failure of the international cooperation with the Chinese government on regulations of arms sales to China. Anglo-American aircraft competition was one of the examples of the Anglo-American competition that prevailed over their cooperation in China. Anglo-American economic competition had an impact on the countries' diplomatic cooperation with the Chinese government. This cooperation was important in the face of the increasing Japanese military expansion in China in the 1930s, especially after the Manchurian incident of 1931.

American air assistance to China between 1929 and 1936 also revealed internal conflicts within the China arms policy of the U.S. government. The arms policy was an important part of the general China policy of the American government during the 1930s. Since China was involved in civil wars, the Department of State exercised strict control over the export to China of arms and munitions of war, including military planes, to prevent domestic violence in China and to strengthen and stabilize the Nationalist government. Without doubt, the Department of State tried to base its China arms policy upon humanitarian consideration.

Other countries, however, did not exercise strict control over the sale of arms and munitions to China. As a consequence, the American government had placed American exporters at a disadvantage. Under the circumstances, the Department of State had to include economic factors in its China arms policy because it did not want American businessmen to lose the potential aircraft market in China. It was difficult for the State Department to formulate its China arms policy. If it regulated the sale of arms and munitions to China rigidly, American economic interests would be hurt, and American businessmen would complain. If the sale policy were relaxed, the Nanjing government would protest because arms and military airplanes would reach local warlords. As a result, the State Department modified its China arms policy from time to time whenever the conditions changed. Sometimes the State Department, influenced more strongly by political motives, weighted its policy in favor of strengthening the Nanjing government; a strong stable Chinese government would help to promote American business interests in China. The final objective of modifying the China arms policy was in support of American economic interests too. In general, the State Department based its arms policy more on realistic business principles than on humanitarian principles; economic considerations won priority over humanitarian ends.

From 1929 to 1936, there was a conflict between the State Department's policy and those of the Commerce and War Departments on the China arms trade. From the standpoint of commercial trade, the Commerce Department promoted the sale of airplanes and air aid to the Chinese because of the benefits for national economic interests, especially during the Great Depression. The Commerce Department most actively aided the Nationalists in securing the services of Americans in an effort to enhance the sale of American aircraft to China. From the standpoint of military considerations, the Department of War favored sales of military airplanes to China also, because the aeronautical industry in the United States had to maintain a surplus productive capacity in peacetime so that the aircraft factories could be equipped quickly for reinforcement and expansion of the U.S. air forces during a war. Therefore, the War Department pursued a policy of encouraging sales of military aircraft to China and other countries in peacetime. But the War Department did not favor sending American air missions to China because direct American involvement with the Nationalist military aviation might promote hostilities between Japan and the United States.

Although the State Department deferred to economic interests in formulating its China arms policy, under certain circumstances it had to consider political and diplomatic consequences as well. The Department did not want Americans involved too heavily in Chinese military aviation because those activities could adversely affect relations between Washington and Tokyo. In addition, sales of planes to South China could also impact relations between Nanjing and Washington. The differing attitudes and policies of the Commerce, War, and State Departments concerning the sale of military airplanes and air assistance to China resulted in differing, sometimes contradictory, actions by the departments. At times, therefore, there was internal conflict within the foreign policies of the U.S. government and a lack of a unified viewpoint by which the various departments' policies were governed.

After the Pearl Harbor attack, America's objective in East Asia during the war was to force Japan to surrender. To realize the goal, the U.S. government strove to keep China in the war against Japan and to mobilize China's military and economic strength to fight the Japanese, using economic and military aid as a major lever. Washington established the China-Burma-India theater and began military cooperation with the Chinese fighting Japan after the eruption of the Pacific War. During the war from 1942 to 1945, the United States rendered air assistance to China by establishing an American air force in China, delivering airplanes to the Chinese Air Force, training the Chinese pilots, and transporting war materials by air to the Chinese. This American air support played a significant role in defeating Japan.

However, U.S. air support, an important part of U.S. aid, was minimal in comparison to U.S. air aid to other countries fighting Nazism. The limited American air aid, becoming one of the conflict points between China and American leaders, created Sino-U.S. diplomatic tension and a crisis of Sino-U.S. military cooperation. Different military goals and strategies between the United States and China, Stilwell's emphasis on the role of ground force and contempt for air power, the construction of the Ledo Road, and the lack of understanding of Chinese military politics, all contributed to the tension. Chiang, a man determined to survive and to take China in a particular direction of his own choosing, resisted Stilwell and other American military leaders when their preoccupations and plans threatened his position. However, if Stilwell and other American military leaders had considered adequately Chinaese political factors and better understood the Chinese culture, including Chinese politics, Chinese military, and personal characteristics of the Chinese leaders, then some differences on air support for China could have been avoided or reduced. The success of the Wedemeyer mission proved this in the last year of the war.

After the Second World War, Chiang Kai-shek had to rely entirely on the United States as his mainstay, using U.S. aid to preserve his power and to maintain Nationalist rule in China. The Chinese Air Force, one of his important tools to consolidate his power, had to depend on American pilots and aircraft. At Chiang's request, the Truman administration continued air aid to the Nationalists immediately after the war in order to help the Nationalists to complete expeditiously the surrender of all Japanese forces in China, to maintain American influence in the Chinese Air Force, and to prevent the Nationalists from seeking Russian air aid again. Huge quantities of aircraft and equipment were delivered and sold to the GMD government. Thousands of Chinese personnel received American air training. The U.S. Army Air Force airlift of a large number of the Nationalist ground forces from West China to East and North China gave the Nationalists military advantages over the Communists before outbreak of the full civil war. American advisers helped the Chinese to improve the area of logistics and maintenance, rendered assistance to the air units which were flying combat sorties, gave advice to transport squadrons in training and operation, and played a role in obtaining spare parts and in sustaining adequate stock levels. The Nationalist Air Force, however, did not play a critical role in the civil war. After the Pacific War, Americans also tried to increase American influence in Chinese commercial aviation. They sold aircraft to the Chinese airlines, helped to establish a new airline in China, reached the Sino-American Air Transport Agreement with the Nationalist government, and attempted to send a civil air mission to China. American influence in commercial aviation in China was limited during the postwar period.

In spite of the limited role of the CAF, the Communists turned the ineffective U.S. air aid to the Nationalists into an effective political weapon against their enemies, complaining and protesting against this aid to the Nationalists. They, however, kept their anti-American propaganda in a low-key fashion while the Marshall mission was active. Washington reduced its air aid to the Nationalists when the civil war started, but the Communist leaders still exploited it to stir up anti-Americanism among the Chinese populace. Their attack on American air aid became increasingly abusive and increasingly unrelated to factual reality in the following years. It was misleading, but proved successful in manipulating the Chinese nationalism to make the United States the enemy of the Chinese. Tensions between the U.S. government and the Chinese Communists increased. Certainly, the CCP-U.S. hostility had several causes, but the Communists' attack on American air aid to the Nationalists was one of the factors responsible for creating it in and after 1949.

The development of Chinese military aviation took years of experimentation and learning since China was trying to adapt quickly to modern military ways. Because of its potential for a great success in modern warfare, military aviation acted as a bridge between the East and West and a stimulant for progress. To unify China by force and to resist Japanese aggression, the Chinese needed Western warplanes, air advisers, aviation training, and aeronautical technology to develop its air force. From 1929 to 1949, Americans helped China establish and develop its modern air force. Without the U.S. government's strong support, private American businessmen were successful in helping the Nationalists to create an air force before the Sino-Japanese war. Although the Nationalist Air Force, to some extent, played an important role in the national unification movement and in fighting the Japanese in the first year of the Sino-Japanese war, the CAF was not powerful and efficient in comparison with other countries' air forces largely because of the continuing power struggle among the Nationalist leaders and the government's financial problems.

After 1940, the U.S. government began to commit itself to the expansion of the CAF in order to harass the Japanese troops in China and to bomb Japanese cities. Under the Lend-Lease, Washington provided air support for China by establishing an American air force in China, assisting the CAF, and airlifting supplies to the Nationalists. Nonetheless, the Chinese Air Force did not play a significant role in the war in comparison to the American air force in China mainly because of limited U.S. government air aid.

After the Sino-Japanese war, the U.S. government continued to provide air aid to the Nationalists by transferring a large number of aircraft to China, airlifting the Nationalist troops to Manchuria, and training the Chinese pilots. But during the civil war, the poor leadership and pilots' low morale accounted for the CAF playing a limited role in fighting the Communists who did not have any air force.

Despite 20 years of American assistance, the Chinese Air Force never became a genuinely effective fighting force. Limited American air aid was one of the reasons, but the most important reason was that American aeronautic technology and advice could not bring about a fundamental change in the backward Chinese military deeply rooted in their own politics, society, and culture. Only the Chinese themselves could find the solutions to their own problems and modernize their military forces.

Bibliography

UNPUBLISHED DOCUMENTS

National Archives, Washington, D.C.

Record Group 18, Records of the Army Air Force
Record Group 38, Records of the Office of the Chief of Naval Operations
Record Group 46, Records of the United States Senate
Record Group 59, General Records of the Department of State
Record Group 94, Records of the Adjutant General's Office
Record Group 107, Records of the Office of the Secretary of War
Record Group 151, Records of the Bureau of Foreign and Domestic Commerce, Department of Commerce.
Record Group 160, Records of Headquarters Armed Service Forces
Record Group 165, Records of the War Department General and Special Staffs
Record Group 218, Records of the U.S. Joint Chiefs of Staff
Record Group 225, Records of Joint Army and Navy Boards and Committees
Record Group 407, Records of the Office of the United States Army

The Federal Record Center, Suitland, Md.

Record Group 84, Records of the Foreign Service Posts of the Department of State
Record Group 237, Records of the Civil Aeronautic Administration

Air Force Historical Research Agency, Maxwell Air Force Base, Alabama

Administrative Records, American Volunteer Group
Correspondence, China Air Service Command
Correspondence Files, Chinese-American Composite Wing
General George E. Stratemeyer's Personal File

General James F. Whisenand Collection
Operational Correspondence, Fourteenth Air Force Policy File, 1940–1945, Fourteenth Air Force
Records on the Chinese Air Force, Army Air Force
Records on Operations in China, Army Air Force

Public Record Office, London

Record Class FO 405, China Confidential Print, Foreign Office Archives
Record Class FO 228, China: Correspondence: Series I 1834 to 1930, Embassy and Consular Archives
Record Class FO 676, China: Correspondence: Series II 1875 to 1956, Embassy and Consular Archives
Record Class FO 371, General Correspondence, Foreign Office Archives

Second Historical Archives of China, Nanjing

Record Group 787, File Number 581, Records of the National Commission on Aviation Affairs
Record Group 787, File Number 645, Records of the First Aircraft Factory

Archives of Guangdong Province, China

Records of the Guangdong Air Force

PUBLISHED GOVERNMENT DOCUMENTS

China

Hanghong shiliao [Historical documents on aviation]. Taipei: Guoshiguan, 1991.
Selected Works of Mao Tse-tung. Peking: Foreign Languages Press, 1961.
Xianzongtong Jianggong shuxin yanlun congji [The collection of President Chiang's speeches and writings]. Taipei: Zhongyang danshi weiyuanhui, 1963.
Zhonggong zhongyang nanjingju: Zhonggong lishiziliao [The Nanjing bureau of the CCP central committee: The CCP history documents]. Beijing: CCP Historical Press, 1990.
Zhongguo shibian: lujunzouzhanshi [China incident: The combat history of the Japanese Army]. Department of History, Defense Institute of the Defense Minister of Japan, translated by Ji Fulin. Beijing: Zhonghua shuju, 1981.
Zhonghua minguo zhongyao shiliao chupian: Duiri kangzhan shiqi [The first edition of the important historical sources of the Republic of China: The period of resisting Japan]. Taipei: Zhongguo guomindang zhongyang weiyuanhui dangshi weiyuanhui, 1976.
Zhonghua minguoshi ziliao conggao [The historical records of the Republic of China]. Beijing: Zhonghua shuju, 1983.
Zhongri waijiao shiliao congbian [A series of records of the Sino-Japan diplomatic history]. Taipei: Zhonghua minguo waijiaoshi xuehui, 1965.
Zhou Enlai Nianpu [Zhou Enlai's chronicle]. Beijing: Renmin chubanshe, 1989, edited by Zhonggong zhongyan wenxien yanjiushi [The Department of Archives and Research, the Central Committee of the CCP].

Japan

Air Operations in the China Theater, 1937–1945. Japanese Monograph No. 76. Prepared by Headquarters, USAFFE and Eighth U.S. Army, 1956.

Ike, Nobutaka, trans. and ed. *Japan's Decision for War: Records of the 1941 Policy Conferences*. Stanford, Calif.: Stanford University Press, 1967.

United Kingdom

Woodward, E. L., and Rohan Butler, eds. *Documents on British Foreign Policy 1919-1939,* Third Series. 10 vols. London: Her Majesty's Stationery Office, 1949–1971.

United Nations

United Nations. *League of Nations Treaty Series, 1921–1922.* New York, 1946.

United States

United States Policy in the Far East: Selected Executive Session Hearings of the Committee, 1943–1950, vol. 7. Washington, D.C., 1976.

U.S. Congress, Senate. Special Committee on Investigation of the Munitions Industry. *Hearings before the Special Committee Investigating the Munitions Industry.* 40 parts. 73d Congress, 2d sess., 1934.

U.S. Congress, Senate. Special Committee on Investigation of the Munitions Industry. *Munitions Industry.* 7 parts. Senate Report 944, 74th Congress, 1st sess., 1935.

U.S Department of State. *Foreign Relations of the United States, 1929–1949.* Washington, D.C.: Government Printing Office, 1946–1970.

U.S. Department of State. *Foreign Relations of the United States, Japan:1931–1941.* Washington, D.C.: Government Printing Office, 1946–1970.

U.S. Department of State. *Press Releases, 1929–1940.* Washington, D.C.: Government Printing Office, 1946–1970.

U.S. Department of State. *Treaties and Other International Acts Series.* Washington, D.C.: Government Printing Office, 1946.

U.S. Department of State. *United States Relations with China, with Special Reference to the Period, 1944–1949.* Washington, D.C., 1949.

U.S. Department of State. Office of Foreign Liquidation. *Report on War Air Furnished by the United States to the USSR.* Washington, D.C.: Department of State, 1945.

MANUSCRIPT COLLECTIONS

Chennault, Claire. Papers. Hoover Institution on War, Revolution, and Peace. Stanford University, California.

Currie, Laughlin. Papers. Hoover Institution.

Hornbeck, Stanley K. Papers. Hoover Institution.

Hull, Cordell. Papers. Department of Manuscripts, Library of Congress. Washington D.C.

Johnson, Nelson T. Papers. Department of Manuscripts, Library of Congress.

Morgenthau, Henry Papers. Franklin D. Roosevelt Library and Museum. Hyde Park, New York.

Pawley, William D. Papers. George C. Marshall Foundation Library. Lexington,

Virginia.
Price, Ernest B. Papers. Hoover Institution.
Soong, Tzu V. Papers. Hoover Institution.
Willauer, Whiting Papers. Firestone Library. Princeton University, New Jersey.
Young, Arthur N. Papers. Hoover Institution.
Zakhartchenko, Constantine L. Papers. Hoover Institution.
Zhou, Zhirou. Kongjun zhandou shilu [The record of the combat history of the Air Force], Department of Asian Collections, Library of Congress.

NEWSPAPERS AND JOURNALS

In English

Aerospace Facts and Figures (Washington, D.C.)
Aircraft Year Book (Washington, D.C.)
The Baltimore Sun
China Business Review
China Handbook, 1937–1945
China Monthly Review (Shanghai)
China Press (Shanghai)
China Trade Monthly Report (Washington, D.C.)
China Weekly Review (Shanghai)
China Year Book (Shanghai)
The Far Eastern Review
Los Angeles Times
New York Times
The North China Herald (Shanghai)
The Times (London)
The Wall Street Journal
The Washington Post

In Chinese

Hanghong xuexiao yuekan [Aviation School Monthly]
Hangkong zhazhi [Aviation Magazine]
Hongjun [Air Force]
Hongjun yuekan [Air Force monthly]
Junshi zhazhi [Military Magazine]
Renmin ribao [People's Daily]
Zhongguo de hongjun [China's Air Force]

BOOKS

In English

Acheson, Dean. *Present at the Creation: My Years in the State Department.* New York: W. W. Norton, 1969.
Anders, Leslie. *The Ledo Road: General Joseph W. Stilwell's Highway to China.* Norman: University of Oklahoma Press, 1965.
Anderson, Irvine. *The Standard Vacuum Oil Company and the United States East Asian Policy.* Princeton, N. J.: Princeton University Press, 1975.

Allen, H. C. *Great Britain and the United States.* New York: St. Martin's Press, 1965.

Atwater, Elton. *American Regulation of Arms Exports.* Washington, D.C.: Carnegie Endowment for International Peace, 1941.

Augelucci, Enzo. *The Rand McNally Encyclopedia of Military Aircraft, 1941–1980.* New York: Military Press, 1983.

Bailey, Thomas Andrew. *Theodore Roosevelt and the Japanese-American Crises: An Account of the International Complications Arising from the Race Problem on the Pacific Coast.* Gloucester, Mass.: P. Smith, 1964.

Barnes, Harry Elmer, ed. *Perpetual War for Perpetual Peace.* Caldwell, Idaho: Caxton Printers, 1953.

Barnett, A. Doak. *China and the Major Powers in East Asia.* Washington, D.C.: Brookings Institution, 1977.

Beard, Charles A. *President Roosevelt and the Coming of the War, 1941.* New Haven, Conn.: Yale University Press, 1948.

———. *American Foreign Policy in the Making, 1932–1940: A Study in Responsibilities.* Hamden, Conn.: Archon Books, 1968.

Bedeski, Robert E. *State-Building in Modern China: The Kuomintang in the Prewar Period.* Berkeley: Center for Chinese Studies Publications, University of California, 1981.

Belden, Jack. *China Shakes the World.* New York: Harper & Brothers, 1949.

Best, Antony. *Britain, Japan and Pearl Harbor: Avoiding War in East Asia, 1936–1941.* London: Routledge, 1995.

Bisson, Thomas A. *America's Far Eastern Policy.* New York: Macmillan, 1945.

Bluestone, Barry. *Aircraft Industry Dynamic: An Analysis of Competition, Capital, and Labor.* Boston: Auburn House, 1981.

Blum, John Morton. *From the Morgenthau Diaries: Years of Crisis, 1928–1938.* Boston: Houghton-Mifflin, 1959.

Bond, Charles, Jr., and Terry Anderson. *A Flying Tiger's Diary.* College Station, Texas: Texas A & M University Press, 1984.

Borg, Dorothy. *The United States and the Far Eastern Crisis of 1933–1938.* Cambridge, Mass.: Harvard University Press, 1964.

———. *American Policy and the Chinese Revolution 1925–1928.* New York: Octagon Books, 1968.

Borg, Dorothy, and Shumpei Okamoto, eds. *Pearl Harbor as History: Japanese-American Relations, 1931–1941.* New York: Columbia University Press, 1973.

Borg, Dorothy, and Waldo Heinrichs, eds. *Uncertain Years: Chinese-American Relations, 1947–1950.* New York: Columbia University Press, 1980.

Botjer, George F. *A Short History of Nationalist China, 1919–1949.* New York: G. P. Putnam's Sons, 1979.

Brisow, O. B., and B. T. Koloskov. *Soviet-Chinese Relations, 1945–1970.* Bloomington: Indiana University Press, 1975.

Brockway, Fenner. *The Bloody Traffic.* London: Victor Gollancz, 1933.

Buhite, Russell D. *Nelson T. Johnson and American Policy toward China 1925–1941.* East Lansing: Michigan State University Press, 1968.

———. *Soviet-American Relations in Northeast Asia, 1945–1954.* Norman: University of Oklahoma Press, 1981.

Burns, Richard D. *The International Trade in Armaments Prior to World War II.* New York: Garland, 1972.

Byrd, Martha. *Chennault: Giving Wings to the Tiger.* Tuscaloosa: University of

Alabama Press, 1987.

Campbell, Charles S. *Special Business Interest and the Open Door Policy.* New Haven, Conn.: Yale University Press, 1951.

Chan, Anthony. *Arming the Chinese: The Western Armaments Trade in Warlord China, 1920–1928.* Vancouver: British Columbia Press, 1982.

Chang, Gordon H. *Friends and Enemies: The United States, China, and the Soviet Union, 1942–1972.* Stanford, Calif.: Stanford University Press, 1990.

Chassin, Lionel Max. *The Communist Conquest of China: A History of the Civil War, 1945–1949.* Translated from the French by Timothy Osato and Louis Gelas. Cambridge, Mass.: Harvard University Press, 1965.

Chen, Jian. *China's Road to the Korean War: The Making of the Sino-American Confrontation.* New York: Columbia University Press, 1994.

Cheng, Hawthorne. *China: After Seven Years of War.* New York: Macmillan, 1945.

Chennault, Anna. *Chennault and the Flying Tigers.* New York: P.S. Eriksson, 1963.

Chennault, Claire. *Way of a Fighter: The Memoirs of Claire Lee Chennault.* New York: G. P. Putnam's Sons, 1949.

Chern, Kenneth S. *Dilemma in China: America's Policy Debate, 1945.* Hamden, Conn.: Archon Books, 1980.

Chi, Hsi-sheng. *Warlord Politics in China, 1919–1928.* Stanford, Calif.: Stanford University Press, 1976.

Cochran, Sherman. *Big Business in China: Sino-Foreign Rivalry in the Cigarette Industry.* Cambridge, Mass.: Harvard University Press, 1980.

Cohen, Warren I. *America's Response to China: A History of Sino-American Relations*, 3d ed. New York: Columbia University Press, 1990.

Cole, Wayne S. *America First: The Battle Against Intervention, 1940-1941.* Madison: University of Wisconsin Press, 1953.

———. *Charles A. Lindbergh and the Battle Against American Intervention in World War II.* New York: Harcourt Brace Jovanovich, 1974.

———. *An Interpretive History of American Foreign Relations.* Revised ed. Homewood, Ill.: Dorsey Press, 1974.

———. *Roosevelt and the Isolationists, 1932–45.* Lincoln: University of Nebraska Press, 1983.

Collier, Basil. *Arms and the Men: The Arms Trade and Governments.* London: H. Hamilton, 1968.

Conroy, Hilary, and Harry Wray, eds. *Pearl Harbor Reexamined: Prologue to the Pacific War.* Honolulu: University of Hawaii Press, 1990.

Cornelius, Wanda. *Ding Hao, America's Air War in China, 1937–1945.* Gretna, La.: Pelican, 1980.

Craven, Wesley, ed. *The Army Air Forces in the World War II,* vol. IV. Chicago: University of Chicago Press, 1950.

———. *The Army Air Forces in the World War II,* vol. II. Chicago: University of Chicago Press, 1953.

———. *The Army Air Forces in the World War II,* vol. VII. Chicago: University of Chicago Press, 1958.

Cunningham, William Glenn. *The Aircraft Industry: A Study in Industrial Location.* Los Angeles: Lorrin L. Morrison, 1951.

Dayer, Roberta Albert. *Bankers and Diplomats in China 1917–1925: The Anglo-American Relationship.* London: Frank Cass, 1981.

Dietrich, Ethel B. *Far Eastern Trade of the United States.* New York: Haddon

Craftsmen, 1940.

Divine, Robert A. *The Illusion of Neutrality.* Chicago: University of Chicago Press, 1962.

Doolittle, James H. "Jimmy." *I Could Never Be So Lucky Again: An Autobiography.* New York: Bantam Books, 1991.

Drummond, Donald. *The Passing of American Neutrality, 1937-1941.* Ann Arbor: University of Michigan Press, 1955.

Eastman, Lloyd E. *The Abortive Revolution: China under Nationalist Rule, 1927–1937.* Cambridge, Mass.: Harvard University Press, 1974.

———. *Seed of Destruction: Nationalist China in War and Revolution, 1937–1949.* Stanford, Calif.: Stanford University Press, 1984.

Elsey, George M. *Roosevelt and China: The White House Story.* Wilmington, Del.: Michael Glazier, 1979.

Endicott, Stephen Lyon. *Diplomacy and Enterprise: British China Policy, 1933–1937.* Vancouver: University of British Columbia Press, 1975.

Fairbank, John King. *The United States and China.* Cambridge, Mass.: Harvard University Press, 1983.

Feis, Herbert. *The Road to Pearl Harbor: The Coming of the War between the United States and Japan.* Princeton, N.J.: Princeton University Press, 1950.

———. *The Chinese Tangle: The American Effort in China from Pearl Harbor to the Marshall Mission.* New York: Atheneum, 1965.

Ferrell, Robert H. *American Diplomacy in the Great Depression: Hoover-Stimson Foreign Policy 1929–1933.* New Haven, Conn.: Yale University Press, 1957.

Field, Frederick. *American Participation in the China Consortium.* Chicago: University of Chicago Press, 1931.

Ford, Daniel. *Flying Tigers: Claire Chennault and the American Volunteer Group.* Washington, D.C.: Smithsonian Institution Press, 1991.

Frank, Lewis A. *The Arms Trade in International Relations.* New York: Praeger, 1969.

Frillman, Paul, and Graham Peck, *China: The Remembered Life.* Boston: Houghton Mifflin, 1968.

Furuya, Keiji. *Chiang Kai-shek: His Life and Times.* New York: St. John's University Press, 1981.

Generalissimo Chiang Kai-shek. *Resistance and Reconstruction: Messages During China's Six Years of War, 1937–1943.* New York: Harper & Brothers, 1943.

Grasso, June M. *Truman's Two-China Policy, 1948–1950.* New York: East Gate Books, 1987.

Grew, Joseph. *Ten Years in Japan: A Contemporary Record Drawn from the Diaries and Private and Official Papers.* New York: Simon and Schuster, 1944.

Gilbert, Martin, ed. *A Century of Conflict, 1850–1950: Essays for A. J. Taylor.* New York: Atheneum , 1966.

Griswold, A. Whitney. *The Far Eastern Policy of the United States.* New Haven, Conn.: Yale University Press, 1938.

Harding, Harry, and Yuan Ming. *Sino-American Relations, 1945–1955: A Joint Reassessment of a Critical Decade.* Wilmington, Del.: Scholarly Resources, 1989.

Harkavy, Robert. *The Arms Trade and International System.* Cambridge, Mass.: Bollinger, 1975.

Hart, Robert A. *The Eccentric Tradition: American Diplomacy in the Far East.* New York: Charles Scribner's Sons, 1976.

Hau, Chi-ming. *Foreign Investment and Economic Development in China, 1840–1937.*

Cambridge, Mass. : Harvard University Press, 1965.

Hayward, Keith. *Government and British Civil Aerospace.* Manchester, England: Manchester University Press, 1983.

Head, William P. *America's China Sojourn: America's Foreign Policy and Its Effects on Sino-American Relations, 1942–1948.* Lanham, Md.: University Press of America, 1983.

Heinrich, Waldo. *Threshold of War: Franklin D. Roosevelt and American Entry into World War II.* New York: Oxford University Press, 1988.

Herzog, James H. *Closing the Open Door: American-Japanese Diplomatic Negotiations, 1936–1941.* Annapolis, Md.: Naval Institute Press, 1973.

Hsu, Long-hsuen. *History of the Sino-Japanese War, 1937–1945.* Taipei: Chungwu, 1985.

Hsu, Immanuel Chung-yueh. *The Rise of Modern China.* Revised ed. New York: Oxford University Press, 1990.

Hunt, Michael. *The Genesis of Chinese Communist Foreign Policy.* New York: Columbia University Press, 1996.

Iriye, Akira. *After Imperialism: The Search for a New Order in the Far East, 1921–1931.* New York: Atheneum, 1965.

———. *Across the Pacific: An Inner History of American–East Asian Relations.* New York: Harcourt, Brace and World, 1967.

———. *The Origins of the Second World War in Asia and the Pacific.* New York: Longman, 1987.

Jiang, Annold Xiangze. *The United States and China.* Chicago: University of Chicago Press, 1988.

Jonas, Manfred. *Isolationism in America.* Ithaca, N. Y.: Cornell University Press, 1966.

Kalyagin, Aleksandr Ya. *Along Alien Roads.* New York: Columbia University, East Asia Institute, 1983.

Kirby, William C. *Germany and Republican China.* Stanford, Calif.: Stanford University Press, 1984.

LaFeber, Walter. *The Clash: A History of U.S.-Japan Relations.* New York: W. W. Norton, 1997.

Langer, William L., and Everett Gleason, *The Challenge to Isolation, 1937–1940.* New York: Harper, 1952.

———. *The Undeclared War, 1940–1941.* New York: Harper, 1953.

Lary, Diana. *Region and Nation: The Kwangsi Clique in Chinese Politics, 1925–1937.* London: Cambridge University Press, 1974.

———. *Warlord Soldiers: Chinese Common Soldiers, 1911–1937.* New York: Cambridge University Press, 1985.

Lattimore, Owen. *China Memoirs: Chiang Kai-shek and the War against Japan.* Tokyo: University of Tokyo Press, 1990.

Lawrence, C. L. *Our National Aviation Program.* New York: Aeronautic Chamber of Commerce of America, 1932.

Leary, William M. *Perilous Missions: Civil Air Transport and CIA Covert Operations in Asia.* Tuscaloosa: University of Alabama Press, 1984.

———. *The Dragon's Wing: The China National Aviation Corporation and the Development of Commercial Aviation in China.* Athens: University of Georgia, 1976.

Levine, Steven I. *Anvil of Victory: The Communist Revolution in Manchuria, 1945–1948.* New York: Columbia University Press, 1987.

Liang, Chin-tung. *General Stilwell in China, 1942–1944: The Full Story.* New York:

St. John's University Press, 1972.

Liao, Kai-lung. *From Yenan To Peking: The Chinese People's War of Liberation from Reconstruction to First Five-Year Plan.* Peking: Foreign Languages Press, 1954.

Liu, F. F. *A Military History of Modern China, 1929-1949.* Princeton, N. J.: Princeton University Press, 1956.

Liu, Kwang-ching. *Anglo-American Steamship Rivalry in China, 1862–1874.* Cambridge, Mass.: Harvard University Press, 1962.

Loh, Pichon P. Y. *The Kuomintang Debacle of 1949: Conquest or Collapse?* Boston: D. C. Heath, 1965.

Louis, Roger. *British Strategy in the Far East, 1919–1939.* Oxford: Clarendon Press, 1971.

Lowe, Peter. *Great Britain and the Origins of the Pacific War: A Study of British Policy in East Asia, 1937–1941.* Oxford: Clarendon Press, 1977.

Lukas, Richard C. *Eagles East: The Army Air Force and the Soviet Union, 1941–1945.* Tallahassee: Florida State University Press, 1970.

Marshall, Richard E., and John E. Wiltz. *The Search for Meaning: Viewpoints in American History.* Philadelphia: Lippincott, 1973.

May, Ernest. *The Truman Administration and China, 1945–1949.* Philadelphia: Lippincott, 1975.

Melby, John F. *The Mandate of Heaven.* Toronto: University of Toronto Press, 1969.

McCormick, Thomas J. *China Market: America's Quest for Informal Empire 1893–1901.* Chicago: Quadrangle Books, 1967.

Millis, Walter, ed. *The Forrestal Diaries.* New York: Viking Press, 1951.

Molesworth, Carl. *Wing to Wing: Air Combat in China, 1943–1945.* New York: Orion Books, 1990.

Morgenthau Diary. Washington, D.C.: U.S. Government Printing Office, 1965.

Myers, Wiliam Starr. *The Foreign Policy of Herbert Hoover: 1929–1933.* New York: Charles Scribner's Sons, 1940.

Nagai, Yon Osuke, and Akira Iriye. *The Origins of the Cold War in Asia.* New York: Columbia University Press, 1977.

Nevins, Allan. *The New Deal and the World Affairs.* New York: United States Publishing Association, 1950.

Nicholas, Herbert. *Britain and the U.S.A.* Baltimore, Md.: Johns Hopkins Press, 1963.

Nish, Ian Hill. *Alliance in Decline: A Study in Anglo-Japanese Relations: 1908–23.* London: Athlone Press, 1972.

Okumiya, Masatake, and Jiro Horikoshi. *Zero!* New York: E. P. Dutton, 1956.

Parrini, Carl. *Heir to Empire: United States Economic Diplomacy, 1916–1923.* Pittsburgh: University of Pittsburgh Press, 1965.

Pepper, Suzanne. *The Civil War in China: The Political Struggle, 1945–1949,* Berkeley: University of California Press, 1980.

Pierre, Andrew J. *Arms Transfer and American Foreign Policy.* New York: Council on Foreign Relations, 1979.

Pollard, Robert T. *China's Foreign Relations: 1917–1931.* New York: Macmillan, Company, 1933.

Prange, Gordon William. *At Dawn We Slept: The Untold Story of Pearl Harbor.* New York: McGraw-Hill, 1981.

———. *Pearl Harbor: The Verdict of History.* New York: McGraw-Hill, 1986.

Pye, Lucian W. *Warlord Politics: Conflict and Coalition in the Modernization of*

Republican China. New York: Praeger, 1971.

Quigley, Harold S. *Far Eastern War, 1937–1941.* Westport, Conn.: Greenwood Press, 1973.

Rae, John B. *Climb to Greatness: The American Aircraft Industry, 1920–1960*, Cambridge, Mass. : The MIT Press, 1969.

Rauch, Basil. *Roosevelt from Munich to Pearl Harbor.* New York: Creative Age, 1950.

Reardon-Anderson, James. *Yenan and the Great Powers: The Origins of the Chinese Communist Foreign Policy, 1944–1946.* New York: Columbia University Press, 1980.

Remer, C. F. *Foreign Investments in China.* New York: Macmillan, 1933.

Reynolds, Quentin. *The Amazing Mr. Doolittle: A Biography of Lieutenant General James H. Doolittle.* New York: Appleton-Century-Crafts, 1953.

Ritchie, Sebastian. *Industry and Air Power: The Expansion of British Military Aircraft Production, 1935–1941.* London: Frank Cass, 1997.

Romanus, Charles F., *Stilwell's Mission to China.* Washington, D.C.: Office of the Chief of Military History, Department of the Army, 1953.

Romanus, Charles F., and Riley Sunderland, *Stilwell's Command Problems.* Washington, D.C.: Office of the Chief of Military History, Department of the Army, 1956.

———. *Time Runs Out in CBI.* Washington, D.C.: Office of the Chief of Military History, Department of the Army, 1959.

Rossinger, Lawrence K. *Deadlock in China.* San Francisco: Institute of Pacific Relations, 1940.

Schaller, Michael. *The U.S. Crusade in China, 1938–1945.* New York: Columbia University Press, 1979.

Shaw, Chonghal Petey. *The Role of the United States in Chinese Civil Conflicts, 1944–1949.* Salt Lake City, Utah: Charles Schlacks, Jr., 1991.

Sheridan, James E. *China in Disintegration: The Republican Era in Chinese History, 1912–1949.* New York: Free Press, 1975.

Sherwood, Robert E. *Roosevelt and Hopkins: An Intimate History.* New York: Harper & Brothers, 1948.

Simonson, G. R. *The History f the American Aircraft Industry: An Anthology.* Cambridge, Mass.: M. I. T. Press, 1968.

Sladkovsky, M. I., ed. *Leninism and Modern China's Problem.* Moscow: Progress, 1972.

Smith, Sara. *The Manchurian Crisis, 1931–1932: A Tragedy in International Relations.* New York: Columbia University Press, 1948.

So, Wai-chor. *The Kuomintang Left in the National Revolution, 1924–1931.* Hong Kong: Oxford University Press, 1991.

Stueck, William. *Road to Confrontation: American Policy toward China and Korea, 1947–1950.* Chapel Hill: University of North Carolina Press, 1981.

Sunderland, Riley, and Charles F. Romanus. *Stilwell's Personal File: China-Burma India, 1942–1944.* Wilmington, Del.: Scholarly Resources, 1976.

Tamagna, Frank M. *Italy's Interests and Politics in the Far East.* New York: International Secretariat, Institute of Pacific Relations, 1941.

Tansill, Charles C. *Back Door to War: The Roosevelt Foreign Policy, 1933–1941.* Chicago: Regnery, 1952.

Thayer, George. *The War Business.* New York: Simon and Schuster, 1969.

The National Cyclopedia of American Biography. Clifton, N.J.: James T. White, 1980.

Thomson, James C., Peter W. Stanley, and John C. Perry. *Sentimental Imperialist: The American Experience in East Asia.* New York: Harper & Row, 1981.

Thomson, James. *While China Faced West.* Cambridge, Mass.: Harvard University Press, 1969.

Thorne, Christopher. *The Limits of Foreign Policy: The West, the League, and the Far Eastern Crisis of 1931–1933.* New York: Putnam's, 1972.

Tien, Fong-cheng. *A History of Sino-Soviet Relations.* Washington, D.C.: Public Affairs Press, 1951.

Toynbee, Arnold J. *Survey of International Affairs, 1937.* London: Oxford University Press, 1938.

Trotter, Ann. *British and East Asia, 1933–1937.* Cambridge, England: Cambridge University, 1975.

Tuchman, Barbara. *Stilwell and the American Experience in China, 1911–1945.* New York: Macmillan, 1971.

Tucker, Nancy B. *Patterns in the Dust: Chinese-American Relations and the Reconstruction Controversy, 1949–1950.* New York: Columbia University Press, 1983.

Utley, Freda. *Last Chance in China.* Indianapolis, Ind.: Bobbs-Merrill, 1947.

Utley, Jonathan G. *Going to War with Japan, 1937–1941.* Knoxville: University of Tennessee Press, 1985.

Valone, Stephen. *"A Policy Calculated to Benefit China": The United States and the China Arms Embargo, 1919–1929.* New York: Greenwood Press, 1991.

Varg, Paul. *The Closing of the Door: Sino-American Relations 1936–1946.* East Lansing: Michigan State University Press, 1968.

Vevier, Charles. *The United States and China 1906–1913: A Study of Finance and Diplomacy.* Brunswick, N. J.: Rutgers University Press, 1955.

Wasserstrom, Jeffrey. *Students Protest in Twentieth-Century China: The View from Shanghai.* Stanford, Calif.: Stanford University Press, 1991.

Wedemeyer, Albert. *Wedemeyer Reports!* New York: Henry Holt, 1958.

Westad, Odd Arne. *Cold War and Resolution: Soviet-American Rivalry and the Origins of the Chinese Civil War, 1944–1946.* New York: Columbia University Press, 1993.

White, Theodore. *The Stilwell Papers.* New York: William Sloane Associates,1948.

Williams, William A. *The Tragedy of American Diplomacy.* New York: Cleveland and New World Publishing, 1959.

Wilson, Joan Hoff. *American Business and Foreign Policy 1920–1933.* Lexington: University Press of Kentucky, 1933.

Wiltz, John E. *In Search of Peace: The Senate Munitions Inquiry, 1934–1936.* Baton Rouge: Louisiana State University Press, 1963.

Wittner, Lawrence S. *Rebels Against War.* New York: Columbia University Press, 1969.

Young, Arthur N. *China and the Helping Hand, 1937–1945.* Cambridge, Mass.: Harvard University Press, 1963.

——— . *China's Nation Building Effort, 1927–1937.* Stanford, Calif.: The Hoover Institution Press, 1971.

In Chinese

Chen, Zunkong. *Lieqiang tui zhongguo de junhuo jinyun, 1919–1929* [China arms embargo, 1919–1929]. Taipei: Institute of Modern History, Academia Sinica, 1983.

Dangdai zhongguo hongjun [China today: Air Force]. Beijing: Zhongguo shehui kexue chubanshe, 1989.

Fu, Qixue. *Zhongguo waijiaoshi* [A diplomatic history of China], 3d ed. Taipei: Shangwu yinshu kuan, 1991.

Georges, Michel. *1909–1949 nian de zhongguo hanghong* [China's Aviation, 1909–1949]. Beijing: Hanghong gongye chubanshe, 1992. Translated by Yang Changxiu.

Guangdong wenshi ziliao [Sources of literature and history of Guangdong]. Guangzhou: Wenshi ziliao chubanshe, 1970.

He, Yingqin. *Heshangjiang kangzhan qiqian junshi baogao* [General He's military report during the Sino-Japanese war]. Taipei: Wenxin chubanshe, 1962.

——. *Rijun jinhua banian kangzhanshi* [The eight-year history of Chinese resistance to Japan's invasion]. Taipei: Guofangbu shizhenju chuban, 1983.

Hongjun kangri zhanshi [The combat history of Chinese Air Force fighting Japan]. Taipei: Intelligence Department, Air Force Headquarters, 1963.

Hongjun zhandou shilu [The records of the combats of the Air Force]. Guangzhou: Lizhun chubanshe, 1938.

Jiang, Changying. *Zhongguo hanghong shi* [The history of Chinese aviation]. Taipei: Zhongguo zhiyi chubanshe, 1993.

Jiang, Weiguo. *Kangri yuwu* [Resisting Japanese invasion]. Taipei: Liming wenhua chubangongsi, 1979.

Jiangzongtong milu [The private history of President Chiang]. Taipei: Central Daily, 1976.

Jiaofei zhanshi [The history of suppressing the bandits]. Taipei: Guofangbu shizhengju, 1967.

Li, Shushan. *Zhongguo hongjun zhanshi* [The fighting history of the Chinese Air Force]. Guangzhou: Guangdong renmin chubanshe, 1996.

Ma, Minfu. *Zhongguo junshi hangkong, 1908–1949* [China's military aviation, 1908–1949]. Beijing: Hanghong gongye chubanshe, 1994.

Shijie hongzhong zouzhan pashinian [The air combats of the world for 80 years]. Shanghai: Shanghai kexue chubanshe, 1988.

Tao, Wenzhao. *Kangri zhanzheng shiqi zhongguo duiwai guanxi* [China's foreign relations during the war against Japan]. Beijing: Zhonggong dangshi chubanshe, 1991.

Tong, Hollington K. *Chiang Kai-shek.* Taipei: China Publishing Company, 1953.

Wang, Tuonian. *Fangong kanluan* [To suppress the rebellion of the Communists]. Taipei: Liming wenhua chubangongsi, 1982.

Wuci weijiao zhanshi [The combat history of the Fifth Campaign]. Taipei: Peace Press, 1967.

Yao, Jin. *Zhongguo hanghong shi* [A history of China's aviation]. Zhengzhou: Daxian chubanshe, 1998.

Zhongguo jindai junshishi lunwenji [A collection of papers on modern Chinese military history]. Beijing: Junshi kexue chubanshe, 1987.

Zhongguo shibian: lujunzouzhanshi [China Incident: The combat history of the Japanese Army]. Beijing: Zhonghua shuju, 1981. Department of History, Defense Institute of the Defense Minister of Japan. Translated by Ji Fulin.

Zhou Enlai Nianpu [Zhou Enlai's chronicle]. Beijng: Renmin chubanshe, 1989.

Zuo, Poyi. *Jiangxi suweiai zhengquan jianli yiqi benghui* [The establishment of the Jiangxi Soviet and its collapse]. Taipei: East Asian Institute of the National Political

University, 1969.

ARTICLES

Bennett, G. "British Policy in the Far East, 1933–1936: Treasury and Foreign Office." *Modern Asian Studies* 26 (1992): 545–68.

Braisted, William R. "The United States and the American China Development Company." *Far Eastern Quarterly* 2 (February 1952): 147–166.

———. "China, the United States Navy, and the Bethlehem Shell Company, 1909–1929." *Business History Review* 42 (Spring 1968): 62–66.

"Central Air Transport Corp. Grows Rapidly" *China Economist* 2 (August 1948): 203.

Chan, Lau Kit Ching. "Britain and the Sino-Japanese War: Arms Traffics to China through Hong Kong, 1937–1941." *Asia Quarterly* 3 (1977): 175–202.

Chang, Jui-te. "Nationalist Army Officers during the Sino-Japanese War, 1937–1945." *Modern Asian Studies* 30, no. 4 (1996): 1033–56.

Chen, Lawrence M. "Aviation in China." *Information Bulletin* (Nanking), 3 (May 1937), issue 12.

Chi, Hsi-sheng "General Stilwell's Approach to China and the Burma Campaign." *Sino- American Relations* 24 (1998): 65–84.

Cohen, Warren I. "The Development of Chinese Communist Policy toward the United States." *Orbit,* II (Spring and Summer 1967).

———. "Who Fought the Japanese in Hunan? Some Views of China's War Effort." *Journal of Asian Studies* 27 (Nov. 1967): 111–115.

Costigliola, Frank C. "Anglo-American Financial Rivalry in the 1920s." *Journal of Economic History* 37 (1977): 911–34.

Garver, John W. "The Origins of the Second United Front: The Comintern and the Chinese Communist Party." *China Quarterly* no. 113 (March 1988).

Haith, Michael E. " 'Vinegar Joe' Stilwell in the CBI." *Military Review* 72, no. 5 (1992): 73–83.

Hao, Yufan, and Zhihai Zhai, "China's Decision to Enter the Korean War: History Revisited." *China Quarterly,* vol. 121 (March 1990).

Harry, Schreiber. "World War I as Entrepreneurial Opportunity: Willard Straight and the American International Cooperation." *Political Science Quarterly* 48 (1969): 486–511.

Hunt, M. H. "American Businessmen in the China Market: Economic Opportunity and Economic Nationalism, 1890–1931." *Business History Review* 43 (1974): 212–233.

Jiang, Wandi. "Old Soldiers Never Die." *Beijing Review,* 42 (November 1991): 33–35.

Jouett, John. "War Planes over China," *Asia* 37 (1937): 827–30.

Kirwin, Harry W. "The Federal Telegraph Company: A Testing of the Open Door." *Pacific Historical Review* 22 (August 1953): 271-86.

Leary, William M. "Wings for China: The Jouett Mission, 1932–1935." *Pacific Historical Review* 38 (1969): 447–62.

———. "Aircraft and Anti-Communists: CAT in Action, 1949–52." *China Quarterly,* no. 52 (1972).

———. "Portrait of an Intelligence Official James McHugh in China, 1937–1942," in *Naval history.* Edited by William B. Gogar. Wilmington, Del.: Scholarly Resources Inc., 1988.

Leary, William, and William Stueck. "The Chennault Plan to Save China: U.S. Containment in Asia and the Origins of the CIA's Aerial Empire, 1949–1950,

Diplomatic History 8 (1984): 349-64.

Lorence, James J. "Business and Reform: The American Asiatic Association and Exclusion Laws." *Pacific Historical Review* 93 (November 1970): 421-438.

Masland, John W. "Commercial Influence Upon American Foreign Policy, 1937–1941." *Pacific Historical Review* 11 (October 1942): 281–99.

Mazurnan, George T. "Our New Gold Goes Adventuring: The American International Corporation in China." *Pacific Historical Review* 43 (1974): 212–233.

Megaw, M. Ruth. "The Scramble for the Pacific Anglo–United States Rivalry in the 1930s." *Historical Studies* 17 (1977): 458–73.

Prouty, Herbert C. "China in the Sky." *Army Air Force Review* (September 1946): 28– 31.

Pugach, Noel H. "Anglo-American Aircraft Competition and the China Arms Embargo, 1919–1921." *Diplomatic History* 2 (1978): 351–72.

Rabe, Stephen J. "Anglo-American Rivalry for Venezuelan Oil, 1919–1929." *Mid-America* 58 (1976): 97–109.

Rosenburg, Emily S. "Anglo-American Economic Rivalry in Brazil during World War I." *Diplomatic History* 2 (1978): 131–52.

Schaller, Michael. "American Air Strategy in China, 1939–1941: The Origins of Clandestine Air Warfare." *American Quarterly* 28 (1976): 2–19.

Sheng, Michael M. "Chinese Communist Policy Toward the United States and the Myth of the 'Lost Chance,' 1948–1950." *Modern Asian Studies* 28 (1994).

Shephard, Johun E. "Warriors and Politics: The Bitter Lesson of Stilwell in China." *Parameters* 19, no. 1 (1989): 61–75.

Sunderland, Riley. "The Secret Embargo." *Pacific History Review* 29, no. 1(1960): 75–80.

Thorne, Christopher. "The Quest for Arms Embargo: Failure in 1933." *Journal of Contemporary History* 5, no. 4 (1970): 129–49.

Turner, William. "Technology or Manpower." *Air University Quarterly Review* 5 (Fall 1952): 3–22.

Wheeler, Gerald E. "Isolated Japan: Anglo-American Diplomatic Cooperation 1917-1936." *Pacific Historical Review* 30, no. 3 (1961): 165–78.

Wu, Shuoqi. "Some Remarks on Joseph Stilwell's Relationship with Jiang Jieshi." *Republican China* 14, no. 2 (1989): 37–55

Zhang, Qifu. "On the Stilwell Incident and the Contradiction Between the United States and Jiang Jieshi," *Republican China* 14, no. 2 (1989): 21–36.

PH.D. DISSERTATIONS AND SPECIAL STUDIES

Chan, Anthony Bernard. "Chinese Warlords and the Western Armaments Trade, 1920–1928." York University, Canada, 1979.

Dayer, Roberta. "Struggle for China: The Anglo-American Relationship, 1917–1925." State University of New York at Buffalo, 1972.

Feaver, John Hansen. "The Truman Administration and China, 1945–1950: The Policy of Restrained Intervention." University of Oklahoma, 1980.

Fox, John P. "The Development of Germany's Far East Policy, 1933–1936." London School of Economics, 1972.

George, Brian Thomas. "The Open Door and the Rise of Chinese Nationalism: American Policy and China,1917–1928." University of Mexico, 1977.

Hoyt, Frederick Bernard. "American in China and the Formation of American Policy

1925–1937." University of Wisconsin, 1971.

Lorence, James John. "The American Asiatic Association,1898–1925: Organized Business and the Myth of the China Market." University of Wisconsin, 1970.

Munson, V. L. "American Merchants of Capital in China: The Second Chinese Banking Consortium." University of Wisconsin, 1968.

Nutter, Thomas Edward. "American Telegraphy and the Open Door Policy in China 1900–1930." University of Missouri-Columbia, 1968.

Payer, Cheryl Ann. "Western Economic Assistance to Nationalist China, 1927–1937: A Comparison with Postwar Aid Progress." Harvard University, 1971.

Pickler, Gordon K. "United States Aid to the Chinese Nationalist Air Force, 1931–1949." Florida State University, 1971.

Russel, Michael B. "American Silver Policy and China, 1933–1936." University of Illinois, 1972.

Seps, Jerry Bernard. "German Military Advisers and Chiang Kai-shek, 1927–1938." University of California, Berkeley, 1972.

Stephen, James Valone. "A Policy of Such Common Interest: The United States and the Diplomacy of the China Arms Embargo." University of Rochester, New York, 1989.

Taylor, Joe G. "Air Interdiction in China in World War II," USAF Historical Studies No. 132. USAF Historical Division, Air University, 1956.

Zhang, Xiaoming. "Toward Arming China: United States Arms Sales and Military Assistance, 1921–1941." University of Iowa, 1994.

Index

About the Author

GUANGQIU XU is Associate Professor of History at Iowa Wesleyan College. He has published numerous articles in journals such as *Modern Asian Studies*, *The Journal of Contemporary History*, and *The Journal of Asian History*. He is currently working on another book on the U.S. Congress and its China policy.

Recent Titles in
Contributions in Military Studies

Native vs. Settler: Ethnic Conflict in Israel/Palestine, Northern Ireland, and South Africa
Thomas G. Mitchell

Battling for Bombers: The U.S. Air Force Fights for Its Modern Strategic Aircraft Programs
Frank P. Donnini

The Formative Influences, Theories, and Campaigns of the Archduke Carl of Austria
Lee Eysturlid

Great Captains of Antiquity
Richard A. Gabriel

Doctrine Under Trial: American Artillery Employment in World War I
Mark E. Grotelueschen

Ordered to Die: A History of the Ottoman Army in the First World War
Edward J. Erickson

Surgeons at War: Medical Arrangements for the Treatment of the Sick and Wounded in the British Army during the late 18th and 19th Centuries
Matthew H. Kaufman

Soviet Military Assistance: An Empirical Perspective
William H. Mott IV

The Battle of Ap Bac, Vietnam: They Did Everything but Learn from It
David M. Toczek

The Emperor's Friend: Marshal Jean Lannes
Margaret Scott Chrisawn

An Uncertain Trumpet: The Evolution of U.S. Army Infantry Doctrine, 1919–1941
Kenneth Finlayson

Death Waits in the "Dark": The Senoi Praaq, Malaysia's Killer Elite
Roy Davis Linville Jumper